Pitman Research Notes in Mathematics Series

Submission of proposals for consideration
Suggestions for publication, in the form of outlines
are invited by the Editorial Board for assessment. I
approach one of the main editors or another memb
citing the relevant AMS subject classifications. Alte
sent directly to the publisher's offices. Refereeing i
and other mathematical authorities in the topic con
world.

Preparation of accepted manuscripts
On acceptance of a proposal, the publisher will sup
preparation of manuscripts in a form suitable for d
reproduction. Specially printed grid sheets are pro
offered by the publisher towards the cost of typing
subject to the publisher's approval, is also acceptal

Illustrations should be prepared by the authors, re
without further improvement. The use of hand-dra
avoided wherever possible, in order to maintain m

The publisher will be pleased to give any guidance
preparation of a typescript, and will be happy to a

Important note
In order to avoid later retyping, intending author
begin final preparation of a typescript before rece
and special paper. In this way it is hoped to prese
the series.

Longman Scientific & Technical
Longman House
Burnt Mill
Harlow, Essex, UK
(tel (0279) 426721)

Titles in this series

52 Wave propagation in viscoelastic media
F Mainardi
53 Nonlinear partial differential equations and their applications: Collège de France Seminar. Volume I
H Brezis and J L Lions
54 Geometry of Coxeter groups
H Hiller
55 Cusps of Gauss mappings
T Banchoff, T Gaffney and C McCrory
56 An approach to algebraic K-theory
A J Berrick
57 Convex analysis and optimization
J-P Aubin and R B Vintner
58 Convex analysis with applications in the differentiation of convex functions
J R Giles
59 Weak and variational methods for moving boundary problems
C M Elliott and J R Ockendon
60 Nonlinear partial differential equations and their applications: Collège de France Seminar. Volume II
H Brezis and J L Lions
61 Singular systems of differential equations II
S L Campbell
62 Rates of convergence in the central limit theorem
Peter Hall
63 Solution of differential equations by means of one-parameter groups
J M Hill
64 Hankel operators on Hilbert space
S C Power
65 Schrödinger-type operators with continuous spectra
M S P Eastham and H Kalf
66 Recent applications of generalized inverses
S L Campbell
67 Riesz and Fredholm theory in Banach algebra
B A Barnes, G J Murphy, M R F Smyth and T T West
68 Evolution equations and their applications
F Kappel and W Schappacher
69 Generalized solutions of Hamilton-Jacobi equations
P L Lions
70 Nonlinear partial differential equations and their applications: Collège de France Seminar. Volume III
H Brezis and J L Lions
71 Spectral theory and wave operators for the Schrödinger equation
A M Berthier
72 Approximation of Hilbert space operators I
D A Herrero
73 Vector valued Nevanlinna Theory
H J W Ziegler
74 Instability, nonexistence and weighted energy methods in fluid dynamics and related theories
B Straughan
75 Local bifurcation and symmetry
A Vanderbauwhede
76 Clifford analysis
F Brackx, R Delanghe and F Sommen
77 Nonlinear equivalence, reduction of PDEs to ODEs and fast convergent numerical methods
E E Rosinger
78 Free boundary problems, theory and applications. Volume I
A Fasano and M Primicerio
79 Free boundary problems, theory and applications. Volume II
A Fasano and M Primicerio
80 Symplectic geometry
A Crumeyrolle and J Grifone
81 An algorithmic analysis of a communication model with retransmission of flawed messages
D M Lucantoni
82 Geometric games and their applications
W H Ruckle
83 Additive groups of rings
S Feigelstock
84 Nonlinear partial differential equations and their applications: Collège de France Seminar. Volume IV
H Brezis and J L Lions
85 Multiplicative functionals on topological algebras
T Husain
86 Hamilton-Jacobi equations in Hilbert spaces
V Barbu and G Da Prato
87 Harmonic maps with symmetry, harmonic morphisms and deformations of metrics
P Baird
88 Similarity solutions of nonlinear partial differential equations
L Dresner
89 Contributions to nonlinear partial differential equations
C Bardos, A Damlamian, J I Díaz and J Hernández
90 Banach and Hilbert spaces of vector-valued functions
J Burbea and P Masani
91 Control and observation of neutral systems
D Salamon
92 Banach bundles, Banach modules and automorphisms of C*-algebras
M J Dupré and R M Gillette
93 Nonlinear partial differential equations and their applications: Collège de France Seminar. Volume V
H Brezis and J L Lions
94 Computer algebra in applied mathematics: an introduction to MACSYMA
R H Rand
95 Advances in nonlinear waves. Volume I
L Debnath
96 FC-groups
M J Tomkinson
97 Topics in relaxation and ellipsoidal methods
M Akgül
98 Analogue of the group algebra for topological semigroups
H Dzinotyiweyi
99 Stochastic functional differential equations
S E A Mohammed

100 Optimal control of variational inequalities
V Barbu
101 Partial differential equations and dynamical systems
W E Fitzgibbon III
102 Approximation of Hilbert space operators. Volume II
C Apostol, L A Fialkow, D A Herrero and D Voiculescu
103 Nondiscrete induction and iterative processes
V Ptak and F-A Potra
104 Analytic functions – growth aspects
O P Juneja and G P Kapoor
105 Theory of Tikhonov regularization for Fredholm equations of the first kind
C W Groetsch
106 Nonlinear partial differential equations and free boundaries. Volume I
J I Díaz
107 Tight and taut immersions of manifolds
T E Cecil and P J Ryan
108 A layering method for viscous, incompressible L_p flows occupying R^n
A Douglis and E B Fabes
109 Nonlinear partial differential equations and their applications: Collège de France Seminar. Volume VI
H Brezis and J L Lions
110 Finite generalized quadrangles
S E Payne and J A Thas
111 Advances in nonlinear waves. Volume II
L Debnath
112 Topics in several complex variables
E Ramírez de Arellano and D Sundararaman
113 Differential equations, flow invariance and applications
N H Pavel
114 Geometrical combinatorics
F C Holroyd and R J Wilson
115 Generators of strongly continuous semigroups
J A van Casteren
116 Growth of algebras and Gelfand–Kirillov dimension
G R Krause and T H Lenagan
117 Theory of bases and cones
P K Kamthan and M Gupta
118 Linear groups and permutations
A R Camina and E A Whelan
119 General Wiener–Hopf factorization methods
F-O Speck
120 Free boundary problems: applications and theory, Volume III
A Bossavit, A Damlamian and M Fremond
121 Free boundary problems: applications and theory, Volume IV
A Bossavit, A Damlamian and M Fremond
122 Nonlinear partial differential equations and their applications: Collège de France Seminar. Volume VII
H Brezis and J L Lions
123 Geometric methods in operator algebras
H Araki and E G Effros
124 Infinite dimensional analysis–stochastic processes
S Albeverio
125 Ennio de Giorgi Colloquium
P Krée
126 Almost-periodic functions in abstract spaces
S Zaidman
127 Nonlinear variational problems
A Marino, L Modica, S Spagnolo and M Degiovanni
128 Second-order systems of partial differential equations in the plane
L K Hua, W Lin and C-Q Wu
129 Asymptotics of high-order ordinary differential equations
R B Paris and A D Wood
130 Stochastic differential equations
R Wu
131 Differential geometry
L A Cordero
132 Nonlinear differential equations
J K Hale and P Martinez-Amores
133 Approximation theory and applications
S P Singh
134 Near-rings and their links with groups
J D P Meldrum
135 Estimating eigenvalues with *a posteriori/a priori* inequalities
J R Kuttler and V G Sigillito
136 Regular semigroups as extensions
F J Pastijn and M Petrich
137 Representations of rank one Lie groups
D H Collingwood
138 Fractional calculus
G F Roach and A C McBride
139 Hamilton's principle in continuum mechanics
A Bedford
140 Numerical analysis
D F Griffiths and G A Watson
141 Semigroups, theory and applications. Volume I
H Brezis, M G Crandall and F Kappel
142 Distribution theorems of L-functions
D Joyner
143 Recent developments in structured continua
D De Kee and P Kaloni
144 Functional analysis and two-point differential operators
J Locker
145 Numerical methods for partial differential equations
S I Hariharan and T H Moulden
146 Completely bounded maps and dilations
V I Paulsen
147 Harmonic analysis on the Heisenberg nilpotent Lie group
W Schempp
148 Contributions to modern calculus of variations
L Cesari
149 Nonlinear parabolic equations: qualitative properties of solutions
L Boccardo and A Tesei
150 From local times to global geometry, control and physics
K D Elworthy

151 A stochastic maximum principle for optimal control of diffusions
U G Haussmann

152 Semigroups, theory and applications. Volume II
H Brezis, M G Crandall and F Kappel

153 A general theory of integration in function spaces
P Muldowney

154 Oakland Conference on partial differential equations and applied mathematics
L R Bragg and J W Dettman

155 Contributions to nonlinear partial differential equations. Volume II
J I Díaz and P L Lions

156 Semigroups of linear operators: an introduction
A C McBride

157 Ordinary and partial differential equations
B D Sleeman and R J Jarvis

158 Hyperbolic equations
F Colombini and M K V Murthy

159 Linear topologies on a ring: an overview
J S Golan

160 Dynamical systems and bifurcation theory
M I Camacho, M J Pacifico and F Takens

161 Branched coverings and algebraic functions
M Namba

162 Perturbation bounds for matrix eigenvalues
R Bhatia

163 Defect minimization in operator equations: theory and applications
R Reemtsen

164 Multidimensional Brownian excursions and potential theory
K Burdzy

165 Viscosity solutions and optimal control
R J Elliott

166 Nonlinear partial differential equations and their applications. Collège de France Seminar. Volume VIII
H Brezis and J L Lions

167 Theory and applications of inverse problems
H Haario

168 Energy stability and convection
G P Galdi and B Straughan

169 Additive groups of rings. Volume II
S Feigelstock

170 Numerical analysis 1987
D F Griffiths and G A Watson

171 Surveys of some recent results in operator theory. Volume I
J B Conway and B B Morrel

172 Amenable Banach algebras
J-P Pier

173 Pseudo-orbits of contact forms
A Bahri

174 Poisson algebras and Poisson manifolds
K H Bhaskara and K Viswanath

175 Maximum principles and eigenvalue problems in partial differential equations
P W Schaefer

176 Mathematical analysis of nonlinear, dynamic processes
K U Grusa

177 Cordes' two-parameter spectral representation theory
D F McGhee and R H Picard

178 Equivariant K-theory for proper actions
N C Phillips

179 Elliptic operators, topology and asymptotic methods
J Roe

180 Nonlinear evolution equations
J K Engelbrecht, V E Fridman and E N Pelinovski

181 Nonlinear partial differential equations and their applications. Collège de France Seminar. Volume IX
H Brezis and J L Lions

182 Critical points at infinity in some variational problems
A Bahri

183 Recent developments in hyperbolic equations
L Cattabriga, F Colombini, M K V Murthy and S Spagnolo

184 Optimization and identification of systems governed by evolution equations on Banach space
N U Ahmed

185 Free boundary problems: theory and applications. Volume I
K H Hoffmann and J Sprekels

186 Free boundary problems: theory and applications. Volume II
K H Hoffmann and J Sprekels

187 An introduction to intersection homology theory
F Kirwan

188 Derivatives, nuclei and dimensions on the frame of torsion theories
J S Golan and H Simmons

189 Theory of reproducing kernels and its applications
S Saitoh

190 Volterra integrodifferential equations in Banach spaces and applications
G Da Prato and M Iannelli

191 Nest algebras
K R Davidson

192 Surveys of some recent results in operator theory. Volume II
J B Conway and B B Morrel

193 Nonlinear variational problems. Volume II
A Marino and M K Murthy

194 Stochastic processes with multidimensional parameter
M E Dozzi

195 Prestressed bodies
D Iesan

196 Hilbert space approach to some classical transforms
R H Picard

197 Stochastic calculus in application
J R Norris

198 Radical theory
B J Gardner

199 The C* – algebras of a class of solvable Lie groups
X Wang

200 Stochastic analysis, path integration and dynamics
D Elworthy
201 Riemannian geometry and holonomy groups
S Salamon
202 Strong asymptotics for extremal errors and polynomials associated with Erdös type weights
D S Lubinsky
203 Optimal control of diffusion processes
V S Borkar
204 Rings, modules and radicals
B J Gardner
205 Two-parameter eigenvalue problems in ordinary differential equations
M Faierman
206 Distributions and analytic functions
R D Carmichael and D Mitrović
207 Semicontinuity, relaxation and integral representation in the calculus of variations
G Buttazzo
208 Recent advances in nonlinear elliptic and parabolic problems
P Bénilan, M Chipot, L Evans and M Pierre
209 Model completions, ring representations and the topology of the Pierce sheaf
A Carson
210 Retarded dynamical systems
G Stepan
211 Function spaces, differential operators and nonlinear analysis
L Paivarinta
212 Analytic function theory of one complex variable
C C Yang, Y Komatu and K Niino
213 Elements of stability of visco-elastic fluids
J Dunwoody
214 Jordan decompositions of generalised vector measures
K D Schmidt
215 A mathematical analysis of bending of plates with transverse shear deformation
C Constanda
216 Ordinary and partial differential equations Vol II
B D Sleeman and R J Jarvis
217 Hilbert modules over function algebras
R G Douglas and V I Paulsen
218 Graph colourings
R Wilson and R Nelson
219 Hardy-type inequalities
A Kufner and B Opic
220 Nonlinear partial differential equations and their applications. College de France Seminar Volume X
H Brezis and J L Lions
221 Workshop on dynamical systems
E Shiels and Z Coelho
222 Geometry and analysis in nonlinear dynamics
H W Broer and F Takens
223 Fluid dynamical aspects of combustion theory
M Onofri and A Tesei
224 Approximation of Hilbert space operators. Volume I. 2nd edition
D Herrero
225 Operator Theory: Proceedings of the 1988 GPOTS–Wabash conference
J B Conway and B B Morrel
226 Local cohomology and localization
J L Bueso Montero, B Torrecillas Jover and A Verschoren
227 Nonlinear waves and dissipative effects
D Fusco and A Jeffrey
228 Numerical analysis. Volume III
D F Griffiths and G A Watson
229 Recent developments in structured continua. Volume III
D De Kee and P Kaloni
230 Boolean methods in interpolation and approximation
F J Delvos and W Schempp
231 Further advances in twistor theory, Volume 1
L J Mason and L P Hughston
232 Further advances in twistor theory, Volume 2
L J Mason and L P Hughston
233 Geometry in the neighborhood of invariant manifolds of maps and flows and linearization
U Kirchgraber and K Palmer
234 Quantales and their applications
K I Rosenthal
235 Integral equations and inverse problems
V Petkov and R Lazarov
236 Pseudo-differential operators
S R Simanca
237 A functional analytic approach to statistical experiments
I M Bomze
238 Quantum mechanics, algebras and distributions
D Dubin and M Hennings
239 Hamilton flows and evolution semigroups
J Gzyl
240 Topics in controlled Markov chains
V S Borkar
241 Invariant manifold theory for hydrodynamic transition
S Sritharan
242 Lectures on the spectrum of $L^2(\Gamma\backslash G)$
F L Williams

M Onofri and A Tesei (Editors)

University of Rome, Italy

Fluid dynamical aspects of combustion theory

Longman Scientific & Technical

Copublished in the United States with
John Wiley & Sons, Inc., New York

Longman Scientific & Technical,
Longman Group UK Limited,
Longman House, Burnt Mill, Harlow,
Essex CM20 2JE, England
and Associated Companies throughout the world.

Copublished in the United States with
John Wiley & Sons, Inc., 605 Third Avenue, New York, NY 10158

First published 1991

AMS Subject Classification: (Main) 76–06, 80–06, 80A25
(Subsidiary) 65M

ISSN 0269-3674

British Library Cataloguing in Publication Data
A catalogue record for this book is
available from the British Library

Library of Congress Cataloging-in-Publication Data
Fluid dynamical aspects of combustion theory / M. Onofri and A. Tesei, editors.
p. cm. --(Pitman research notes in mathematics, ISSN 0269-3674; 223)
ISBN 0-470-21589-5 (Wiley)
1. Combustion. 2. Fluid dynamics. I. Onofri, M. Tesei, A. (Alberto) III. Series: Pitman research notes in mathematics series; 223.
QD516.F593 1990
541.3′61--dc20 89-13183
CIP

Printed and bound in Great Britain
by Biddles Ltd, Guildford and King's Lynn

Contents

5 - Numerical Combustion

6 - Heterogeneous Combustion

List of Contributors

Borghi Roland,
Université de Rouen, Faculté des Sciences et des Techniques
B.P. 118 - 76134 Mont St Aignan Cedex, France

Boris Jay P.,
Laboratory for Computational Physics and Fluid Dynamic, Naval Research Laboratory
Washington, D.C. 20375 - USA

Brauner Claude Michel,
U.F.R. Mathematiques & Informatique, Universitè Bordeaux I
351 Cours de la Libèration, 33405 Talence Cedex - France

Bray Kenneth N.C.,
Engineering Department, Cambridge University
Trumpington Street, Cambridge, CB2 1PZ, England

Bruno Claudio,
Dipartimento di Meccanica e Aeronautica, Università di Roma "La Sapienza"
Via Eudossiana, 18 - 00184 Roma, Italy

Buckmaster John D.,
Dept. Aeronautical and Astronautical Engineering, University of Illinois,
101, Transportation Building, Urbana, IL 61801, USA

Clavin Paul,
Laboratoire de Recherche en Combustion, URA 1117 Universitè Aix-Marseille I/CNRS
Av. Escadrille Normandie Niemen, S252, Centre Saint Jerome, 1339 Marseille , Cedex 13, France

Di Blasi Colomba,
Dipartimento di Ingegneria Chimica, Università di Napoli
Piazzale V. Tecchio, 80125 Napoli, Italy

Dold J. William,
School of Mathematics, University of Bristol
University Walk, Bristol BS8 1TW, United Kingdom

Frankel Michael L.,
Department of Mathematical Sciences, Indiana University-Purdue University
Indianapolis, Indiana 46205, USA

Ghoniem Ahmed F.,
Massachusetts Institute of Technology, Mechanical Engineering Dept.
Cambridge, MA 02139 - USA

Kapila Ash K.,
Department of Mathematical Sciences, Rensselaer Polytechnic Institute
Troy, New York 12180–3590, USA

Larrouturou Bernard,
INRIA, Unité de Recherche Sophia-Antipolis
2004, Route des Lucioles, B.P. 109 - 06561 Valbonne Cedex, France

Libby Paul A.,
Dept. of Applied Mechanics and Engineering Science, B-010
University of California San Diego, La Jolla, CA 92093, USA

Linan Amable,
Escuela Tecnica Superior de Ingenieros Aeronauticos, Universidad Politecnica de Madrid
Plaza Cardenal Cisneros, 3 - 28040 Madrid

Onofri Marcello,
Dipartimento di Meccanica e Aeronautica, Università di Roma "La Sapienza"
Via Eudossiana, 18 - 00184 Roma, Italy

Oran Elaine S.,
Laboratory for Computational Physics and Fluid Dynamic, Naval Research Laboratory
Washington, D.C. 20375 - USA

Peters Norbert,
Institut für Technische Mechanik, RWTH Aachen
Templergraben 64, D-5100 Aachen, Federal Republik of Germany

Roytburd Victor,
Department of Mathematical Sciences, Rensselaer Polytechnic Institute
Troy, New York 12180–3590, USA

Schmidt-Lainé Claudine,
CNRS et Ecole Normale Superieure de Lyon
46, Alléé d'Italie, 69364 Lyon Cedex 07 - France

Sivashinsky Gregory I.,
School of Mathematical Sciences, Tel-Aviv University
Ramat-Aviv, Tel-Aviv 69978, Israel

Tesei Alberto,
Ist. Applicazioni del Calcolo "M. Picone"
V.le del Policlinico, 137 - 00161 Roma, Italy

Warnatz Jurgen,
Institut für Technische Verbrennung, Universität Stuttgart
7000 Stuttgart, Germany

Williams Forman A.,
Dept. of Applied Mechanics and Engineering Science, B-010
University of California San Diego, La Jolla, CA 92093, USA

Preface

Research in combustion science is becoming one of the most challenging fields of advanced engineering. This happens by two main reasons, respectively connected with its scientific and practical interest.

First, combustion is a subject that exhibits both fundamental and applied characters and includes theoretical and experimental aspects, thus spanning from engineering applications to theoretical modelling. Therefore it involves some of the most advanced research fields: gasdynamics, turbulence, thermochemistry and chemical kinetics. At the same time, it offers stimulating problems in numerical analysis and applied mathematics.

Secondly, the interest connected with many industrial applications, e.g., the need for accurate prediction of fluid flows in aerospace applications, or the demand for control of pollutant emissions has driven the push for more and more advanced methods to predict complex fluid fields involving chemical reactions. In this context the introduction of large scale computers makes it possible to carry out cumbersome calculations to solve more sophisticated models, with more complicated reaction mechanisms, for longer times, thus resulting in an attractive challenge to theoretical capabilities.

As a matter of fact, the interaction of complex fluid motion with the thermodynamic changes associated with chemical reactions is a difficult problem, especially when unsteady phenomena have to be analyzed. Nevertheless, in recent years the theoretical basis of combustion phenomena and the capability of numerical prediction of the relevant flow fields have been impressively developed. Only eight years ago in the preface of their book "Theory of Laminar Flames", Buckmaster and Ludford wrote that "Existing combustion books are primarily phenomenological in the sense that ...mathematical modelling is employed (with) "ad hoc" irrational approximation, the emphasis being on the explanation of existing esperimental results". Today, only a few years later, mathematical methods based on asymptotic expansions have allowed a remarkable growth in the knowledge of the fluid dynamical aspects of combustion; moreover, a growing contribution is coming from numerical and qualitative analysis of detailed simulation models, by taking full advantage of recent developments achieved in Computational Fluid Dynamics (CFD).

In particular, much attention has been recently devoted to the solution of turbulent compressible flows; flows with finite rate chemistry and supersonic combustion have received also renewed attention due to the development of aerospace applications, and large efforts have been spent for a deeper understanding of reaction mechanisms and on developing reduced reaction models. At the same time, mathematical approaches based on asymptotic methods have been growing, providing new insight into detonation phenomena and flame behaviour.

As a consequence of such large and rapid growth of these studies, it is really difficult for a researcher, even familiar with the background knowledge in turbulence and aerothermochemistry, to get a correct perspective about the current state of the art of combustion science, and about actual capabilities to predict reacting flowfields of interest to industrial applications.

An attempt to provide a systematic presentation of the present state of combustion science was made in 1989 at the Istituto per le Applicazioni del Calcolo "M. Picone" of the Italian National Research Council (CNR), during the Special Year devoted to the "Fluid Dynamical Aspects of Combustion Theory".

A prominent result of that activity was the organization of six Lecture Series, each a week long, given by outstanding researchers in combustion. The main goal of these seminars was to encourage university and industrial researchers in mathematics and engineering to enter the field, introducing them to suitable theoretical instruments. Thus, rather than a presentation of very specialized topics, the speakers were asked to provide the audience with the fundamentals of combustion theory, their recent contribution, and motivation for future research. In particular, some speakers were asked to give introductory mini-courses, other to set their research within the broad scope of the combustion literature, and to illustrate their own approach and motivations.

The result, in our opinion, was beyond expectation. In fact, comments of the partecipants were so encouraging that we decided to collect the contributions to have a written version of this experience.

The present volume represents the final result of the above mentioned activity. It is an introduction to combustion theory, and it may be suitable for graduate courses in applied mathematics, mechanics and aerospace or chemical engineering. Moreover, it should represent a basis for researchers interested in entering the combustion field.

Nevertheless, the reader should be aware that this work has not been conceived as a complete presentation of combustion science. There are topics excluded, though important for practical applications (e.g., solid propellant combustion, high Mach number combustion, etc.) or novel approaches connected with the use of new sophisticated numerical techniques. Moreover, no specific attention has been given to experiments, that represents a third major field of research in combustion. As a matter of fact, it should be noticed that the growing importance of experimental techniques and the recent development of combustion flow diagnostics need a separate tutorial treatment.

The topics presented in the book require an appropriate background in fluid dynamics and turbulence; moreover, in giving adeguate emphasis to different specialized research fields, not enough space was left to the fundamentals of aerothermochemistry. To this end the reader may resort to some graduate textbooks, such as "Combustion Fundamentals" by R. Strehlow or to more synthetic works such as "The Mathematics of Combustion" by J.D. Buckmaster (ed.), in which detailed references for deeper study can be found. This book is complementary to many others on combustion, and indications to this effect are given in the list of references of each contribute paper. In any case, general references are books such as "Combustion Theory" by F.A. Williams, "Turbulent Reacting Flows" by P.A. Libby and F.A. Williams (eds.), "Detonation" by

W. Fickett and W.C. Davis, and "Theory of Laminar Flames" by J.D. Buckmaster and G.S.S. Ludford.

This volume is organized as follows. In the introduction details about the argument discussed, the reason of their choice and their organization are reported. Chapter 1 discusses flame theory, showing different methods to analize the behaviour of laminar premixed and nonpremixed flames, while Chapter 2 refers about the modelling of turbulent reacting flows. Chapter 3 deals with detonation theory and Chapter 4 with chemical kinetic problems. Numerical methods are discussed in Chapter 5, while two significant examples of heterogeneous combustion are shown in Chapter 6.

The activities of the Special Year 1989 were made possible by the financial support of the Institute "M. Picone", due to special funds delivered by CNR; contributions from the "Scuola di Ingegneria Aerospaziale" and the "Dipartimento di Meccanica e Aeronautica" of the University of Rome "La Sapienza" are also gratefully acknowledged. Many researchers at these istitutions gave their patient and precious contribution. Mrs. G. Castellini and Mrs. G. Silveri of the Institute " M. Picone" gave a constant and efficient support in collecting the material of this volume and in the practical organization of the Special Year.

Rome, April 1990

Marcello Onofri
Alberto Tesei

Introduction

The aim of this volume is to provide a survey of studies on the fluid dynamical aspects of combustion, including theoretical modelling and numerical calculation of flames, detonations and turbulent reacting flows in both homogeneous and heterogeneous systems.

Nevertheless, combustion science is too large and multifaceted a field to be amenable to a complete description. Theoretical knowledge of the field has grown rapidly in recent years as a result of a large number of studies following different methodological approaches. These increased theoretical capabilities, in turn, have stimulated the development of numerical methods to simulate complex combusting flows typical of engineering applications. In fact, the current points of view are so many and so varied that the contributions collected in this volume can only aim at presenting some of the more interesting approaches and their contribution to the development of theoretical understanding of combustion phenomena.

Following one possible key to a reading of combustion literature, the papers presented here cover three main areas: *i*) analytical approaches that treat simplified models of the governing equations, *ii*) numerical approaches that employ detailed modelling of turbulent reacting flows, and *iii*) studies oriented towards practical applications, e.g., those concerned with chemical reaction mechanisms, or development of suitable numerical techniques.

The first area, particularly suited for the analysis of laminar flames and detonations, employs simplified models derived via asymptotics. The introduction of physically sound approximations, such as the hydrodynamic limit and the assumptions of low Mach number, large Damköhler number, or large activation energy, can lead to substantial reductions of analytical complexity. The hydrodynamic limit, for instance, treats the flame as a hydrodynamic discontinuity, propagating at a speed whose dependence on the kinetics and transport parameters can be determined by means of an asymptotic analysis of the structure of the discontinuity. The low Mach number approximation assumes that the kinetic energy of the flow is small when compared to the static entalpy, leading to equations in which pressure gradients are negligible and from which acoustic phenomena have been filtered out. For diffusion flames, the limit of large Damköhler number implies that reactants cannot coexist except in a thin flame sheet, thereby dividing the flow into regions in which either the fuel or the oxidant is absent.

Models based on activation energy asymptotics have been adopted widely. These models rely on the extreme sensitivity of the reaction rate to changes in temperature, as represented by Arrhenius kinetics. For temperature below a critical value the reaction rate is negligible, while for temperatures above critical the reactants are either in chemical equilibrium (for nonpremixed systems) or the deficient reactant is completely consumed (in premixed systems). In either case, reaction is again confined to thin zones, or flame sheets, while the remaining field can be treated in terms of equilibrium/frozen regimes. Within the flame sheet itself simplifications are possible, making it easier the to model the reaction zone.

The large activation-energy limit has already increased our understanding of flame propagation and combustion instability, and it is expected that analyses based on this limit will still play prominent roles in the future, continuing to provide new insights. Fundamentals of this approach are reported in Chapter 1 of "The Mathematic of Combustion" by J.D. Buckmaster, and in "Theory of Laminar Flames" by the same author and G.S.S. Ludford.

The second area follows a different philosophy, based on equation models that provide the full, time dependent description of turbulent reacting flows, including a detailed modelling of chemical kinetics.

The direct numerical integration of these equations is one of the present and future fields of research in turbulence. However, these solutions require a suitable resolution in space and time, and result in a tremendous amount of CPU time, varying at least as $Re^{2.75}$. These limitations on solving large scale problems result in solution of idealized models that are thus of fundamental rather than applied interest.

A common alternative is to model the turbulence considering its nature as a stochastic process, which may be described by probabilistic approaches. Among the many methods available, one of the most popular approaches is to resort to the classic 'moment methods' to compute the fluid dynamic variables, coupled with approaches based on probability-density-functions (pdf) to obtain closure models for the thermochemical variables.

The conservation equations are averaged to obtain a spatial distribution of statistical quantities: each variable is split into mean and fluctuating components, and closure models are used to determine the new unknowns. In this procedure two points are critical: the criteria of the averaging of the equations and the closure models.

As far as averaging is concerned, it should be pointed out that the density correlations must be carefully considered, as they undoubtedly influence the turbulent reacting flow behaviour. In this sense the use of mass–weighted averaging, commonly denoted as Favre averaging, is strongly indicated for these flows, since it results in fewer terms in the equations than the conventional averaging adopted for constant density flows.

As to the closure problems, modelling of the turbulent transport (i.e., Reynolds stresses and fluxes) and of the chemical source terms are required. For the fluid mechanical aspects, classical descriptions (like k–ϵ modelling, etc.) are commonly used, while more sophisticated models are required to deal with some phenomena (i.e., countergradient diffusion). Regarding the chemical aspects of premixed systems, a useful treatment is to define in space and time functions which assume values in the range zero to one, as the system passes from reactant to products, and whose statistical behaviour is described by probability density functions of a progress variable. In nonpremixed combustion, the species can be computed by the 'conserved scalar' method.

A recent interesting feature of this approach, introduced to model the chemical aspects of turbulent reacting flows, is the "laminar flamelets model". Following this method, the chemical reactions are supposed to be confined to surfaces, considered as laminar flamelets stretched by the turbulent flow. The combustion process will be the result of the behaviour of such flamelets, that can be separately studied in such a way that the stretching and possible extinction can be accounted for.

A good presentation of turbulent combustion modelling is provided in "Turbulent Reacting Flows" by P.A. Libby and F.A. Williams (eds.).

The third group of studies includes a wide spectrum of approaches, resulting from the large number of practical applications in the combustion field.

An important role is covered by the improvements performed in developing both detailed and reduced chemical reaction mechanisms. The detailed knowledge of combustion mechanisms is important for the understanding and control of practical combustion processes like pollutant formation, or induced ignition and auto-ignition or engine knock. However, for many approaches the possibility of obtaining a reduced reaction mechanism is often crucial when many reactions are involved.

Another considerable group of studies is the result of efforts devoted to develop computational methods solving specific phenomena occurring in combusting flows. In particular most numerical simulations have aimed at analyzing both the fundamental mechanisms in detonation shock waves and the complex interchange between turbulence and combustion, including interactions such as vortex-flame, separating flow-flame, and shear layer-combustion.

Finally, studies are reported concerning specific treatments of heterogeneous combustion processes.

An introduction to chemically detailed combustion modelling is given in "Combustion Chemistry" by W.C. Gardiner (ed.), while a survey of many fundamental aspects of numerical combustion is given by E. Oran and J. Boris in "Numerical Simulation of Reacting Flow".

The papers presented in this volume relate to studies performed by all of the above mentioned approaches.

Chapter 1 deals with flame theory, including premixed and non-premixed laminar flames, and turbulent flames in the flamelet regime. Representative studies based on the use of asymptotic expansions are reported in the papers of Liñan, who considers nonpremixed flames, and Frankel and Sivashinsky, Clavin, Brauner and Schmidt-Lainè, who analyze premixed flames.

Linan's paper summarizes the basic theoretical aspects of the analysis of nonpremixed flames and shows how powerful asymptotics can be in providing an understanding of flame behaviour. The low Mach number limit and infinite Damköhler number model are cleverly exploited in order to analyze the structure of diffusion flames. Then the large activation energy limit is used to describe the occurrence of triple flame structures, including both premixed and diffusion flames. Finally diffusion flames are analyzed in strained mixing layers, where ignition and extintion phenomena may take place.

An analysis of flame propagation based on the hydrodynamic limit is carried out by

Frankel and Sivashinsky. In their paper, the equations for surface dynamics of slightly perturbed plane flame fronts, obtained by means of multiple scale asymptotic analysis, are extended to generalized coordinate frames. This formulation allows consideration of the evolution of curved fronts with freely expanding smooth surfaces, and of specific effects such as the stretch and self turbulization of the front.

An extension of the hydrodynamic approach in the flamelet regime of the propagation of premixed turbulent flames is discussed in Clavin's paper, providing the background for wrinkled and curved fronts dynamics, and the definition of flamelet regime. Then he discusses the cusped flame regime in weakly turbulent flows and the corrugated flamelet regime in high intensity turbulence. Following a semi-phenomenological approach, flame properties are analyzed. In particular, flame propagation is considered to occur between two limiting values: that limited by the combustion of vortex tongues and that controlled by contamination.

Brauner and Schmidt-Lainè discuss some difficulties occurring in the mathematical treatment of laminar premixed flames by asymptotic analysis. Two examples are considered. First in a nonlinear differential system modelling a two step reaction, a distinguished limit is considered, corresponding to fast radicals recombination. The purpose is to determine the burning rate as function of the two reaction rates. Secondly the nonlinear stability analysis of the travelling waves in Equidiffusional Flames is performed considering disturbances perturbing only temperature, not enthalpy, thus resulting in a one-phase boundary problem that can be considered representative of many other situations.

In **Chapter 2** surveys of the modelling of turbulent reacting flows are given by Williams that summaries the current status, by Borghi that gives account of the studies concerning nonpremixed flames, and by Bray and Libby that consider premixed flames.

A general discussion about the modelling of these flows, with a special emphasis on premixed turbulent combustion, is presented by Williams. He refers to current approaches, giving a summary of the most interesting subjects of research. Moreover he discusses the future prospects in this field of studies.

Borghi's paper describes the fundamental physical aspects involved in turbulent flames and in turbulent combustion chambers. Then the paper refers to the state of the art in the numerical modelling of nonpremixed turbulent flames. Two categories of flames are separately considered: with *i)* fast and *ii)* moderately fast chemistry. For the first category the applications of pdf statistical models are analyzed. For the second three different approaches are discussed: *a)* the method of using the joint multidimensional pdf equation for the reactive species, *b)* the flamelet models, *c)* the Lagrangian models that represent the evolution of fluid particles over their life and trajectory within the reacting medium. Assessements of the models and of their range of validity are reported.

The paper by Bray and Libby refers to developments of the turbulence modelling of premixed flames based on the Bray-Moss (BM) and the Bray-Moss-Libby (BML) theories, that deal, respectively, with the modelling of thermochemistry and aerothermochemistry of premixed turbulent combustion.

Applications of the BM model considering a classical gradient transport calculation of planar flames gave predictions not in accord with experiments, due to the shortcomings of fluid mechanic treatment. The BML model was introduced to circumvent this problem. It involves a more suitable description of the mean turbulent fluxes. Instead of using the gradient approximation, this correlation can be written in terms of conditioned averaged velocity components $\overline{u}_{i_r}$ and $\overline{u}_{i_p}$, the mean velocity within reactants and products respectively. Following this approach they show that even countergradient diffusion (i.e., diffusion process opposite in sign to the species gradient) can satisfactorily be accounted for. The last part of the paper deals with laminar flamelet models, with considerations of the mean reaction rate.

Chapter 3 is devoted to detonation theory. Fundamentals are given in the introductory paper by Kapila. Then papers by Dold, Roytburd, and Brauner et al. (Buckmaster, Dold, and Schmidt-Lainè) respectively deal with detonation instability, unsteady behaviour of an emerging detonation, and stability of detonations with respect to two dimensional perturbations.

Kapila introduces the mathematical model describing fast reacting flows and reviews the main features of steadily propagating detonation waves. In the second part of the paper he reports on two topics of current research: an analysis of the early stage of detonation initiation, and the development of an asymptotic theory for propagation of curved detonation fronts. These topics receive a detailed treatment in the papers by Dold and Roytburd, provide two examples of applications of asymptotic theory to detonations, and point out the advantages and limits of this method.

Dold's paper analyzes the way in which a chemically reacting medium can ignite to produce a detonation wave. Recent studies and new interpretations are reviewed. The fundamental features of steady and unsteady behaviour of combustion waves are analyzed by means of a one-dimensional flow model and a simple one-step chemical reaction mechanism, with a large activation temperature. Both weak detonations, the unsteady formation of a shock wave and the development of strong detonation have been considered.

Roytburd considers the dynamically unstable behaviour of travelling waves: small perturbations of the wave structure grow in time resulting in wrinkling of the originally planar front, and in changing of the wave speed. He discusses the most significant contributions in the study of instabilities of planar detonation waves. Linear and weakly nonlinear analyses of one and two-dimensional stability are reviewed. The approach relies on the ideas of weakly nonlinear geometrical optics. The one-dimensional case is described and the main differences introduced by the extension to two dimensions are indicated.

The contribution by Brauner et al. analyzes the evolution of perturbations to a detonation wave in the limit of large activation energy. Under this assumption the detonation wave structure reduces to a square-wave model, in which conditions in the induction zone are uniform and the reaction zone is a discontinuity. The linear stability of this structure is analyzed with respect to slowly varying corrugations of the leading shock of the detonation. The effect of the curvature of the leading shock on the induction zone is modelled by a stream tube divergence term. The results of the linear analysis are complemented by a nonlinear stability analysis of an evolution equation for the shock displacement derived under the above approximations.

Chapter 4 deals with chemical kinetic problems and involves significant studies by Warnatz and Peters. Warnatz describes the developments of detailed chemical kinetics mechanisms and of strained flame libraries, necessary for the description of turbulent combustion by means of probabilistic approaches, while Peters discusses the reduction of complex reaction mechanisms by the introduction of various assumptions including that of a steady state approximation.

Warnatz considers hydrocarbon combustion for ignition problems, by means of a system with as many as 1500 reactions among 150 species. This results in a very large set of nonlinear stiff partial differential equations. By means of sensitivity analysis to designate and eliminate unimportant reactions, he obtains one mechanism involving a large set of kinetic data for the elementary reactions. Representative results obtained by using this mechanism are successfully compared with experimental data for laminar premixed flames at high temperature, and for induced-ignition and auto-ignition at low temperatures. Finally the turbulent mixing of a fuel jet with air is considered with a detailed kinetics mechanism consisting of 72 elementary reactions and 15 species. The turbulent flame computation is based on the flamelet model, considering a burning laminar flamelet library containing 15 stretched flames profiles.

Peters discusses an approach based on a reduction of complex reaction mechanisms by introducing steady state assumptions for the intermediate species produced in the reaction chain. For homogeneous systems this assumption is well founded for species produced by slow reactions and consumed by fast reactions, so that their concentrations remain small and essentially steady. The paper reports the principles to be observed in reducing detailed reaction mechanisms to a smaller number of global reactions. Then it summarizes recent results relevant to nonpremixed methane-air flames and to lean premixed methane and lean-to-rich premixed methane, methanol and propane flames. The results have also been compared with numerical calculations based on full kinetics schemes.

In **Chapter 5** numerical methods to predict combustion phenomena are discussed. Mathematical aspects of solution of combustion problems performed by both analytical and numerical approaches are considered by Larrouturou, while significant examples of computations of flames, detonations and complex flowfields are reported in the papers by Ghoniem and Oran and Boris.

The aim of Larrouturou's paper is to show examples of how mathematical methods are used to solve analytically flame models and to provide support to the numerical analysis of combusting flows. In the first section two examples are reported: a travelling wave solution of a premixed flame propagation and the analysis of the cold boundary difficulty, which arises when investigating the propagation of a steady planar flame in infinite tube. In the second section the mathematical aspects of some numerical schemes are considered in three examples. The first reports the analysis of convergence of a class of schemes for the numerical approximation of the unsteady isobaric 1-D model; in the second the steady planar flame problem, posed in a truncated domain, is studied in order to show that the truncated solution exponentially converges toward the steady solution; the last example discusses how to assure mass fraction positivity when computing multicomponent flows by upwind schemes based on approximate Riemann solvers, such as the Roe and Osher schemes.

Oran and Boris present a general review of reactive flow modelling and of the numerical methods developed over the last years, also reporting selected computations of detonations, flames, and complex flows, with an emphasis on the new physical understandings gained from them. In particular, the computations of detonations consider weak and strong ignition behind reflected shocks, and the analysis of the formation both of structures involving interaction between shocks and of unreacted pockets of material behind the fronts. Calculations of premixed laminar flames are performed in order to understand the relative importance of the different transport processes to the burning velocity and to determine their structure and stability. Solutions of complex flowfields are considered to show how acoustic interactions are responsible for large scale effects in the flow structures, and to analyze the interpenetration of two fluids of different density as a way to understand the mechanisms which control compressible flow turbulence. Finally new computer architectures are discussed in order to illustrate their impact on the numerical techniques and the great enhancement that should be achieved by their use in the calculation of combusting flows.

The fundamentals of vortex methods and some successful applications to reacting flows are reviewed in Ghoniem's paper. An interesting feature of these methods is that they are inherently Lagrangian, naturally adaptive, with the derivatives of the primary variables, vorticity and scalar gradients, represented by discrete elements that move along particle trajectories. In the first part of the paper the formulation and the basic concepts are reported. In the following part results are reviewed for three different reacting flows: a three-dimensional reacting shear layer with no heat release, a two-dimensional jet with heat release, and a premixed flame stabilized in a separating flow produced by a sudden expansion in a channel, as in practical dump combustors.

Chapter 6 of this volume discusses two significant examples of heterogeneous combustion. Gas-surface interaction including catalytic effects is described by C. Bruno; this phenomenon plays a substantial role in many engineering applications like furnaces, engines, and reentry problems in external aerodynamics of aerospace shuttles. The second topic, presented by C. Di Blasi, analyzes combustion processes in pool fires. In this case interest in the topic comes from its practical applications for pollution control and in safety science to prevent catastrophic effects of leaks or spill in the storage of flammable liquid or liquified gas.

Both articles deal with flows in contact with active interfaces where combustion is supported by interfacial phenomena. Di Blasi treats flame spreading over liquid fuel pools where the energy feedback from the hot gas vaporizes the cold fuel, creates a rolling vortex by creating a surface tension gradient, and ultimately drives flame front propagation. Bruno, on the contrary, considers a gas-solid interface that is motionless and with zero net mass transfer; at the same time, the surface is an active energy source, feeding both solid and gas in proportions depending on the physics of the problem. For chemical recombination of gas particles most of the heat is transferred to the solid, while for catalytic combustion it is the gas that receives most of the energy, and eventually may react homogeneously.

1 Flame theory

A LINAN

The structure of diffusion flames

1. INTRODUCTION

These notes are devoted to the analysis of reacting flows in systems where the reactants are initially unpremixed. We shall deal here with the combustion reactions, encountered very often in applications, which take place in the gas phase, between a fuel and an oxidizer, (typically the oxygen of the air). The combustion reactions are exothermic, so that we shall see large temperature differences in the flow field, and the reaction rates are very sensitively dependent on temperature. Then in regions of the flow field where the temperature is low the reactants will mix without chemical reactions, while in regions of high temperature the flow will be in chemical equilibrium. When the reaction can be modelled by an overall irreversible reaction between the fuel and the oxygen, the chemical equilibrium condition implies that one of the reactants has locally zero concentration, because it has been previously completely consumed by the chemical reaction. The chemical reaction has taken place in thin reaction layers in the form of rich or lean premixed flames, when they separate regions of frozen flow from regions of equilibrium flow, or diffusion flames, when they separate a region without fuel from a region without oxidizer.

Diffusion flames are formed in large variety of natural and industrial systems. We find diffusion flames in fires, in industrial furnaces, in Jet and Diesel engine combustion chambers, and also in the combustion chamber of Rocket engines.

Very often the fuel is initially in solid or liquid form, so that, for example, the solid must gasify before mixing and chemical reaction with the oxygen may take place in the gas phase. The liquid fuels are injected in the combustion chamber in the form of jets that break, or atomize, to form a spray of small droplets that vaporize to generate the fuel vapors that mix and react with the ambient air.

In the following, we are forced to restrict our discussion to some theoretical aspects of the analysis of combustion in unpremixed systems; so that we will limit our considerations to gaseous systems undergoing chemical reactions that can be modelled by an overall irreversible Arrhemius reaction.

For a more complete description of the combustion processes, the reader may consult the other Lecture Notes in this Course, and, for example the books by Williams (1985) and Zeldovich et al. (1984), the chapter on Combustion in the book of Frank-Kamenetskii (1969) and also the book by Buckmaster and Ludford (1981) on the asymptotic analysis of Combustion.

In the following we shall begin in Section II with the formulation of the conservations equations, and a consideration of the characteristic scales and parameters introduced by the initial and boundary conditions.

Section III will be devoted to a description of the limiting case of very fast reactions, when they take place in thin layers, or reaction sheets at a diffusion controlled rate.

In Section IV we will introduce the large activation energy analysis of diffusion flames, where we shall encounter the triple flame structure mentioned earlier in the Introduction. Section V will be devoted to the analysis of diffusion flames in strained mixing layers where we encounter the concepts of ignition and extinction of diffusion flames.

2. FORMULATION

We shall give below the conservation equations that describe the process of gas phase mixing and chemical reaction between a fuel and an oxidizer, for example the oxygen of the air. The reaction will be modelled by an overall reaction of the form

$$F + \nu_0 0_2 \rightarrow (1 + \nu_0)P + (q) \tag{1}$$

where a mass ν_0 of Oxygen is consumed per unit mass of fuel to yield a mass $(1 + \nu_0)$ of products and a thermal energy q.

Let $\vec{v}$, ρ, p, T be the velocity, density, pressure and temperature of the fluid; and let Y_F, Y_0 and Y_p be the mass fractions of the fuel, oxygen and products (i.e. the ratio of their densities to the fluid density). These variables are functions of the time t and spatial coordinate $\vec{x}$, given by the following system of conservation equations, see Williams (1985),

$$\frac{\partial \rho}{\partial t} + \nabla \cdot (\rho \vec{v}) = 0 \tag{2}$$

$$\frac{\partial}{\partial t}(p\vec{v}) + \nabla \cdot (\rho \vec{v}\vec{v}) = \rho \vec{g} - \nabla p + \nabla \cdot \tau' \tag{3}$$

$$\frac{\partial}{\partial t}(\rho Y_\alpha) + \nabla \cdot (\rho \vec{v} Y_\alpha) = \nabla \cdot (\rho D_\alpha \nabla Y_\alpha) + w_\alpha \tag{4}$$

$$\frac{\partial}{\partial t}(\rho c_p T) + \nabla \cdot (\rho \vec{v} c_p T) = \nabla \cdot (k \nabla T) - q w_F \tag{5}$$

together with the equation of state

$$p = \rho T R / M \tag{6}$$

where R is the universal gas constant and M is the mean molecular mass that we shall, for simplicity consider constant. The specific heat at constant pressure c_p, appearing in Eq. (5) will also be considered to be constant.

The energy conservation Eq. (5) has been written in the simplified form appropriate for the analysis of combustion in flows with low Mach numbers in an open atmosphere. In these flows the spatial variations in pressure, determined by the momentum Eq. (3) which includes the gravity force $\vec{g}$, are small compared with the pressure itself, and these may be neglected in Eq. (6), where p may be approximated by a constant typical value p_a of the pressure in the flow field. In Eqs. (3)-(5) the viscous stress tensor τ' and the diffusion and heat conduction fluxes are given by the well known laws of Navier-Poisson, Fick and

Fourier in terms of the gradients of velocity, mass fractions and temperature. D_α is the diffusivity of the species α, and D_T is the thermal diffusivity. The ratio $D_T/D_\alpha = L_\alpha$, the Lewis number of the species α, will be assumed to be constant, while ρD_T will be considered a known function of T. Eq. (4) is applicable to any species in the fluid mixture. The reaction term $W_\alpha = 0$ for any inert species (like N_2), and

$$W_0/\nu_0 = -W_p/(1+\nu_0) = W_F \tag{7}$$

where we shall use for W_F the Arrhenius overall expression

$$W_F = -\rho B e^{-E/RT} Y_0 Y_F \tag{8}$$

where B is a frequency factor, and E the activation energy of the reaction.

With this form, Eq. (8), of the rate expression, and the assumption that the mean molecular mass M is constant, we only need two species conservation equations (4), for the fuel and oxygen mass fractions Y_F and Y_0 for the description of the flow field.

When solving the system of conservation equations (2)-(8) we must add as initial conditions the values of the mass fractions of the species, together with the initial temperature and velocity distributions. We must also give the temperature and velocity of the fluid at the boundary, together with the condition that the derivatives of the mass fractions of the species normal to the boundary is zero at a non-reacting boundary.

We shall not attempt to write these boundary conditions in more detailed form, but we must indicate that they provide the characteristic times, velocities and lengths of the flow.

As an example, we consider a jet of fuel coming out into ambient air, through an injector of radius a. We must specify the temperature T_0, fuel mass fraction Y_{FO} and velocity distribution upstream in the injector; this velocity distribution is characterized by a mean velocity U. The air may be stagnant far from the injector, with a temperature T_a (that for simplicity in the presentation we shall consider equal to T_0) and an oxygen mass fraction Y_{00}. The product and oxygen concentration upstream in the fuel feed stream will be assumed to be zero, while the product and fuel concentration in the ambient air will also be zero. The derivatives of the mass fractions normal to the injector wall are zero, while the temperature of the fluid at the interface with the injector must be equal to that of the injector; in the extreme case where the injector walls are highly conducting this temperature will be T_0; while in the opposite case we will consider the wall to be adiabatic and the normal derivative of T at the wall surface will be zero. If the injector velocity, for example, changes with time, a characteristic time is introduced by the time t_0 required to change the velocity by a factor of order 1.

The boundary conditions thus introduce characteristic values U, a and t_0 for the velocity, length and time. These can be used for a naive estimate of the order of magnitude of the different terms in the conservation equations. Thus, we give below, the order of the terms in the conservation equation for the fuel mass fraction

$$\frac{\partial Y_F}{\partial t} + \overline{v} \cdot \nabla Y_F - \rho^{-1} \nabla \cdot (\rho D_F \nabla Y_F) = -Be^{-E/RT} Y_0 Y_F \tag{9}$$

$$t_0^{-1} \qquad U/a \qquad D_F/a^2 \qquad t_{ch}^{-1}$$

if we consider Y_{FO} and Y_{00} to be of order unity, and consider that, at a typical point of the flow field at a distance of order a from the injector, Y_F and Y_0 are of order unity and undergo changes of order unity in times of order t_0, and when, at a given time, we move a distance of order a. The characteristic chemical time t_{ch} is of the order of the inverse of the reaction constant $B \exp(-E/RT)$.

When the times t_0, a^2/D_F and t_{ch} that characterize the terms representing the local accumulation, the diffusion and the chemical reaction are compared with the residence time a/U that measures the convective term we obtain the parameters

$St = a/Ut_0 =$ Strouhal number;

$Pe = aU/D_F =$ Peclet number;

$Da = a/Ut_{ch} =$ Damköhler number.

These parameters measure the relative importance of the different terms in the conservation equation for the fuel. In gases the Lewis numbers and the Prandtl number, ratio of the kinematic viscosity ν to the thermal diffusivity, are of order unity, so that the Peclet and Reynolds numbers are of the same order. Thus the same parameters measure, also, the relative importance of the different terms in the other conservation equations.

If the Strouhal number is small compared with unity we introduce the quasi-steady approximation: neglecting the unsteady, accumulation terms, in the equations.

If the Reynolds or Peclet number were small compared with unity, we could neglect the convective term relative to the diffusive term; this is the case when analysing the vaporization and combustion of "small" fuel droplets moving at "small" relative velocities in air.

In many applications in Combustion the characteristic values of the Reynolds numbers are large compared with unity. In this case the diffusion term can be neglected in most of the flow field; then mixing of the reactants is confined to thin mixing layers, of thickness δ_m such that

$$\delta_m^2/D_F = a/U \; . \tag{10}$$

The characteristic diffusion time across the layers is of the order of the residence time. The chemical reaction is obviously confined to these layers.

The mixing layers at high Reynolds numbers are also, typically, thin vorticity layers, introduced in the analysis of fluid flows by Helmholtz in 1868 and shown by Kelvin in 1871 to be unstable to small disturbances; see Drazin and Reid (1981). Due to the Helmholtz-Kelvin instabilities, the flow in the mixing layers becomes turbulent: the vorticity first concentrates in discrete vortices that grow in size by pairing (in a non periodic way), and are also subject to three-dimensional instabilities that produce vortices of smaller and smaller scale.

The mixing layers are strongly corrugated by the unsteady, three-dimensional, turbulent flow, so that mixing is very much enhanced by the turbulence. For this reason, in order to reduce the size of the combustion chambers, or increase their output, these are designed for the flow to be turbulent. For the description of the fluctuating temperature, velocity and concentration fields we can use the conservation equations written above, with the terms representing the local accumulation, convection and diffusion of the same order in the regions where molecular mixing and reaction take place; the characteristic times associated with accumulation, convective transport and diffusion are of the same order t_λ for the eddies of size λ, of typical velocity V_λ given according to the Kolmogorov by the relations

$$V_\lambda^3/\lambda = V^3/l \;, \; t_\lambda = \lambda/V_\lambda \tag{11}$$

in terms of the eddy size λ and characteristic velocity V and size l of the large eddies. These relations can be used for eddy sizes smaller than l and larger than the Kolmogorov eddy size l_k, given by (11) and the condition $V_k l_k/\nu = 1$. For the flow with velocity U coming out of an injector of size a, $l \sim a$ and $V \sim U$.

The importance of the reaction term relative to the convective transport term (the most of important of the transport terms in high Reynolds number flows) is measured by the Damköhler number:

$$D_a = aU^{-1}Be^{-E/RT} \tag{12}$$

whose value may change across the flow field due to the changes in the Arrhenius exponential with temperature. In a turbulent flow the residence time a/U, should be replaced by $(\lambda/a)^{2/3}a/U$, when measuring the importance of the reaction term in the thin mixing layers surrounding an eddy of size λ.

When the Damköhler number is everywhere in the flow field $Da \ll 1$, the reactants can not coexist in the flow field. If the order of magnitude of the terms in the left hand side of Eq. (9) is U/a and $Be^{-E/RT}$ is $O(t_{ch}^{-1})$, then we may conclude that

$$Y_F Y_0 = 0(t_{ch}U/a) = 0(D_a^{-1}) \;.$$

This implies, in the limit $D_a \to \infty$, that Eq. (9) simplifies to

$$Y_F = 0 \quad \text{or} \quad Y_0 = 0 \;. \tag{13}$$

The reactants can not coexist; the flowfield is divided in two domains Ω_F and Ω_0 separated by a reaction sheet Σ_f. In the fuel domain Ω_F, $Y_0 = 0$ the oxygen concentration is zero. In the oxygen domain Ω_0, $Y_F = 0$. Both reactants have zero concentration at the surface Σ_f, where the chemical reaction takes place.

3. THE BURKE-SCHUMANN ANALYSIS OF DIFFUSION FLAMES

In 1928 Burke and Schumann showed how the structure of diffusion flames could be described in the limiting case of very fast reactions. The Burke-Schumann (B-S) description corresponds to the limit $D_a \to \infty$, when the reactants can not coexist.

The temperature and concentration profiles are shown schematically in Fig. 1 for a given cross-section AA′. The profiles show the temperature and concentration distributions as continuous but with discontinuous derivatives at the flame sheet Σ_f. The discontinuities are associated with a Dirac delta form of the reaction terms in the conservation equations

$$\frac{\partial Y_F}{\partial t} + \vec{v} \cdot \nabla Y_F - (\rho L_F)^{-1} \nabla \cdot (\rho (D_T \nabla Y_F) = W_F/\rho \tag{14}$$

$$\frac{\partial Y_0}{\partial t} + \vec{v} \cdot \nabla Y_0 - (\rho L_0)^{-1} \nabla \cdot (\rho D_T \nabla Y_0) = \nu_0 W_F/\rho \tag{15}$$

$$\frac{\partial T}{\partial t} + \vec{v} \cdot \nabla T - \rho^{-1} \nabla \cdot (\rho D_T \nabla T) = -(q/c_p) W_F/\rho \tag{16}$$

for the concentrations and temperature.

The Dirac delta character of the non-dimensional reaction term $-\frac{a}{U} W_F/\rho = (a/U) B e^{-E/RT} Y_F Y_0 = D_a Y_F Y_0$ is associated with the limit $Da \to \infty$, which leads to a zero value of W_F outside the flame sheet and an infinite value of W_F at the flame sheet.

From the asymptotic structure of the solution, represented in Fig. 1, we can anticipate that in the thin reaction zone the large reaction terms are balanced by the diffusion terms normal to the flame sheet. That is

$$-\frac{1}{\rho}\frac{\partial}{\partial n}\{\rho D_T \frac{\partial}{\partial n}\left(\frac{Y_E}{L_F}\right)\} = -B e^{-E/RT} Y_F Y_0 \tag{17}$$

$$-\frac{1}{\rho}\frac{\partial}{\partial n}\{\rho D_T \frac{\partial}{\partial n}\left(\frac{Y_0}{L_0}\right)\} = -\nu_0 B e^{-E/RT} Y_F Y_0 \tag{18}$$

$$-\frac{1}{\rho}\frac{\partial}{\partial n}\{\rho D_T \frac{\partial}{\partial n}(T)\} = \frac{q}{c_p} B e^{-E/RT} Y_F Y_0 \tag{19}$$

where n is the outer normal distance to the flame sheet. The reaction terms can be eliminated from these equations, and then integrated across the flame sheet to yield

$$-\rho D_T \frac{\partial}{\partial n}\left(\frac{Y_F}{L_F} - \frac{Y_0}{\nu_0 L_0}\right) = m''_F = m''_0/\nu_0 \tag{20}$$

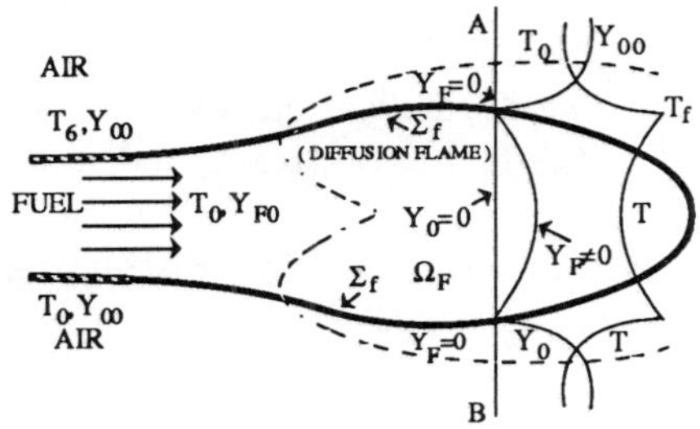

Figure 1. The temperature and concentration profiles of diffusion flames in the Burke-Schumann limit.

$$\left(\rho D_T \frac{\partial T}{\partial n}\right)_{-\infty} - \left(\rho D_T \frac{\partial T}{\partial n}\right)_{+\infty} = q m_F''/c_p \tag{21}$$

when we take into account that $Y_F = 0$ at $n \to \infty$ and $Y_0 = 0$ at $n \to -\infty$.

In these relations m_F'', m_0'' and qm_F'' are, respectively, the rate of mass consumption of fuel and oxygen, and the rate of heat release per unit flame surface and time. Thus, the reactants reach the flame sheet, from opposite sides, by diffusion at stoichiometric proportions; the heat released by the reaction, qm_F'' per unit flame surface and time, is conducted away from the flame sheet.

Thus for the Burke-Schumann description of the diffusion flame, we should solve Eqs. (14) and (16), with $W_F = 0$ and $Y_0 = 0$ in Ω_F (the fuel side of the flow field), and Eqs. (15) and (16), with $W_F = 0$ and $Y_F = 0$ in Ω_0 (the oxygen side of the flow field). The flame sheet Σ_f must be determined as a free boundary problem by requiring that Y_F, Y_0 and T are continuous at the flame sheet and satisfy there the jump conditions given by Eqs. (20) and (21). Burke and Schumann (1928) indicated how this problem could be solved when the Lewis numbers of the fuel and oxygen are equal to unity (the equidiffusional approximation): in this case the linear combinations $(Y_F - Y_0/\nu_0)$ and $Y_F + c_pT/q$ diffuse like passive scalars. These so called, Schvab-Zeldovich, coupling functions have been widely used with the equidiffusional approximation. See Williams (1985).

We generalize here this approach to the general case of non-unity Lewis numbers. See Liñán (1989) for the details of the analysis.

Let us define first the generalized mixture fraction variable

$$Z = (S\hat{Y}_F + 1 - \hat{Y}_0)/(S+1) \tag{22}$$

where

$$\hat{Y}_F = Y_F/Y_{FO}\ ,\ \hat{Y}_0 = Y_0/Y_{00} \qquad \text{and} \qquad S = \nu_0 Y_{FO} L_0 / Y_{00} L_F \tag{23}$$

is a generalized air fuel stoichiometric ratio. The mixture fraction variable Z is a normalized coupling function, $Y_F/L_F + (Y_{O0} - Y_O)/\nu_O L_O$, which according to Eq. (20) diffuses without reaction across the flame sheet. The normalization, by division by its value on the fuel feed stream, insures that $Z = 1$ in the fuel feed stream, and $Z = 0$ in the ambient air.

A conservation equation for Z can be obtained by substracting from Eq. (14), Eq. (15) divided by ν_O. We thus obtain the equation

$$L_e(Z)\rho\left(\frac{\partial Z}{\partial t} + \vec{v}\cdot\nabla Z\right) - \nabla\cdot(\rho D_T \nabla Z) = 0 \tag{24}$$

if we take into account that, according to the (B-S) equilibrium conditions, Eq. (13),

$$Y_O = 0 \quad \text{for} \quad Z > Z_s \quad \text{and} \quad Y_F = 0 \quad \text{for} \quad Z < Z_s \tag{25}$$

with, Z_s, the stoichiometric value of the mixture fraction, defined by

$$Z_s = 1/(s+1) = (1 + \nu_O Y_{FO} L_O / Y_{O0} L_F)^{-1} \tag{26}$$

and $L_e(Z)$, an effective Lewis number, defined by

$$L_e = L_F \quad \text{for} \quad Z > Z_s \quad \text{and} \quad L_e = L_O \quad \text{for} \quad Z < Z_s\,. \tag{27}$$

Eq. (24) is a non-linear conservation equation for the modified mixture fraction variable Z, which, together with Eq. (25) may used to replace Eqs. (14) and (15). In the particular case when the Lewis numbers are equal to unity, Z reduces to the widely used form of the mixture fraction variable Z, described by the linear conservation equation (24) with $L_F = L_O = 1$.

The flame surface is given by the level surface

$$Z(\overline{x}, t) = Z_s \tag{28}$$

of the mixture fraction variable.

A similar conservation equation free from the reaction term can be obtained from Eqs. (14)-(16), to replace Eq. (16) when analysing the structure of diffusion flames in the (B-S) limit of infinite Damköhler numbers. Thus if we add Eq. (14) to Eq. (15) multiplied by S/ν_O to Eq. (16) multiplied by $(1+S)c_p/q$, we obtain, if Eq. (25) is taken into account, the following conservation equation

$$N(Z)\rho\left(\frac{\partial Z}{\partial t} + \vec{v}\cdot\nabla Z\right) + \rho\left(\frac{\partial H}{\partial t} + \vec{v}\cdot\nabla H\right) - \nabla\cdot(\rho D_T \nabla H) \tag{29}$$

for the excess enthalpy H, defined by

$$H = \hat{Y}_F + \hat{Y}_O - 1 + (1+S)L_F c_p (T - T_O)/q Y_{FO} \tag{30}$$

where

$$N(Z) = (L_F - 1)(S+1)/S \quad \text{if} \quad Z > Z_s \tag{31a}$$

$$N(Z) = (1 - L_0)S + 1 \quad \text{if} \quad Z < Z_s \, . \tag{31b}$$

The excess enthalpy function H is equal to zero in the fuel feed stream, and also in the ambient unperturbed air side of the diffusion flame.

The temperature and concentrations will then be given in terms of Z and H by the relations.

$$H + (1 + 1/S)(1 - Z) = (1 + S)L_F c_p(T - T_0)/qY_{FO} \tag{32a}$$

when

$$\hat{Y}_0 = 0 \, , \; S\hat{Y}_F = (S+1)Z - 1 \geq 0$$

and

$$H + (1 + S)Z = (1 + S)L_F c_p(T - T_0)/qY_{FO} \tag{32b}$$

when

$$\hat{Y}_F = 0 \, , \quad \hat{Y}_0 = 1 - (1 + S)Z \geq 0 \, .$$

For unity Lewis numbers, $N = 0$. If, in addition, the walls of the injector are adiabatic the solution of Eq. (29) gives $H \equiv 0$, and the flame sheet temperature T_f takes the adiabatic flame temperature value T_e,

$$T_e = T_0 + qY_{FO}/c_p(1 + S) \, . \tag{33}$$

For other values of the Lewis numbers, the flame sheet temperature takes the value

$$T_f = T_0 + (H_f + 1)qY_{FO}/c_p L_F(1 + S) \tag{34}$$

where H_f, the value of H at $Z = Z_s$, is no-zero due to the so called preferential diffusion effects.

In summary, when describing the diffusion flame structure in the (B-S) limit of infinite Damköhler numbers, the conservation equations (14)-(16), which include, in their right hand sides, Dirac delta functions of initially unknown strength placed at an unknown flame surface, may be replaced by the equilibrium conditions (32) and the non-linear transport equations (24) and (29) for Z and H. These equations can be written in conservation form

$$\frac{\partial}{\partial t}(\rho G) + \nabla \cdot (\rho \vec{v} G) - \nabla \cdot (\rho D_T \nabla Z) = 0 \tag{35}$$

$$\frac{\partial}{\partial t}\{\rho(K + H)\} + \nabla \cdot \{\rho \vec{v}(K + H)\} - \nabla \cdot (\rho D_T \nabla H) = 0 \tag{36}$$

in terms of the functions $G(Z)$ and $K(Z)$, defined by

$$G = (Z - Z_s)L_e(Z) \quad \text{and} \quad K = (Z - Z_s)N(Z) \, . \tag{37}$$

This description of the diffusion controlled combustion process will be particularly useful in the modelling of turbulent combustion flows, or in their direct numerical simulation.

Finally, we shall indicate that if we are interested in the description of the reaction zone structure for large but finite values of the Damköhler number, we should notice that the reaction is confined to a thin layer placed along the flame sheet surface, given by $Z = Z_s$ with the (B-S) description. Outside the reaction layer the temperature and concentrations are well described by the (B-S) solution.

In the reaction layer, as first shown by Liñán (1961) and later by Fendell (1965), the conservation equations reduce to Eqs. (17) and (19), representing a reactive-diffusive balance. These equations can be additionally simplified because the differences of T from T_f are so small at large Damköhler numbers that the Arrhenius factor can be replaced by $\exp(-E/RT_f)$, and in addition, ρ and D_T by their (B-S) flame sheet values ρ_f and D_{Tf}. Then the conservation equations take the form

$$\frac{\partial^2 Y_F}{\partial n^2} = L_F D_{Tf}^{-1} B e^{-E/RT_f} Y_F Y_O \tag{38}$$

$$\frac{\partial}{\partial n}(Y_O/\nu_O L_O - Y_F/L_F) = m_F''/\rho_f D_{Tf} \tag{39}$$

$$\frac{\partial^2 H}{\partial n^2} = 0 \tag{40}$$

where Eq. (39) and (40), with H defined by Eq. (30), are combinations of Eqs. (17) and (19). Eqs. (38)-(40) must be solved with the boundary conditions obtained by matching with the outer $(B-S)$ solutions.

Then

$$H = H_f + nH_{nf} \tag{41}$$

where $H_{nf} = \partial H/\partial n$, evaluated at the flame sheet $Z = Z_s$ using the $(B-S)$ solution.

Similarly, Eq. (39) can be integrated to yield

$$Y_O/\nu_O L_O - Y_F/L_F = nm_F''/\rho_f D_{Tf} \tag{42}$$

where

$$L_F m_F''/\rho_f D_{Tf} = -Y_{FO}(1 + 1/S)Z_{nf} \tag{43}$$

and

$$Z_{nf} = \partial Z/\partial n \quad \text{at} \quad Z = Z_s \, .$$

Eq. (38) can be used with Eq. (42), and the boundary conditions

$$Y_F = 0 \quad \text{at} \quad n \to \infty \, , \; Y_O = 0 \quad \text{at} \quad n \to -\infty \tag{44}$$

to calculate $Y_F(n)$ and $Y_O(n)$; the temperature will then be given by Eq. (41).

The equation for Y_F takes the normalized form

$$d^2y/d\eta^2 = y(y+\eta) \tag{45}$$

and the boundary conditions

$$y = 0 \quad \text{at} \quad \eta \to \infty \; , \; y + \eta = 0 \quad \text{at} \quad \eta \to -\infty \tag{46}$$

with the appropriate scaling of the variables Y_F and n. The resulting value of Y_F in the reaction zone is of order

$$Y_{Fr} \sim D_a^{-1/3} = \{Bd^2 D_{T_f}^{-1} \exp(-E/RT_f)\}^{-1/3} \tag{47}$$

where $1/d$ is the $(B-S)$ value, at the flame sheet, of the gradient of the fuel mass fraction, given by the right hand side of Eq. (43).

The corresponding changes, from T_f, of the temperature in the reaction zone are of order $(T_e - T_0)D_a^{-1/3}$. When these changes become, for more moderate values of the Damköhler number, of order RT_f^2/E, the Arrhenius factor can not be approximated by a constant factor, and we encounter the extinction regime described by Liñán (1974).

Let us analyse, as an example, the combustion of a spherical fuel pocket in air. For simplicity in the presentation, we shall carry out the analysis using the thermal-diffusive approximation; in this approximation the velocities induced by the thermal expansion are neglected, or equivalently the density is considered constant. In addition, we shall consider the case in which the Lewis numbers are equal to 1.

For the initial conditions, at $t = 0$, we take

$$T = T_0 \; , \; Y_F = Y_{FO} \; , \; Y_0 = 0 \quad \text{at} \quad r < a \tag{48a}$$

$$T = T_0 \; , \; Y_F = 0 \; , \; Y_0 = Y_{00} \quad \text{at} \quad r > a \; . \tag{48b}$$

Then in the $(B-S)$ limit of infinite Damköhler numbers, the temperature and concentration distributions are given by Eqs. (32) with $H \equiv 0$ and Z given by the equation

$$\frac{\partial Z}{\partial \tau} - \frac{1}{x^2}\frac{\partial}{\partial x}\left(x^2 \frac{\partial Z}{\partial x}\right) = 0 \tag{49}$$

where $x = r/a$ and $\tau = tD_F/a^2$.

This equation has to be solved, for $\tau = 0$, with the initial conditions

$$Z = 1 \quad \text{for} \quad x < 1 \; , \qquad Z = 0 \quad \text{for} \quad x > 1 \; . \tag{50}$$

The solution is

$$Z = \frac{1}{2}erfc\left(\frac{1+x}{2\sqrt{\tau}}\right) + \frac{1}{2}erfc\left(\frac{1-x}{2\sqrt{\tau}}\right) - \frac{\sqrt{\tau}}{x\sqrt{\pi}}\left(e^{-\frac{(1-x)^2}{4\tau}} - e^{-\frac{(1+x)^2}{4\tau}}\right) \; . \tag{51}$$

The flame sheet is given by $Z(x,\tau) = Z_s$. It disappears at $\tau = \tau_f$ given by

$$Z_s = erfc(1/2\sqrt{\tau_f}) - (\pi\tau_f)^{-1/2}\exp(-1/4\tau_f) \ . \tag{52}$$

4. LARGE ACTIVATION ENERGY ASYMPTOTICS

We shall indicate in this Section how the large activation energy, $E/RT_0 \gg 1$, of the exothermic, $q/c_pT_0 \sim 1$, combustion reactions determines the structure of diffusion flames.

Due to the exothermicity of the combustion reaction we will find temperatures in the flow field ranging from T_0 to T_e. (In this discussion we consider for simplicity the case $L_F = L_0 = 1$). Due to the large activation energies, the characteristic reaction times t_{cho} and t_{che} at the initial and flame temperature satisfy the inequality

$$t_{cho} = B^{-1}e^{E/RT_0} \gg t_{che} = B^{-1}e^{E/RT_e} \ . \tag{53}$$

Then, the Damköhler number

$$D_a = au^{-1}Be^{-E/RT} \tag{54}$$

changes very much across the flowfield.

Let us define a cross-over temperature T_c given by

$$au^{-1}Be^{-E/RT_c} = 1 \ . \tag{55}$$

Then if $T_c < T_0$, the Damköhler number will be $D_a \gg 1$, everywhere in the flow field, and the combustion process will be diffusion-controlled well described by the Burke-Shumann limit. When $T_c > T_e$, the chemical reaction will be frozen everywhere.

In the important case $T_0 < T_c < T_e$ we will find equilibrium flow with either $Y_F = 0$ or $Y_0 = 0$ in regions of temperature higher than T_c, while the reactions will be frozen in regions where $T < T_c$.

The regions of equilibrium flow are separated from the regions of frozen mixing by thin reaction sheets of the premixed flame type, where one of the reactants is completely consumed, with only partial consumption of the other. The fuel free region is separated from the oxygen free region by a diffusion flame sheet. See Fig. 2.

In order to understand why this is so, let us write the conservation equation (9) for the fuel in the form

$$-(a/u)\left\{\frac{\partial Y_F}{\partial t} + \overline{v}\cdot\nabla Y_F - \rho^{-1}\nabla\cdot(\rho D_F \nabla Y_F)\right\} = e^{\frac{E}{RT_c}\frac{T-T_c}{T}}Y_FY_0 \ . \tag{56}$$

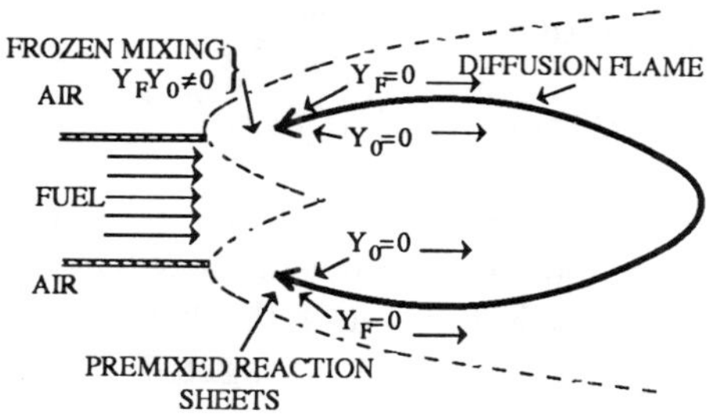

Figure 2. The triple flame structure of diffusion flames.

While we may expect each of the transport terms in the left hand side of the equation to be 0(1), the reaction term in the right hand side will be infinitely small for $T < T_c$ if $E/RT_c \to \infty$, leading to frozen flow. For $T > T_c$, the Arrhenius factor $\exp\{(E/RT_c)(T-T_c)/T\}$ will become infinitely large when $E/RT_c \to \infty$, so that the factor $Y_F Y_0$ must be infinitely small because the two members in Eq. (56) must balance.

The reaction term in the limit, $E/RT_c \to \infty$, of infinite activation energies becomes a Dirac delta function placed at a reaction sheet, where $T = T_c$; the reaction is balanced at the premixed flame sheet by the term representing diffusion normal to the sheet. At the flame sheet the temperature and concentrations are continuous, with either $Y_F = 0$ or $Y_0 = 0$ at the flame sheet and also in the equilibrium side. There are, however, jumps of the temperature and concentration gradients related by the stoichiometric coefficients of the reaction (1).

The reaction term is also of the Dirac delta type in the diffusion flame sheet that separates the equilibrium regions. Along the diffusion flame sheet the temperature rises from T_c, the value at its leading edge where it meets the premixed flames, to T_e well downstream, where the cooling effect of the injector walls is no longer felt, and thereby $H_f = 0$. If the injector walls are good conductors of heat their temperature will be close to T_0; if heat from the triple flame system reaches the injector walls by heat conduction H will be negative in the triple flame region, so that the flame sheet temperature may be $T_c < T_e$ at its leading edge. See Liñán (1988).

Upstream heat conduction is essential for the diffusion flame to be anchored close to the injector lips, with a leading edge temperature $T_c < T_e$. The triple flame region, in flows with large Reynolds numbers, lies in the Navier-Stokes region at the core of the triple deck region formed at the lip of the injector; see Daniels (1977). It turns out that the characteristic residence time in this flame attachement region, to be used in the evaluation means of Eq. (55), of the flame temperature T_c, is precisely a/U, if the fuel flow velocity profile at the injector exit is parabolic. Otherwise, the injector radius a should be replaced in Eq. (55) by the fuel flow boundary layer thickness at the injector exit.

Notice that, according to Eq. (55), T_c will grow with the flow velocity U. When this velocity grows to a value U_L such that

$$aU^{-1}\exp(-E/RT) = 1\ , \tag{57}$$

the triple flame will lift-off from the injector lip. It will move downstream to a lift-off height where, in a rough way of speaking, its upstream velocity equals the local flow velocity. The determination of the lift-off height is difficult if the flow is turbulent, when the triple flame will move in the strongly strained and corrugated "laminar" mixing layers. See Peters and Williams (1983) for an attempt to evaluate the lift-off height using percolation ideas. The flame spread is not possible in regions where a large enough fraction of the corrugated laminar mixing layers can not sustain a diffusion flame, because they are subject to excessive strain.

5. DIFFUSION FLAMES IN STRAINED MIXING LAYERS

In turbulent flows the thin mixing layers between the fuel and air are distorted and strained by the flow field. A positive strain leads to a growth of the area covered by the fluid particles in the mixing layer and enhances the mixing process. The strain rate A is the inverse of a time that replaces the residence time a/U in the definition of the Damköhler number. For the diffusion flame to exist the strain rate can not exceed a critical extinction value A_E.

We shall give below a short description of the ignition and extinction processes in strained diffusions flames. A detailed account of ignition and extinction in diffusion flames may be found in Williams (1985); for applications of these ideas to turbulent diffusion flames, see Peters (1986).

For simplicity in the presentation, we shall use the thermal diffusive approximation in the description of the velocity field. We shall describe a planar mixing layer between two counter-flowing, two-dimensional, streams of fuel and oxidizer. The velocity field that we consider is the potential field, given by

$$u = 0\ , \quad v = -Ay\ , \quad w = Az \tag{58}$$

with a constant strain rate A.

The conservation equations for the temperature and mass fractions of the fuel and oxygen have similarity solutions with T, Y_F, Y_0 functions of the time t, the coordinate y, transverse to the mixing layer and the coordinate x in the direction perpendicular to the strain. If, for simplicity, we limit our discussion to the case of unit Lewis numbers with a constant thermal diffusivity, the conservation equations take the form

$$L(\alpha) = W_F/\rho \tag{59}$$

with

$$L(\alpha) = \frac{\partial \alpha}{\partial t} - Ay\frac{\partial \alpha}{\partial y} - D_T\frac{\partial^2 \alpha}{\partial y^2} - D_T\frac{\partial^2 \alpha}{\partial x^2} \tag{60}$$

$\alpha = \{Y_F, Y_0/\nu_0,\ c_pT/q\}$ and

$$W_F/\rho = -Be^{-E/RT}Y_0Y_F \tag{61}$$

to be solved with the boundary conditions

$$y \to \infty : T - T_0 = Y_0 - Y_{00} = Y_F = 0 \tag{62a}$$

$$y \to -\infty : T - T_0 = Y_0 = Y_F - Y_{FO} = 0 \tag{62b}$$

and initial conditions

$$t = 0 : T = T_I(y,x)\ ,\ Y_F = Y_{FI}(y,x)\ ,\ Y_0 = Y_{OI}(y,x)\ . \tag{63}$$

It is easy to show that, independently of the initial conditions the mixture fraction variable Z and the excess temperature H, defined by Eqs. (22) and (30), which satisfy the boundary conditions $Z = H = 0$ at $y \to \infty$ and $Z - 1 = H = 0$ at $y \to -\infty$, are given for large times by

$$H = 0\ ,\ Z = \frac{1}{2}erfc(y/\sqrt{2D_T/A})\ . \tag{64}$$

We shall, for simplicity, consider that the initial concentrations and temperature are such that (64) also holds at $t = 0$.

The system of equations (61)-(64) can thus be reduced to the solution of the following equation

$$\frac{\partial \theta}{\partial t} - y\frac{\partial \theta}{\partial y} - \frac{\partial^2 \theta}{\partial y^2} - \frac{\partial^2 \theta}{\partial x^2} = D_0\hat{Y}_F\hat{Y}_0e^{\beta_0\theta/(1+\alpha_0\theta)} \tag{65}$$

for $\theta = c_p(T - T_0)/qY_{FO}$ in terms of the coordinates x, y and time t, scaled here with the thickness $\delta_m = \sqrt{D_T/A}$ of the mixing layer and the time A^{-1}.

Here the D_0 is a Damköhler number

$$D_0 = A^{-1}BY_{00}\exp(-E/RT_0) \tag{66}$$

and

$$\alpha_0 = qY_{FO}/c_pT_0\ ,\ \beta_0 = (E/RT_0)(qY_{FO}/c_pT_0)\ . \tag{67}$$

Eq. (66) has to be solved with the boundary conditions

$$\theta = 0 \quad \text{for} \quad y \to \pm\infty\ . \tag{68}$$

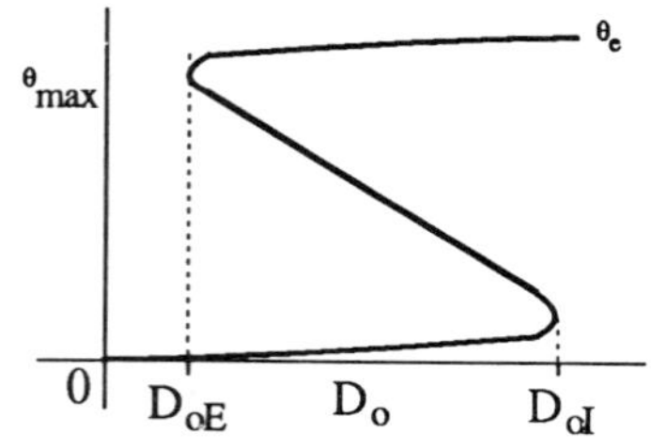

Figure 3. The maximum temperature in the mixing layer as a function of the Damköhler number.

The reduced mass fractions $\hat{Y}_F$ and $\hat{Y}_0$ are given by the relations

$$\hat{Y}_F + \theta = 1 - \hat{Y}_0 - S\theta = Z = \frac{1}{2} erfc(y/\sqrt{2}) . \tag{69}$$

We should add an initial condition

$$t = 0 : \quad \theta = \theta_I(y, x) . \tag{70}$$

5.1 One dimensional steady solutions

There are one dimensional steady solutions of this problem, in the form $\theta = \theta(y)$ given by the equation

$$y\frac{\partial \theta}{\partial y} + \frac{\partial^2 \theta}{\partial y^2} = -D_a \hat{Y}_F \hat{Y}_0 e^{\beta_0 \theta/(1+\alpha_0 \theta)} \tag{71}$$

with $\theta = 0$ at $y \to \pm\infty$.

This problem has multiple solutions for values of the Zeldovich number β_0, larger than a critical value $\beta_c(\alpha_0, S)$. These are sketched in Fig. 3.

Liñán (1974) has carried our an asymptotic description of the solution of Eq. (71), for large values of the Zeldovich number, to calculate the ignition and extinction conditions, and the interval (D_{OE}, D_{OI}) of multiplicity of solutions.

We simply indicate here that for large Damköhler numbers, the maximum temperature in the nearly frozen solution for $D_0 < D_{OI}$ differs from O only by an amount of order $1/\beta_0$

at most. In the upper, diffusion controlled solution, for $D_0 > D_{OE}$, θ_{max} is close to the adiabatic flame temperature θ_e corresponding to the $(B-S)$ solution of infinite Damköhler numbers. In this $(B-S)$ limit:

$$\hat{Y}_0 = 0 \quad \text{for} \quad Z > Z_s \ , \ \hat{Y}_F = 0 \quad \text{for} \quad Z < Z_s \tag{72}$$

and the flame sheet is located at $y = y_f$, given by

$$Z_s = \frac{1}{2} erfc(y_f/\sqrt{2}) \ , \tag{73}$$

and there the flame temperature takes the value $\theta_e = Z_s = 1/(S+1)$.

The intermediate, unstable, branch has reaction sheet temperatures θ_f, given approximately, by the following balance

$$1 = D_0 \exp\{\beta_0 \theta_f/(1 + \alpha_0 \theta_f)\} \tag{74}$$

obtained from a rough order of magnitude estimate of the terms in Eq. (71), leaving aside a power factor of β_0.

When D_0 is such that the temperature θ_f given by (74) equals 0, the resulting value of $D_{OE} = 1$ is a rough estimate of D_{OI}. Similarly the Damköhler number given by Eq. (74) with $\theta_f = \theta_e$ is a rough estimate of the extinction value of D_0.

5.2 Ignition and Extinction Fronts

In the Damköhler number interval between the extinction and ignition values, the nearly frozen solutions and the near equilibrium solutions may coexist in the mixing layer; we could encounter the first one at large negative values of x, and the near-equilibrium solution at large positive values of x, joined by a transition front as shown in Fig. 4.

Eq. (65) has travelling wave solutions of the form $\theta = \theta(x', y)$, with $x' = x + U_p t$, joining the two solutions. They are given by the equation

$$U_F \theta_{x'} - y\theta_y - \theta_{yy} - \theta_{x'x'} = D_0 \hat{Y}_F \hat{Y}_0 \exp\{\beta_0 \theta/(1 + \alpha_0 \theta)\} \tag{75}$$

to be solved with the boundary conditions

$$\theta = 0 \quad \text{for} \quad y \to \pm\infty \tag{76}$$

and

$$\theta = \theta_l(y) \quad \text{for} \quad x' \to -\infty \tag{77}$$
$$\theta = \theta_u(y) \quad \text{for} \quad x' \to \infty \tag{78}$$

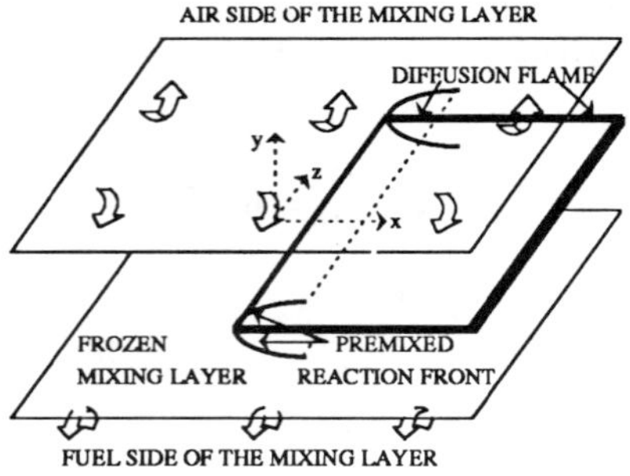

Figure 4. Schematic description of an ignition or extinction front in a strained mixing layer.

where θ_l and θ_u are the lower (nearly frozen) and upper (near-equilibrium) solution of Eq. (71). U_F is the velocity of propagation of the front, an eigenvalue of the problem,

$$U_F = U_F(D_0, \beta_0, \alpha_0, S) . \tag{79}$$

The resulting value of U_F is positive for ignition fronts, for values of $D_0 > D_0^c$, and U will be negative for values of $D_0 < D_0^c$, corresponding to extinction fronts.

A considerable effort is being currently devoted to the analysis of ignition and extinction fronts; see, for example, the initial work of Dold (1987) and Liñán (1988). These fronts may play an important role in the lift-off and blow-off characteristics of turbulent diffusion flames. The thermal expansion effects, left out in the analysis of these fronts given here, will significantly alter their propagation velocity.

ACKNOWLEDGMENT. This work has been partially sponsored by the Spanish CICYT.

REFERENCES

[1] BUCKMASTER, J.D. and LUDFORD, G.S.S., "*Theory of Laminar Flames*" Cambridge Univ. Press (1981).

[2] BURKE, S.P. and SCHUMMANN, T.E.W., "*Diffusion Flames*", Ind-Eng. Chem. **20**, 998-1004 (1928).

[3] DANIELS, P.G., "*Viscous Mixing in a Trailing Edge*", Q. Jl. Mech. Appl. Math. **30**, 319-342 (1977).

[4] DOLD, J.W., "*Flame Propagation in a Non-Uniform Mixture: Analysis of a Slowly Propagating Triple Flame*", Bristol Univ. Report n. AM-87-02 (1987).

[5] DRAZIN, P.G. and REID W.H., "*Hydrodynamic Stability*", Cambridge Univ. Press (1981).

[6] FENDELL, F.E., "*Ignition and Extinction on Combustion of Initially Unmixed Reactants*", J. Fluid Mech., **21**, 291-303, (1965).

[7] LIÑAN, A., "*On the Internal Structure of Laminar Diffusion Flames*", OSR/EOAR, TN 62-69. Astia Doc. n. 273069, INTA. Madrid (1961).

[8] LIÑAN, A., "*The Asymptotic Structure of Counter-Flow Diffusion Flames for Large Activation Energies*", Acta Astronáutica, **1**, 1007-1035, (1974).

[9] LIÑAN, A., "*Diffusion Flame Attachment and Flame Propagation along Mixing Layers*", in Mathematical Modelling in Combustion and Related Topics. Ed. C.M. Brauner and C. Schmidt-Lainé, pp. 151-154, Martinus Nijhoff Publ. (1988).

[10] LIÑAN, A., "*The Analysis of Diffusion Flames for non-Unity Lewis Numbers*", submitted for publication, (1989).

[11] PETERS, N., "*Laminar Flamelet Concepts in Turbulent Combustion*", 21st Symp. (Int.) on Combustion, 1231-1250, The Combustion Institute (1986).

[12] PETERS, N. and WILLIAMS, F.A., "*The Lift-off Characteristics of Turbulent Jet Diffusion Flames*", AIAA, J **21**, 423-429, (1983).

[13] WILLIAMS, F.A., "*Combustion Theory*", 2nd Edition, Benajmín Cummings, (1985).

[14] ZELDOVICH, YA, B., BARENBLATT, G.I., LIBROVICH, V.B. and MAKHVILADZE, G.M., "*Mathematical Theory of Combustion and Explosions*", Consultants Bureau, (1984).

A. Liñán

Escuela Técnica Superior de Ingenieros Aeronáuticos
Universidad Politécnica de Madrid

M L FRANKEL AND G I SIVASHINKSY

Surface evolution equations modelling intrinsic dynamics of premixed flames

1. INTRODUCTION

It is well known (cf. [1]), that due to a strong temperature dependence of the reaction rate the zone of the essential chemical activity in the flame structure becomes extremely thin, and the dynamics of premixed flames may be studied through that generated by a free interface problem. It is the location of the free interface imitating the reaction zone that is usually referred to as the flame front.

Adopting the simplest possible assumptions concerning the chemical kinetics and the transport properties of the system, the appropriately nondimensionalized set of the pertinent conservation laws may be written as follows:

Heat

$$\rho\Theta_t + \rho\mathbf{u}\cdot\nabla\Theta = \nabla^2\Theta + Q\delta_F\ . \tag{1.1}$$

Diffusion

$$\rho C_t + \rho\mathbf{u}\cdot\nabla C = \frac{1}{Le}\nabla^2 C - Q\delta_F \tag{1.2}$$

State

$$\rho = \frac{1}{1-\gamma+\gamma\Theta} \tag{1.3}$$

Continuity

$$\rho_t + \nabla\cdot\rho\mathbf{u} = 0 \tag{1.4}$$

Momentum

$$\mathbf{u}_t + (\mathbf{u}\cdot\nabla)\mathbf{u} = -\frac{1}{\rho}\nabla p + \frac{Pr}{\rho}A(\mathbf{u}) \tag{1.5}$$

where

$$[A(\mathbf{u})]_i = \frac{\partial}{\partial x_k}\left(\frac{\partial u_i}{\partial x_k} + \frac{\partial u_k}{\partial x_i} - \frac{2}{3}\delta_{ik}\frac{\partial u_m}{\partial x_m}\right),$$

$Q = exp[\frac{1}{2}\beta(\Theta-1)]$, $\delta_F = |\nabla F|\delta(F)$ is the surface delta function localized along the interface (flame front) $F(\mathbf{r},t) = 0$ separating the fresh combustible mixture ($F < 0$) from the burned matter ($F > 0$) (Figure 1).

The nondimensional quantities in Eqs.(1.1)-(1.5) are defined as follows:

$\Theta = (T - T_u)/(T_b - T_u)$ is the reduced temperature expressed through the the temperature T and the temperatures of the unburned T_u and the burned T_b gas far away from the flame front;

C is the concentration of the deficient reactant referred to the initial concentration at the "cold" end C_u ;

ρ is the gas density referred to that of the unburned gas ρ_u ;

the velocity of the gas $\mathbf{u} = (u_1, u_2, u_3)$ is referred to the speed of the planar steady combustion wave with respect to the burned gas U_b ;

the pressure p is referred to $\rho_b U_b^2$;

the coordinates x, y, z are expressed in the units of the the typical width of the thermal-diffusive structure of the flame $l_{th} = D_{th}/U_b$, where D_{th} is the thermal diffusivity of the gas;

Le is the the Lewis number defined as the ratio of the thermal diffusivity D_{th} to the molecular diffusivity D_{mol} ;

Pr is the Prandtl number;

β is the Zeldovich number: $\beta = E(T_b - T_u)/RT_b^2$, where E is the activation energy of the chemical reaction and R is the universal gas constant.

Far ahead of the flame front the reduced temperature Θ is zero, while the concentration C is unity. Far behind the front the temperature is equal to the adiabatic temperature of the combustion products, i.e. $\Theta = 1$. In the burned gas region $(F < 0)$ the deficient reactant is completely consumed: $C \equiv 0$.

In this formulation the normal speed of the planar flame wave (U_s) is regarded as a prescribed physico-chemical parameter of the system.

For all its conceptual simplicity the premixed flame exhibits an unusually rich spectrum of phenomena, ranging from self-turbulization of the flame to its self-extinction. The equations (1.1)-(1.5), however, are obviously too difficult to be expected to directly yield any relevant information on the complex dynamical patterns of the system even considering large scale numerical simulations. The central role of the theory, therefore, becomes extraction and elucidation of the features that can be described in a fashion sufficiently simple to be tractable and insightful on the one hand, while accurate enough to be close[1] to the essential dynamical pattern created by the original model (1.1)-(1.5) on the other hand.

[1] If not asymptotically, then at least in some qualitative sense.

Since the flame fronts, whose dynamics is in the focus of attention of the flame theory, are represented in the above model (1.1)-(1.5) by the free interface separating the burned and the unburned regions, it is highly desirable to be able to lower the dimensionality of the problem, in some satisfying way, in order to obtain a closed equation for the free interface only. Such reduced equations are, as a rule, very instructive physically and, clearly, constitute a much more attractive object for numerical experimentation as well as, in many respects, analytical treatment than the original free boundary problem.

For a slightly perturbed planar steady flame such a procedure may indeed be carried out by the means of a quite rational multiple-scale asymptotic analysis. As a result one obtains a weakly nonlinear evolution equation [2] for a small perturbation ϕ of the plane flame front propagating downward $z = \phi(x, y, t) - t$:

$$\phi_t + \frac{1}{2}(\nabla\phi)^2 + \epsilon\nabla^2\phi + 4\nabla^4\phi = \frac{\gamma}{2}I[\phi] \tag{1.6}$$

where

$$I[\phi] = \frac{1}{2\pi}\int_{-\infty}^{\infty}\int_{-\infty}^{\infty} |\mathbf{k}|\phi(\mathbf{x}', t)e^{i\mathbf{k}(\mathbf{x}'-\mathbf{x})}d\mathbf{x}'d\mathbf{k}\ , \quad \mathbf{x} = (x, y)\ . \tag{1.7}$$

The parameters $\epsilon = (1/2)\beta(1 - Le) - 1$ and γ in Eq.(1.6) are assumed to be small.

Eq.(1.6) received much attention recently due to its ability to generate chaotically fluctuating solutions, specifically for $\gamma = 0$, which corresponds to the so-called constant density model of combustion consisting of the reaction-diffusion Eqs.(1.1)-(1.2) which are decoupled from the hydrodynamical part of the problem. This equation proved to be relevant to a variety of physical systems (cf. [2-8]). Perhaps the simplest free boundary problem (as opposed to free interface) that leads to Eq.(1.6) consists of a single heat equation and curvature dependent boundary conditions [9].

The problem is much more difficult and yet much more realistic physically when the flame configuration is not a weak perturbation of a plane but is essentially curved as, for instance, a freely expanding closed smooth surface diffeomorphic to the sphere. The expanding flames are known to exhibit a number of specific effects, including those due to the so-called stretch [10] that may probably delay the self-turbulization of the front [11]. It is, therefore, of considerable interest to have a modified version of Eq.(1.6) that is not linked to a specific coordinate frame. Such an equation would also reflect the important invariance of the original problem (1.1)-(1.5) with respect to translations and rotations of the coordinate frame.

The present paper offers a brief account of some recent developments on this subject.

2. GEOMETRICALLY-INVARIANT EQUATION MODELING THE THERMAL-DIFFUSIVE INSTABILITY OF FLAME FRONTS

One of the basic patterns of behavior that is produced solely by the thermal-diffusive structure of the flame and the chemical kinetics: the cellular-turbulent evolution corresponding to the so-called thermal-diffusive instability of flame propagation. The cellular-turbulent dynamics is manifested in the onset of a cellular pattern and self-turbulence of the flame front.[2] This behavior occurs in the thermal-diffusive model of gas combustion with the δ-function approximation of the chemical reaction which corresponds to Problem (1.1)-(1.5) with $\mathbf{u} \equiv 0,\ \rho \equiv 1$ (constant density model). We obtain, therefore, the following problem:

Heat:

$$\Theta_t = \nabla^2\Theta + Q\delta_F \tag{2.1}$$

Diffusion

$$C_t = \frac{1}{Le}\nabla^2 C - Q\delta_F \tag{2.2}$$

One can easily find a unique (up to translation and rotation) basic solution of Eqs. (2.1),(2.2) represented by a plane boundary propagating with constant speed: $F_b = -t$ if the boundary is defined by $z = F(x, y, t)$.

Let us briefly discuss the ideas leading to Eq.(1.6) (for $\gamma = 0$) [2] which governs the thermal-diffusive dynamics of small perturbations of the plane fronts near the onset of instability. Linear stability analyses performed about the basic solution near the instability threshold: $\epsilon = \alpha - 1 \ll 1$ results in the following approximate dispersion relationship:

$$\omega = \epsilon k^2 - 4k^4 \tag{2.3}$$

where the infinitesimally small perturbation of the boundary has the form: $F - F_b = A\exp(\omega t - i\mathbf{k}\cdot\mathbf{x})$, the parameter α is defined as $\alpha = (1/2)\beta(1 - Le)$, while $k^2 \sim \epsilon$ and $\omega \sim \epsilon^2$ for the unstable modes. This separation of scales dictates the following rescaling of variables:

$$\xi = \sqrt{\epsilon}x, \quad \tau = \epsilon^2 t, \quad f = F - F_b = \epsilon\phi, \quad \Theta - \Theta_b = \epsilon^2\theta, \quad C - C_b = \epsilon^2 c\,. \tag{2.4}$$

The choice of scale for the front perturbation f, however, cannot be made on the basis of the linear analyses but is rather motivated by the observation that if the normal velocity remains unperturbed: $V = F_t/\sqrt{1+F_x^2} = -1$, the spatial distortion of the flame front requires the following balance:

$$V + 1 \approx \epsilon^2 f_\tau + \frac{1}{2}\epsilon f_\xi^2 = 0\ , \tag{2.5}$$

and the scale for $\Theta - \Theta_b,\ C - C_b$ can then be obtained from the linearized problem.

[2]Generally speaking there is another, distinctly different dynamical pattern that results from the thermal-diffusive instability for a different range of the parameters. It occurs in the gas as well as in the solid fuel combustion and manifests itself in the onset of velocity oscillations of the plane front or waves propagating along the interface. The spinning dynamics can also be modelled by a geometrically invariant equation [12].

Introduction of the scaled variables (2.4) leads to rather elementary calculations for the solution sought as an asymptotic expansion up to ϵ: $\phi = \phi_0 + \epsilon\phi_1 + \ldots$, $\theta = \theta_0 + \epsilon\theta_1 + \ldots$, and leads in the principal order of ϕ to [2]:

$$\phi_\tau + \frac{1}{2}(\nabla\phi)^2 + \epsilon\nabla^2\phi + 4\nabla^4\phi = 0 . \tag{2.6}$$

The spatial derivatives in Eq.(2.6) should be understood in terms of the stretched variables.

The idea leading to derivation of the coordinate-free equation that is capable of imitating the dynamics of arbitrarily shaped flame sheet configurations is as follows [13],[14]. Suppose that it was observed, say from numerical simulation of the model problem (2.1)-(2.2), that even for the values of parameters relatively far from the stability threshold the principal curvatures of the boundary κ_1, κ_2 and the perturbation of normal velocity against that of the plane steady front $v = V+1$ remain small, or, in other words, that the dynamics generated by the free boundary problem is quasi-plane and quasi-steady.

We assume, therefore, that

$$\kappa_1 \sim \kappa_2 \sim v \sim \delta \ll 1 \tag{2.7}$$

for some artificially introduced small parameter δ unrelated to the physical parameters of the problem. Then in terms of the stretched space and time variables: $\xi = \delta x$, $\eta = \delta y$, $\phi = \delta F$, $\tau = \delta\ t$, one can find an asymptotic solution of (2.1)-(2.2) in powers of δ resulting, in particular, in the following expansion terms for the normal velocity:

$$V_0 = -1, \quad V_1 = -\epsilon K_0, \quad V_2 = -\epsilon K_1 + C_1 K_0^2 + C_2 H_0 , \tag{2.8}$$

$$V_3 = -\epsilon K_2 + 2C_1 K_0 K_1 + C_2 H_1 + C_3 K_0^3 + C_4 K_0 H_0 - (\alpha^3 + 3\alpha)\Delta K_0,$$

where K_i, H_i are the expansion terms of the rescaled mean K and Gaussian H curvatures, C_i are some coefficients expressed through α.

Adding the above terms multiplied by appropriate powers of δ: $V = \Sigma\delta^i V_i$, and restoring the original nonscaled variables one can synthesize the following equation (up to δ^3) for the front dynamics:

$$V = -1 - \epsilon\kappa + P(\kappa, q) - (\alpha^3 + 3\alpha)\Delta_s\kappa , \tag{2.9}$$

where $\kappa = \kappa_1 + \kappa_2 = -div\mathbf{n}$ is the (double) mean curvature of the flame surface, $q = \kappa_1\kappa_2$ is the Gaussian curvature, P is a polynomial $\sim \kappa_i^2$ with coefficients expressed through α and Δ_s is the invariant Laplace-Beltrami operator (surface laplacian).

The above surface evolution equation has a unique (up to translation and rotation) plane solution $v = 0$ for any values of the parameters. For small perturbation of the plane solution the approximate dispersion relationship (2.5) is obviously recovered. Near the stability threshold $\epsilon \ll 1$ it is easy to show that

$$\kappa \sim \epsilon^2, \quad \Delta_s\kappa \sim v \sim \epsilon^3, \quad q \sim \epsilon^4 , \tag{2.10}$$

and Eq.(2.9) is reduced (up to ϵ^3) to the following compact relationship

$$V = -1 - \epsilon\kappa - 4\Delta_s\kappa . \tag{2.11}$$

In 2-D Eq.(2.11) becomes:

$$V = -1 - \epsilon\kappa - 4\frac{\partial^2 \kappa}{\partial s^2}\ , \tag{2.12}$$

where s is the arc-length of the 2-D front (curve).

If we express all the values in terms of rescaled perturbation function ϕ and expand up to ϵ^3, Eq.(2.11) is farther reduced to its weak counterpart (2.6). Thus, Eq.(2.6) is an approximate coordinate form "imprint" of the geometrically invariant Eq.(2.11).

The equation of surface dynamics (2.11) represents a very natural modification of the classical Huygens principle and can probably be obtained as a Taylor expansion from a general form of the dependence of normal velocity on the principal curvatures and their derivatives with respect to the "surface coordinates" if rotational invariance is additionally assumed. This leads us to the view upon the equation as representing a certain generic type of surface evolution manifested qualitatively in spontaneous "self-cellularization" and self-turbulence for different ranges of the governing parameter. At the same time solutions of Eqs.(2.9) and (2.11) should retain and even improve the smoothness of the initial configuration due to their locally parabolic nature.

Comparison between Eq.(2.6) and its invariant counterpart (2.11) allows to observe the following termwise correspondence:

$$f_t \leftrightarrow v, \quad \nabla^2 f \leftrightarrow \kappa, \quad \nabla^4 f \leftrightarrow \Delta_s \kappa\ , \tag{2.13}$$

where $v = V+1$ is the deviation of the normal velocity from that of the basic plane solution. It is noteworthy, that should we possess this information beforehand and rewrote the linear "original" of the dispersion equation (2.3): $f_t = -\nabla^2 f - 4\nabla^4 f$ in invariant terms, we would quite painlessly obtain Eq.(2.11) and, by the same token, Eq.(2.6) as its coordinate form expansion. This observation probably contains a hint to a certain formalism that may be helpful in designing geometrically invariant evolution equations.

3. GEOMETRICALLY INVARIANT EQUATION MODELING THE DARRIEUS-LANDAU FRONT DYNAMICS IN POTENTIAL FLOWS

Eq.(1.6) was derived for near-critical values of parameters and reflects the development of both the thermal-diffusive instability and the hydrodynamic instability in the flame front. If $\epsilon < 0$ the thermal-diffusive structure has a stabilizing effect upon the front and its dynamics is essentially described by a degenerated version [2] of Eq.(1.6):

$$\phi_t + \frac{1}{2}(\nabla\phi)^2 + \epsilon\nabla^2\phi = \frac{\gamma}{2}I[\phi]. \tag{3.1}$$

In the long wavelength limit the spatial resolution becomes "insensitive" to the thermal-diffusive structure and Eq.(3.1) can be further reduced to [2]:

$$\phi_t\ +\ \frac{1}{2}(\nabla\phi)^2\ =\ \frac{\gamma}{2}I[\phi] \tag{3.2}$$

Therefore, the last equation describes the dynamics of a slightly perturbed plane flame front under the conditions of the thermal-expansion induced instability only. Eq. (3.2) was also derived directly [15] using formal asymptotic expansion in $\gamma \ll 1$ within the framework of a purely hydrodynamic (Darreus-Landau) model of flame propagation.

In Section 2 we saw that the invariant equation (2.11) corresponds to Eq.(1.6) with zero right hand side ($\gamma = 0$). In this context it is natural to ask the following question: what is the geometrical, coordinate-free version of the integral term in Eqs.(1.6), (3.1) and (3.2). In other words, what is the geometrically invariant form of Eq.(3.2) that would model the evolution of curved flame fronts as opposed to small perturbations of the plane ones under the conditions of thermal-expansion induced instability?

Analyzing computations leading to the "near plane" equations (1.6), (3.1) and (3.2) one can observe (cf. [2],[3],[15]) that in the principal order with respect to $\gamma \ll 1$ the gas flow is irrotational both ahead and behind the flame front, which shows that Landau instability is not necessarily related to generation of vorticity within the flame structure. It is natural then to study a model of the evolution of arbitrarily shaped fronts within a purely irrotational flow [16].

Let us therefore consider the Darreus-Landau model in which the flame is regarded as a surface of density discontinuity propagating at a constant velocity relative to the burned gas. The flow is assumed to be incompressible and non-viscous. We shall further assume that the flow is irrotational in both the unburned (-) and the burned (+) gas:

$$\nabla \cdot \mathbf{u} = 0, \qquad \nabla \times \mathbf{u} = 0 \, . \tag{3.3}$$

At the interface the flow is subject to the conditions of continuity, preservation of matter and constant velocity of the front with respect to the gas:

$$\begin{gathered} (\mathbf{u}^- - \mathbf{u}^+) \cdot \boldsymbol{\tau} = 0, \\ \rho^-(\mathbf{u}^- \cdot \mathbf{n} - V) = \rho^+(\mathbf{u}^+ \cdot \mathbf{n} - V) \, , \\ \mathbf{u}^+ \cdot \mathbf{n} - V = 1, \end{gathered} \tag{3.4}$$

where $\rho^+ = 1$ and $\rho^- = (1-\gamma)^{-1}$ are the densities of the burned and the unburned gas respectively, and $\boldsymbol{\tau}$ is the unit vector tangent to the flame front.

The solution of the problem (3.3),(3.4) is obtained as a single layer potential w: $\nabla w = \mathbf{u}$ with constant charge distribution $\gamma/2\pi$ (we consider the 2-D case first):

$$w(\mathbf{r}, t) \;=\; \frac{\gamma}{2\pi} \int_C \ln |\mathbf{r} - \xi| dl_\xi \tag{3.5}$$

where C is the curve (in 2-D case) corresponding to the instantaneous position of the flame front. The dynamics of the front can be found from the last of the conditions (3.4) that overdefines the problem [16]:

$$V(\mathbf{r}, t) \;=\; -1 \,+\, \mathbf{u}^+ \cdot \mathbf{n} \, , \tag{3.6}$$

where

$$\mathbf{u}^+ \cdot \mathbf{n} \;=\; \frac{\gamma}{2}(1 \;+\; \frac{1}{\pi}\int\limits_C \frac{(\mathbf{r}-\xi)\cdot\mathbf{n}}{|\mathbf{r}-\xi|^2}dl_\xi),$$

and $\mathbf{n}$ in the integral is the normal to the instantaneous position of the flame front $C(t)$ at $\mathbf{r}$.

Thus, we obtain a surface dynamics equation that defines the instantaneous normal propagation velocity until, possibly, the front develops a cusp. From Eq (3.6) it is obvious that the integrals are regular for a sufficiently smooth boundary, one can also easily recognize the normal component of the (burned) gas velocity represented by the superposition of uniformly distributed sources along the boundary C.

Eq.(3.6) can be reduced to the near plane version (3.2) for $\gamma \ll 1$ with some specific choice of the contour C when it approaches a slightly perturbed straight line [16]. This provides the interpretation of the integral term in Eqs.(1.6), (3.1) and (3.2) which was exactly our objective

The derivation of Eq.(3.6) is absolutely rigorous and quite simple, once the assumption of the potential character of the flow is made. On the other hand this assumption is not restrictive in the case of small γ where Eqs.(1.6),(3.1) and (3.2) were derived. That is the invariant equation is at least "as correct" as this equation, and additionally reflects some desirable features of the physical problem such as zero velocity of the gas at infinity, and its invariance.

The front dynamics equation (3.6) is obviously invariant with respect to translations and rotations of the coordinate frame as it is expressed in terms of the front velocity in the local normal direction and the distances between the points on the front. The equation can be immediately transferred into three dimensions using Newton's potential instead of the logarithmic one [16]:

$$V(\mathbf{r},t) = 1 + \mathbf{u}^+ \cdot \mathbf{n}, \quad \mathbf{u}^+ \cdot \mathbf{n} = \frac{\gamma}{2}(1 + \frac{1}{2\pi}\int\limits_S \frac{(\mathbf{r}-\xi)\cdot\mathbf{n}}{|\mathbf{r}-\xi|^3}dS_\xi)\;. \tag{3.7}$$

In spite of a rather simple form of Eqs.(3.6) and (3.7) it is not easy to find any explicit solutions. One important illustration, however, is readily provided by an expanding circle (sphere) where the integral can be easily evaluated

$$\int\limits_C \frac{(\mathbf{r}-\xi)\cdot\mathbf{n}}{|\mathbf{r}-\xi|^2}dl_\xi \;=\; -\pi \tag{3.8}$$

to yield $V \equiv -1$, which is obviously the case since, in view of the symmetry, the burned matter is immobilized. The same result is formally produced after evaluation of the integral in the case of the initially ignited plane that generates two plane fronts running away from each other with $V \equiv -1$, but rigorously speaking this is incorrect since the solution is valid only for closed fronts.

4. GEOMETRICALLY INVARIANT EQUATION MODELING THE FLAME FRONT DYNAMICS AS DENSITY DISCONTINUITY IN POTENTIAL FLOWS

Although at this stage we cannot present any results of numerical simulation of Eq.(3.6), in view of the similarity between the weak (3.1) and the invariant form we can expect the development of cusps pointing into the burned matter as numerical simulation of Eq.(3.2) for small ϵ shows [21].It is of considerable interest, therefore, to synthesize the models discussed above to furnish Eqs.(3.6) and (3.7) with appropriate geometrically invariant dissipative terms reflecting the dynamics of the thermal-diffusive structure of the flame that would ensure the regular character of the underlying flame front evolution. Such a synthesis can be carried out within the framework of a hydrodynamical model that again treats the flame as a surface of density discontinuity propagating in an irrotational gas flow with the velocity, however, now dependent on the solution of a constant density thermal-diffusive model that yields the principal part of the interaction within the flame structure.

Let us consider the following model of the flame:

Heat:

$$\rho(\Theta_t + \mathbf{u}\cdot\nabla\Theta) = \nabla^2\Theta + Q\delta_F \ . \tag{4.1}$$

Diffusion:

$$\rho(C_t + \mathbf{u}\cdot\nabla C) = \frac{1}{Le}\nabla^2 C - Q\delta_F \tag{4.2}$$

Incompressible and irrotational flow:

$$\nabla\cdot\mathbf{u} = 0, \qquad \nabla\times\mathbf{u} = 0 \ . \tag{4.3}$$

Continuity of the tangential component across the front:

$$(\mathbf{u}^- - \mathbf{u}^+)\cdot\boldsymbol{\tau} = 0 \ . \tag{4.4}$$

Continuity of the mass flow across the front:

$$\rho^-(\mathbf{u}^-\cdot\mathbf{n} - V) = \rho^+(\mathbf{u}^+\cdot\mathbf{n} - V) \ , \tag{4.5}$$

where again $\rho^+ = 1$ and $\rho^- = (1-\gamma)^{-1}$ are the densities of the burned and the unburned gas respectively, and $Q = exp[\frac{1}{2}\beta(\Theta - 1)]$, $\delta_F = |\nabla F|\delta(F)$.

However, in contrast to the Darreus-Landau model here

$$\mathbf{u}^+\cdot\mathbf{n} - V = U \quad (\neq 1) \ . \tag{4.6}$$

Hence, also the jump of the normal velocity component is not constant

$$\mathbf{u}^+\cdot\mathbf{n} - \mathbf{u}^-\cdot\mathbf{n} = \gamma U \quad (\neq \gamma) \ . \tag{4.7}$$

Nevertheless, for small γ and ϵ this deviation of U from unity is not consequential and can be disregarded in the principal order. As a result one obtains the simplest possible synthesis of Eqs. (2.11) and (3.7):

$$V = -1 - \epsilon\kappa - 4\Delta_s\kappa + \mathbf{u}^+ \cdot \mathbf{n} \ , \tag{4.8}$$

where $\mathbf{u}^+ \cdot \mathbf{n}$ is defined as in Eqs.(3.6) or (3.7) for 2-D and 3-d respectively [16].

It is worth mentioning that for $\epsilon < 0$, when the thermal-diffusive structure of the flame is intrinsically stable, the geometrically invariant equation modeling the flame dynamics is [16]:

$$V = -1 - \epsilon\kappa + \mathbf{u}^+ \cdot \mathbf{n} \tag{4.9}$$

Eqs. (4.8) and (4.9) can also be formally reduced to their weak near-plane counterparts (1.6) and (3.1) respectively for some specific choices of the closed fronts if one adopts appropriate rescalings of variables under which Eqs. (1.6) and (3.1) were derived.

Although the motivation for the purely irrotational flame models came from the observations made for $\gamma \ll 1$ it is of some interest to see what happens when this requirement is relaxed. Here one expects Eq.(4.9) to undergo a rather curious modification with κ being replaced by the so-called "stretch" of the flame Σ defined by [10][3]

$$\Sigma = \kappa(1 - \mathbf{u}^+ \cdot \mathbf{n}) - \mathbf{n} \cdot \nabla \times \mathbf{u}^+ \times \mathbf{n} \tag{4.10}$$

where $\mathbf{u}^+ \cdot \mathbf{n}$ is now determined from the solution of the problem (4.3),(4.4),(4.7) with U defined as (cf. [10],[17],[18]) $U = 1 + \epsilon\Sigma$.

5. CONCLUDING REMARKS

As it was remarked in Introduction the complete flame problem (1.1)-(1.5), which itself is a simplification of a more realistic model with the distributed Arrhenius kinetics, seems far too demanding at this point to be effectively treated either analytically or numerically. The wealth of dynamical phenomena, on the other hand, generated by the flame propagation makes combustion a natural source for simplified problems based on various physically justifiable idealizations that would sacrifice the completeness of the picture in favor of a rectified part that is still rich enough to deserve our close attention.

In the latter context, however, it is remarkable how fruitful a source of rich mathematical models the theory of premixed gas combustion has been and continues to be. It is also remarkable that these models seem to often-times represent a certain generic mechanism meaningful outside the combustion[4].

There may be several reason for this generic nature of some models originated from the premixed combustion. One of them may be related to the fact that the flame propagation

[3]In 2-D case $-\mathbf{n} \cdot \nabla \times \mathbf{u}^+ \times \mathbf{n} = d(\mathbf{u}^+ \cdot \tau)/ds$ where s is the arc-length along the front.

[4]An attempt to present the material from a slightly more general point of view is made in [20].

is strongly reminiscent of the motion of an interface separating two different phases of a substance undergoing a phase transition (cf.[19]). The process is "self-propelled" in the sense that no external driving force is required for the interface (flame front) motion, which obviously results in its invariance with respect to translations and rotations of the coordinate frame as a reflection of homogeneity and isotropy of the space.

At the same time the mode of combustion that is in the focus of our interest, although labeled as slow combustion (as opposed to detonation) is in fact rather fast in comparison to typical phase transition problems. The last feature is responsible for a good and in a certain sense uniform localization of the thermal-diffusive structure which in turn leads to the existence of the steadily propagating plane wave of combustion.

The existence of the plane flame and the quasi-plane quasi-steady character of the flame in general are extremely important. These properties allow the asymptotic treatment of reaction-diffusion part of the problem in scales separated in the transversal and the tangential to the flame front directions, and, consequently, on a still larger, "insensitive" to the thermal-diffusive structure, scale the treatment of the flame as a hydrodynamic discontinuity.

The above discussion summarizes the physical properties of the free flame propagation that result in the simplified models and, ultimately, in the equations of surface dynamics some of which were described in the previous sections. As for the latter they represent, in our view, a step ahead in our understanding of certain aspects of the flame dynamics, and yet a serious mathematical challenge in their own right.

A systematic study of surface dynamics as a separate subject is, to the best of our knowledge, just beginning to unfold. Although we seem to understand certain properties of the geometrically invariant equations, such as local existence and smoothness of the solutions, relying upon our experience with partial differential equations, the most interesting questions concerning the overall qualitative picture for prolonged periods of time should be formulated in a global and coordinate free form. For instance, the question of global existence starting, say, from some smooth closed initial configuration should be supplemented by investigation of a possibility of self-intersection, connectedness, preservation of some topological characteristics etc..

It is clear, therefore, that the surface dynamics governed by the above type equations leads in effect to an entirely new mathematical subject in which differential equations are closely intertwined with differential geometry and, possibly, other mathematical disciplines. At the moment, however, we cannot make intelligent predictions concerning long-time behavior for the solutions of the above evolution equations. This brings us to the necessity of an extensive numerical experimentation with the above equations of surface dynamics.

It is easy to predict, however, that we should encounter some serious mathematical difficulties in the course of numerical experiments as well. One of them concerns representation of the evolving surface in an efficient way. For instance, triangulation of the surface with subsequent projection of the parts onto appropriate coordinate planes, representing the surface evolution equation in coordinate form and solving them separately with simultaneous matching would hardly be a realistic procedure in view of extremely complex dynamics, such as turbulent-type behavior, expected from the solutions. If, on the other hand, we use an appropriate global parametrization, the translation of the invariant evolution equations to the parametric terms may lead to rather lengthy and complex expressions including highly unpleasant nonlinearities. Additionally, the parametrizations that were chosen with a reasonably uniform grid for the initial configurations would become increasingly unsuitable as the surface evolves, and one should then face the problem of sustaining the accuracy of the computation.

Finally, there is one more serious difficulty to overcome. It concerns the adequacy of the visualization of the results of numerical simulation of the surface evolution governed by the above geometrically invariant equations which may be no less important than the computation itself.

Thus, the surface evolution equations modeling dynamical patterns of the flame propagation represent a serious mathematical challenge yet to be answered. We believe, however, that studying this rich material, both: analytically and numerically, can be highly rewarding.

REFERENCES

[1] J. Buckmaster, G.S. Ludford, *Theory of laminar flames*, Cambridge University Press, New York, 1982.

[2] G.I. Sivashinsky, Acta Astronaut., **34** (1977), pp. 1177-1206.

[3] G.I. Sivashinsky, Ann. Rev. Fluid Mech. **15** (1983), pp. 179-199.

[4] P. Maneville, in *Propagation in systems far from equilibrium*, J. E. Westfrec et al. eds., Springer-Verlag, Berlin, 1988.

[5] P.C. Hohenberg and B.I. Shraiman, Physica D **37** (1989), pp. 109-115.

[6] O. Thual, U. Frisch and M. Henon, J. Physique, **46** (1985), pp. 1485-94.

[7] G. Joulin, J. Physique **50**, pp. 1069-72.

[8] S. Gutman and G.I. Sivashinsky, *The cellular nature of hydrodynamic flame instability*, Physica D, in press.

[9] M.L. Frankel, Math. Mod. and Num. Analysis, **23** (1989), pp. 283-291.

[10] M. Matalon, Combust. Sci. Technol., **31**, pp. 169-181.

[11] A.G. Istratov and V.B. Librovich, Astronaut Acta, [**14** (1969), pp. 453-467.

[12] M.L. Frankel, Phys. Let. A, **140** (1989), No.7,8, pp. 405-410.

[13] M.L. Frankel and G.I. Sivashinsky, J. Physique **48** (1987), pp. 25-28.

[14] M.L. Frankel and G.I. Sivashinsky, Physica D **30** (1988), pp. 28-42.

[15] G.I. Sivashinsky and P. Clavin, J. Physique, **48**, (1987), pp. 193-198.

[16] M.L. Frankel, *An equation of surface dynamics modeling the flame fronts as density discontinuities in potential flows*, Phys. Fluids, submitted.

[17] M. Matalon and B.J. Matkowsky, J. Fluid. Mech., **124** (1982), pp. 239-259.

[18] P. CLAVIN and G. JOULIN, J. Physique Lett, **1** (1983), pp. 1-12.

[19] M.L. FRANKEL, Phys. Let. A, **128** (1988), pp. 57-60.

[20] M.L. FRANKEL, *Free boundary problems and dynamical geometry associated with flames*, in Dynamical Issues in Combustion Theory, Springer-Verlag, 1990, to appear.

[21] D. MICHELSON and G.I. SIVASHINSKY, Acta Astronaut., **34** (1977), pp. 1207-1221.

Michael L. Frankel

Department of Mathematical Sciences, Indiana University - Purdue University at Indianapolis, Indianapolis, Indiana 46205, USA. The first author (M.L.F.) was partially supported by the Air Force Office of Scientific Research and the National Science Foundation under NSF Grant No. CBT-8905838.

Gregory I. Sivashinsky

The Levich Institute for Physico-Chemical Hydrodynamics, The City College of New York, New York 10031, USA,
and
School of Mathematical Sciences, Tel-Aviv University, Ramat-Aviv, Tel-Aviv 69978, ISRAEL. The second author (G.I.S.) was partially supported by the U.S. Department of Energy under Grant No. DE-FG02-88ER13822 and by the National Science Foundation under Grant No. CTS-8910903.

P CLAVIN

Turbulent flame propagation

1. INTRODUCTION

This paper is devoted to the flamelet regime of the propagation of premixed turbulent flames. The background of the dynamics of wrinkled flames and of curved fronts is recalled in the two first sections. The flamelet regime is defined in §III. The cusped flame regime of an unstable flame front propagating in a weakly turbulent flow is presented in §IV. This regime is an interesting example where the thermal diffusive mechanisms inside the flame are coupled with hydrodynamics of the far flow field providing a selection of large size cells on the front by fluctuations at much smaller length scales in the incoming turbulent flow. The corrugated flamelet regime for a flame propagating in a high intensity turbulent flow is presented in §V which is independent of the preceding sections. Here, the attention is focused on the kinematic problem of a front propagating with a constant normal burning velocity in a prescribed turbulent flow. In the absence of a general method to solve this difficult problem, a semi phenomenological approach is used to improve the understanding of this topic. Two limiting mechanisms for the flame propagation are identified. One is controlled by the combustion time of the fresh mixture engulfed in the burned gases by the turbulent mixing. In the absence of an ignition threshold, the other one is controlled by the fastest contamination mechanism of the turbulent flow. These two limiting regimes are found to merge in the case of a flame propagating in a fully developed turbulent flow represented by the Kolmogorov cascade. Turbulent flame speed, flame thickness as well as some geometrical properties of the flame front are obtained in this case. This paper is an extension of the work presented in a lecture given at the Cargèse Institute on Disorder and Mixing (see ref.17 in the text).

2. DYNAMICS OF WRINKLED FLAMES

Because of the gas expansion across a premixed gas flame, the density of the burned gases is smaller than the density in the fresh mixture ($\rho_b < \rho_u$) and the streamlines crossing a tilted flame front are deflected (see figure 1). Thus, the flame dynamics of wrinkled fronts involves hydrodynamical phenomena which develops on a characteristic length scale of the order of magnitude of the wavelength of the wrinkles Λ which are assumed to be much larger than the flame thickness d ($\varepsilon \equiv d/\Lambda \ll 1$). The governing equations include the energy and species balance, the mass conservation, the Navier-Stokes equations and the ideal gas law which can be written as $\rho T = Ct$ in the small Mach number limit. The variations of density, temperature and mass fraction are localised inside the flame thickness. In the limit of small ε, the flame thickness d becomes negligibly small compare to the hydrodynamical length Λ and the flame can be considered as a surface of discontinuity. In this limit, the problem may be treated by a multiscale analysis[1,2,3] and reduces to an hydrodynamical free boundary problem for incompressible flows with different densities in the upstream and downstream sides of the flame. The boundary conditions satisfied by the flowfields at the flame front are obtained as the result of the analysis of the local structure of the wrinkled flame. The complete solution of the flame dynamics is obtained by solving the hydrodynamical problem with the local relations mentioned above.

2.1. Local laws

The normal burning velocity U_n is defined in the limit $\varepsilon \to 0$ as the normal component of the relative velocity of the unburned gases crossing the flame front. Let $x = \alpha(y,t)$ denote the equation of the flame front (written in the two dimensional case for simplicity), kinematic considerations yield:

$$U_n = \{ u - \partial\alpha/\partial t - v\partial\alpha/\partial y\}/\{1 + (\partial\alpha/\partial y)^2\}^{1/2} \qquad \text{(1-a)}$$

where u and v are the values at the front of the x and y components of the velocity of the upstream flow of fresh mixture (see figure 1).

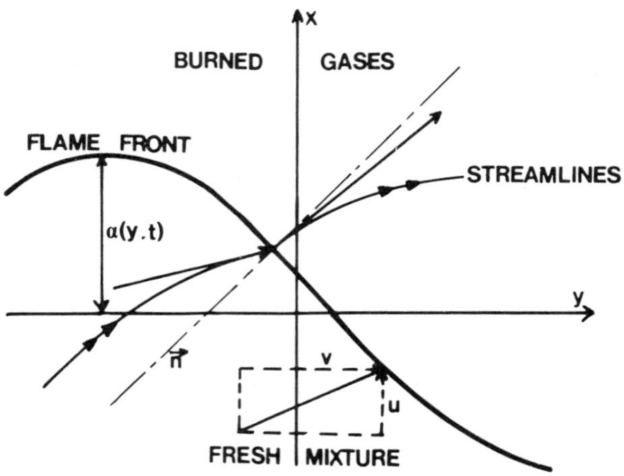

Figure 1: In the limit $\varepsilon \to 0$, the flame appears as a surface of hydrodynamical discontinuity (flame front).

Let S be the area of the flame surface, the combustion rate is expressed as:

$$\int_S \rho_u U_n d\sigma \qquad \text{(1-b)}$$

where $d\sigma$ is the differential elementary area of the flame front. For $\varepsilon = 0$ one has $U_n = U_L$. As anticipated by the phenomenological analysis of Markstein[4] and Karlowitz[5], the flame speed is modified ($U_n \neq U_L$) by front curvature and flow inhomogeneities and these corrections are obtained at the following orders of a perturbation analysis of the wrinkled flame structure in power of ε. Such a study[3] was carried out in the framework of an asymptotic expansion in the limit $\beta \to \infty$. At the first non trivial order, the local analysis of the flame structure yields[3]

$$(U_n/U_L - 1) = Ma \{ d/R + \tau_t \mathbf{n} . \nabla\mathbf{u} . \mathbf{n} \} + O(\varepsilon^2) \qquad \text{(1-c)}$$

where U_L is the laminar flame speed and τ_t the transit time $\tau_t = d/U_L$. R is the mean curvature of the reaction sheet defined to be positive when the front is concave towards the fresh mixture and **n** is the unit vector normal to the reaction sheet oriented towards the fresh gases. $\nabla\mathbf{u}$ is the extrapolation to the reaction sheet of the rate of strain tensor of the upstream flow. The first term in the r.h.s. of Eq. (1-c), describes the curvature effect. The second is the stretch effect. The dimensionless coefficient Ma, called the Markstein number, caracterizes the intensity of stretching and curvature effects. Its value is found as a result of the analysis[1] and depends on the diffusive and reactive properties of the mixture. It is predicted to be positive in ordinary cases[1]. Eq. (1-c) can also be written as[3]:

$$(U_n/U_L - 1) = \text{Ma}\ \{\sigma^{-1}\tau_t\,(d\sigma/dt)\} + O\,(\varepsilon^2) \qquad (2)$$

where σ is the elementary area of the flame surface and the nondimensional rate of stretch is of order ε, $\tau_t\sigma^{-1}(d\sigma/dt) = O(\varepsilon)$. Nonlinear effects generalising (2) may lead to an extinction mechanism when the rate of stretch $\sigma^{-1}(d\sigma/dt)$ becomes of the same order of magnitude as the inverse of the transit time τ_t^{-1}.

2.2. Hydrodynamical effects

Eqs. (1-a) and (1-c) put together yield an equation for the motion of the flame front. But because of the hydrodynamical effects which are produced by the gas expansion, this equation has not a closed form. More precisely, if u_e and v_e are the components of the flow velocity in the absence of the flame, one has :

$$u = u_e + u_i, \qquad v = v_e + v_i, \qquad (3)$$

where u_i and v_i are the components of the flow velocity which are induced upstream by the gas expansion of the burnt gases. By solving the corresponding hydrodynamical problem in which the flame is considered as a surface of discontinuity located at the reaction sheet, the induced flow u_i and v_i is computed for small ε as a functional of the dynamical characteristics of the flame front $x = \alpha(y,t)$. Such a calculation was first carried out indepently by Darrieus (1938) and Landau[6] (1944) in the linear approximation of a front weakly wrinkled around a planar shape and in the limiting case $\varepsilon = 0$ where the effects associated with the flame thickness are completely neglected, $U_n = U_L$. As a result, a planar flame front standing perpendicular to a uniform flow ($u_e = U_L$, $v_e = 0$) is found to be unstable for all wavelengths Λ ($\Lambda = 2\pi/K$) with a linear growth σ of the form:

$$\sigma = a_1(\gamma)\, U_L\, K, \qquad (4)$$

where K is the modulus of the wave vector, $K = 2\pi/\Lambda$. The nondimensional coefficient $a_1(\gamma)$ has the same sign as the gas expansion coefficient $\gamma = (\rho_u - \rho_b)/\rho_u \approx (T_b - T_u)/T_b$. Working in perturbation of ε, relations such as Eq. (1-c) must be used, and the first correction in non zero thickness of the wrinkled flame front ($U_n \neq U_L$) introduces a K^2 term in the r.h.s. of Eq. (4) with a coefficient depending on the reactive and diffusive properties of the reactive mixture. When this coefficient is positive, the corresponding effects stabilise the short wavelengths and the so called Michelson-Sivashinsky[7] (M.S.) model equation for the front dynamic has been derived in the limit of a weak gas expansion $\gamma \ll 1$. At the leading order in a γ expansion this equation takes the following form :

$$\partial\alpha/\partial t = \mathbb{L}\,(\alpha) - |\nabla\alpha|^2/2 \qquad (5)$$

where the linear differential operator $\mathbb{L}$ can be written in the Fourier representation in a conveniently nondimensional form as:

$$\mathbb{L}_k = k - k^2, \qquad (6)$$

where k ($\approx$ Kd) is the modulus of the reduced wavevector scaled in such a way that k=1 is the marginal case and k=1/2 corresponds to the most unstable wavevector. $|\nabla\alpha|^2/2$ is the nonlinear geometrical effect resulting from the tilting of the front.

3. CURVED FRONTS

Flames in channel exhibits a stationary curved front with a size R much larger than the most unstable wavelength of the planar front (k = 1/2, see Eq.6). An approximate steady state solution of the corresponding curved front and a simple explanation of its anomalous stability was presented by Zel'dovich *et al.*[8]. As sketched in Fig.2, the analysis consists in evaluating the growth rate of a localised perturbation initiated close to the tip, stretched and carried along the front by the tangential component of the flow. During its travel, such a perturbation is submitted locally to a linear growth rate similar to the one in the planar case (see Eq.6) but with a varying wavelength which is increasing under the flow stretch. The final amplitude is the result of a time integration in the Lagrangian frame of the perturbation. The most amplified perturbations are initiated near the tip with an initial size of the order of magnitude of the marginal wavelength Λ_m of the dispersion relation of the planar front which corresponds to k_m=1 when it is written in the a nondimensional form (6). The total amplification Γ, can be computed[8] by a W.K.B. method for stationary fronts of curvature radius R much larger than $\Lambda_m$as those observed in experiments. For a general review of this question and more recent extensions see references[14,15,16].

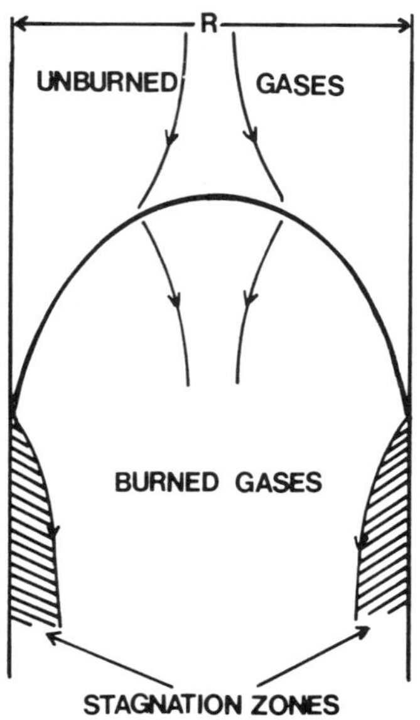

Figure 2. Sketch of the streamlines associated with a curved flame front propagating in a tube. The gas flow in the fresh mixture presents the form of a local stagnation point flow at the tip. Local perturbations which are initiated near the tip are swept toward the wings and stretched by the induced flow. Stagnation zones appear in the burned gases close to the walls.

In curved flames, the leading order of the WKB expansion of Γ is found to be determined by the vicinity of the tip whose the configuration is similar to a stagnation point flow and where the tangential velocity is small and the transit time is long enough to let the instability grow. The total amplification turns out to be of the order of R/Λ_m which corresponds to the Reynolds number $Re=RU_L/\nu$ based on the size of the channel R (Λ_m is proportional to the flame thickness $d \approx D_T/U_L$ and $D_T \approx \nu$, so that $RU_L/\nu = R/\Lambda_m$)

$$\Gamma \approx Re \quad \text{with} \quad Re \approx R/\Lambda_m , \tag{7-a}$$

where by definition of one has

$$A_f/A_i \equiv \exp\Gamma, \tag{7-b}$$

where A_f is the final amplitude reached in the wing of a perturbation initiated close to the tip with an initial amplitude A_i. A stability criterion is roughly obtained by limiting the admissible final amplitude A_f to the radius of the tube. For a given amplitude A_i of the perturbation of the front produced at the tip by the external noise, this gives an estimate of the maximum size of the tube R^* in which a single cell can propagate:

$$(R^*/\Lambda_m) - \text{Log}\,(R^*/\Lambda_m) = \text{Log}\,(\Lambda_m/A_i). \tag{8}$$

When $R>R^*$ the flame front breaks up in at least two cells. The present theory is limited to $A_i << \Lambda_m$ and R^* much larger than Λ_m (a typical order of magnitude is $\Lambda_m \approx 0.3$ cm).The second term in the l.h.s. of Eq.(8) is of a following order in the expansion.

4. THE FLAMELET REGIME OF TURBULENT COMBUSTION.

4.1. *The flamelet regime*

The flamelet regime holds when all the length scales of turbulence are larger than the laminar flame thickness ($l >> d$) i.e. $d < l_K$ where l_K is the Kolmogorov scale. The flame may be considered as a surface of zero thickness in the flow. In order to predict the overall turbulent combustion rate, one has to solve a free boundary problem of an hydrodynamical nature, an to relate the statistical properties of the flame front to those of the incoming turbulent flow (u_e, v_e). In the ideal case of an homogeneous and isotropic initial turbulent flow in an infinite domain without boundary, one may look for a travelling wave solution, one-dimensional and steady on an average and travelling with a constant turbulent flame speed U_T relatively to the mean uniform flow. The very existence of such a solution is questionable. The mean volumetric rate per unit cross sectional area at which the fresh mixture is burned is

$$<U_n \Sigma>, \tag{9-a}$$

the bracket denotes the space average, U_n is the normal flame velocity (1-a) and $\Sigma = S/S_o$ the flame area per unit cross-sectional area $\Sigma = S/S_o$. If the mean position of the flame surface is well defined and planar, the turbulent flame speed may be defined properly by time averaging of (9-a) to give, when this last quantity is constant in time,

$$U_T = <U_n \Sigma>. \tag{9-b}$$

In the limiting cases where $d/l \to 0$ for all wavelengths of the turbulent flow, the deviation of U_n from U_L described by (2) is negligible and one has :

$$U_T = U_L<\Sigma> \tag{9-c}$$

The problem which is left is to relate $<\Sigma>$ to $\underline{u}_e$. Even in the approximation (9-c), the local effects (2), even weak $O(\varepsilon)$, may be important to determine the flame surface and thus U_T through the dependence of $<\Sigma>$ with $\underline{u}_e$. The existence of Eq.(9-c) is not sufficient for insuring a steady state, one has also to verify that the thickness d_T of the turbulent brush has a limit for $t \rightarrow \infty$. This is far from being a trivial question in particular for the ideal case of a turbulent flame propagating in an unbounded domain. Even in the approximation of a zero gas expansion $\rho_u = \rho_b$, for which the initial turbulence is not modified by the flame front ($\underline{u}_i = 0$), all these questions are presently open.

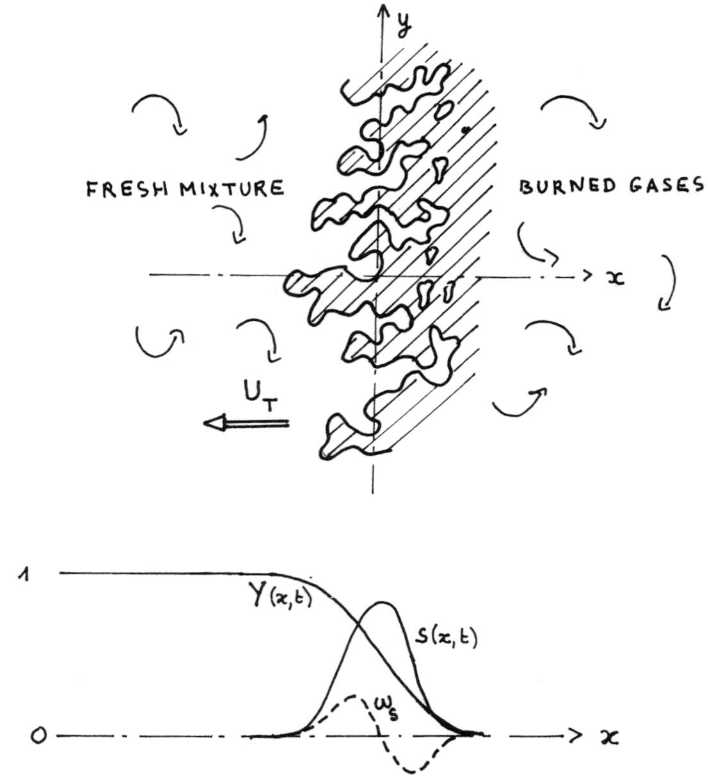

Fig. 3. Sketch of the turbulent flame brush in the flamelet regime.

4.2. Wrinkled flames

The problem is simpler in the wrinkled flame regime which is a particular case of the flamelet regime corresponding to a small intensity of turbulence $u' < U_L$. In this case (9-c) yields :

$$U_T = U_L \langle (1 + |\nabla \alpha|^2)^{1/2} \rangle. \tag{10}$$

This turbulent flame problem has been completely solved in the linear approximation[9] for stable fronts when earth gravity is the stability mechanism. In the unstable case, one must take into account the cellular structures and formations of cusps. We will come back to this case in the next section. Skipping these difficulties for the moment, Eq.(10) may be used in the wrinkled flame regime to evaluate the turbulent flame speed U_T in terms of the velocity fluctuation of the resulting turbulent flow u' (= $u'_e + u_i$) at the front $x = \alpha(y,z,t)$. Retaining only the linear term of Eq. (1-a) written in the referential flame where the mean position of the flame is at rest, $\partial\alpha/\partial t = u'$, the order of magnitude of $|\nabla\alpha|^2$ may be estimated in the framework of a Taylor hypothesis of a frozen field and of an isotropy approximation as

$$|\nabla\alpha|^2 \approx 2(u'/U_T)^2. \qquad (11\text{-a})$$

This yields [13] :

$$U_T/U_L \approx 1 + (u'/U_L)^2, \qquad (11\text{-b})$$

which is valid for a small turbulent intensity ($u'/U_L <1$). Extending (11-a) to high values of the turbulent intensity ($u'>U_L$), Eq. (10) leads to a linear dependence of the turbulent flame speed U_T/U_L on the turbulent intensity u'/U_L

$$U_T \approx u'. \qquad (11\text{-c})$$

This extrapolated result is not well justified here because it concerns a combustion regime outside the approximations of wrinkled flames. Such a result will be shown in §V to be relevant for the corrugated flamelet regime at a sufficiently high turbulent intensity.

5. THE CUSPED FLAME REGIME

When an unstable premixed flame propagates into a weakly turbulent flow ($u_e' \ll U_L$), irregular and chaotic cellular patterns appear on the flame front as a result of an interaction of the external noise constituted by the fluctuations in the velocity of the turbulent flow with the Darrieus Landau instability described by Eqs.(4-6). This regime which we will called the "cusped flame regime" is indeed the generic case of the turbulent wrinkled flame regime.

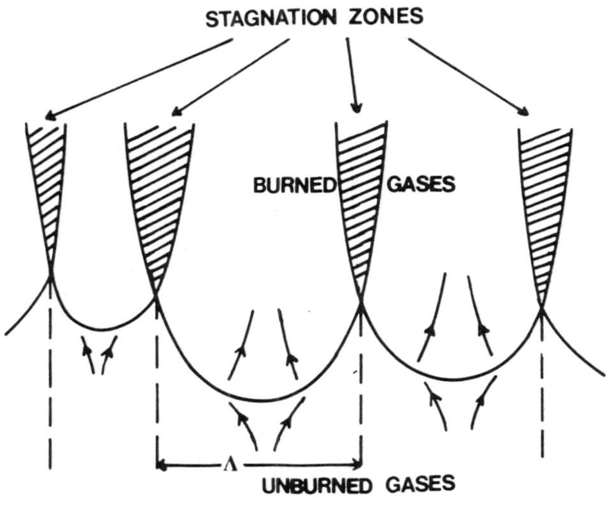

Figure 4. A cusped flame front is constituted of cells separated by sharp edges. Each cell can be considered as the pattern represented in Fig.4 for the propagation in a channel.

Consider for simplicity a free flame front propagating in the direction normal to its mean position which is assumed to be planar. Consider also for simplicity a two dimensional geometry (see Fig.4). On such a front, each cell can be considered as a curved front observed in channel and studied by Zel'dovich *et al.* [8]. The analogy is clearly exhibited when one mentally substitutes the symmetry axis of the cusps by the walls of the channel. To each of these cells is associated an induced flow u_i the amplitude of which is of the order of magnitude of U_L. At the tip of the cells, this flow has a stagnation point. When curvature effects are neglected at the cusps, stagnation zones must be introduced in the burnt gases. The question we want to address now is the selection of the size of these cells in a weakly turbulent flow. In the absence of external noise, cells with a size Λ much larger than the marginally stable wavelength Λ_m which is of the same order of magnitude as the flame thickness, are still stable according to Eq. (7). In this case, the size of the cells is determined by the dimension R of the vessel. As suggested by the numerical calculation of the M.S. equation[7,10], the basin of attraction of the largest possible cell is expected to be very large and, as a rule, this solution is selected in a long time limit for ordinary initial conditions. When the flame propagates in a weakly turbulent flow characterised by $u'_e < U_L$, the mean size of the cells $\langle\Lambda\rangle$ is controlled by the incoming turbulence playing the role of an external noise. In a straightforward manner , one obtains from Eq. (8) the following implicit relation for $\langle\Lambda\rangle$:

$$(<\Lambda>/\Lambda_m>) - \mathrm{Log}(<\Lambda>/\Lambda_m) = \mathrm{Log}\,(\Lambda_m/\alpha_e') \qquad (12\text{-a})$$

where α_e' is the amplitude of the fluctuation of the flame position at the tip produced by u'_e. Eq (12-a) can be considered as the result of a "mean field theory". Notice that $\alpha_e' = \Lambda_m$ is a critical value of Eq. (12-a) corresponding to $\langle\Lambda\rangle = \Lambda_m$: there is no more solution for $\alpha'_e > \Lambda_m$. Eq.(23a) is justified only for $\alpha'_e << \Lambda_m$ and is restricted to the most unstable initial perturbations with a wavelength equal to Λ_m. According to the results presented in §II, one must focus the attention only upon those vortices with a size Λ_m which are the most amplified. As $U_L > u'_e$, the front perturbation due to these vortices is determined by the time of interaction Λ_m/U_L between the vortex and the front in such a way that

$$\alpha'_e = \Lambda_m u'_e/U_L, \qquad (12\text{-b})$$

so that

$$(<\Lambda>/\Lambda_m) - \mathrm{Log}(<\Lambda>/\Lambda_m) = \mathrm{Log}\,(U_L/u'_e) \qquad (12\text{-c})$$

which determines the mean size $\langle\Lambda\rangle$ of the cells in terms of the turbulent intensity. More precisely u'_e is the amplitude of the incoming turbulent flow corresponding to the velocity fluctuations around the length scale $l = \Lambda_m$ .The second term in the l.h.s. of (12-c) is of second order in the limit $u'_e/U_L \to 0$ for which $<\Lambda>/\Lambda_m \to \infty$. Notice that the solution depends upon the curvature effects (2) through Λ_m. The situation described by (12-c) is rather unusual. Small scale structures Λ_m in the incoming turbulent flow of fresh mixture, are found to select the cell size which are much larger and, thus, to control the large scale structures $<\Lambda>$ in the burnt gas flow. Using the relation between the flame speed and the tube radius (cell size) of the approximate solution[8] for steady curved fronts propagating in a channel, the turbulent flame speed is obtained from the mean cell size $<\Lambda>$ given by Eq. (12-c). Notice also that Eq. (12c) delimits unambiguously the cusped flame regime by the relation $u'_e = U_L$: there is no more solution for $<\Lambda>$ when $u'_e > U_L$ and the mean cell size reaches its minimum value $<\Lambda> = \Lambda_m$ at $u'_e = U_L$. Numerical studies of (5) triggered by an external force have confirmed the above predictions[10].

6. THE CORRUGATED FLAMELET REGIME

6.1. *Position of the problem*

The regime of corrugated flamelets in turbulent flows occurs in the flamelet regime ($l >> d$) for a turbulent intensity sufficiently high, $u' >> U_L$. Fast eddies with $u' >> U_L$ roll up the front and can produce pockets (see fig. 3). The shape of the flame front results from the convolution of the effects of all these eddies. In such a situation, one may hope that hydrodynamical effects are less important than for the wrinkled flame regime and analysis based only on kinematics considerations ($u' \approx u_e$) could be of some interest. Even in this approximation, this regime is very difficult to analyse. The kinematic problem of the evolution of a front propagating with a normal velocity U_n in a prescribed turbulent flow $\underline{u}_e(\underline{r},t)$ is governed by the following field equation[18]:

$$\partial G/\partial t + \underline{u}_e \cdot \underline{\nabla} G = U_n |\nabla G| \qquad (13)$$

where $G(\underline{r},t) = Ct$ is the equation of the flame front. In the same conditions as those described at the beginning of §4, it has been shown[18] that $\langle\Sigma\rangle = \langle|\nabla G|\rangle$ holds for particular initial conditions. Despite recent works[19,20], the general solution of the stochastic equation (13) remains unknown and the problem of the evolution of a propagating front in a prescribed turbulent flow is far from being understood. In the absence of a general method to solve (13), we will present in the following a more phenomenological approach which is helpful for a better understanding of this difficult topic.

Without limiting mechanisms, the area Σ of a passive interface convected by a turbulent flow would increase indefinitely in time under the turbulent mixing process and Eq. (9-c) would not correspond to a stationary solution. The limiting mechanism lies on the non passive character associated with the normal flame speed $U_n \approx U_L \neq 0$. The basic idea is that the turbulent flow corrugates the flame front to engulf tongues or pockets of fresh mixture (see fig.3) with a characteristic thickness λ and a burning time $\tau_b = \lambda/U_L$. Without laminar combustion ($U_L=0$), the thickness $\lambda(t)$ of these tongues would decrease continuously in time to achieve the turbulent mixing process between the fresh and burned mixtures. But, for sufficiently small $\lambda(t)$, the decreasing rate U_L/λ by the laminar combustion becomes higher than the scale reduction rate by the turbulence. For a strong turbulent intensity, $u' >> U_L$, this combustion effect is negligible on the geometry of the front in the fresh mixture side where λ being large, the increase of Σ by the turbulence is dominant, but it becomes important at small λ in the burned gases where the laminar combustion leads to a consumption rate of Σ. A typical profile of the production rate of flame area across the flame brush, $\omega_s(x)$, is plotted in fig. 3 ($\omega_s < 0$ corresponds to a consumption). Model equations for the turbulent flame structure will be presented later on, see Eqs. (19 b-c). The laminar burning rate being the competitive process against the area increase and the basic mechanism responsible for the combustion, the limit $U_L/u' \rightarrow 0$ could well be singular. Semi phenomenological analysis of this problem are presented below.

6.2. *Flame vortex interaction*

The interaction of a flame front with vortices is an instructive example for the interplay of scales reduction, increase of flame surface by turbulent stretch and surface consumption by laminar combustion. Let us first consider the interaction of an initially planar and laminar premixed flame crossing a single eddy of size l and of a rotating speed u' with the axis perpendicular to the direction of propagation. Only order of magnitude estimates are presented here for the case $u' >> U_L$. For a detailed analysis see reference[11]. Considering a vortex of a size l with its centre on the flame sheet (see figure 5) and neglecting the motion of the centre relative to the front ($u'>>U_L$), the number of turns after a time t is proportional to

$$N \equiv u't/l. \qquad (14\text{-a})$$

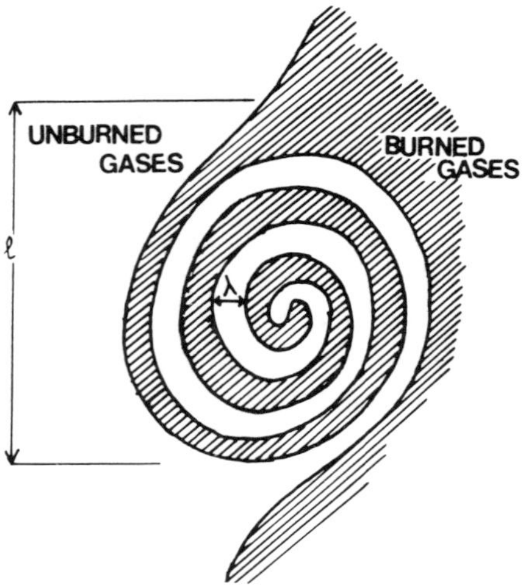

Figure 5. Sketch of the interaction between a flame and a vortex in the case $u' >> U_L$.

The corresponding thickness λ and burning time τ_b of the tongues of fresh mixture engulfed during the time t are respectively

$$\lambda \approx l/N = l^2/u't\ ; \qquad (14\text{-b})$$

and

$$\tau_b \approx \lambda/U_L = l^2(u'U_Lt)^{-1}. \qquad (14\text{-c})$$

In the limit $(u'/U_L) >>1$, the rollup is limited by the combustion of the tongues and proceeds as long as $t < \tau_b$. The fresh mixture involved in the vortex is completely burned when t is of order of τ_b to give

$$\tau_b \approx l/(u'U_L)^{1/2} \qquad (14\text{-d})$$

with a characteristic thickness at which laminar combustion becomes dominant given by:

$$\lambda \approx l(U_L/u')^{1/2}. \qquad (14\text{-e})$$

Notice that the combustion rate presents a maximum at the end of the combustion time of the eddy. Such phenomenon may be important for the generation of sound by a turbulent flame.

6.3.<u>Propagation mechanisms</u>

The flame propagation through an array of vortices such as the Taylor flow[12] is an instructive example. The combustion time τ_b to burn completely a vortex does not necessarily control the flame speed. This is the role of the contamination time τ_e of the flame front between two neighbouring vortices. But because of ignition thresholds which are frequently involved in combustion, the effective contamination time could be longer for the flame than for a passive scalar. As a consequence, τ_e may well depend on the state of the burning structure. A limiting case in which the flame propagation might be controlled by the combustion time τ_b instead of τ_e, corresponds to $\tau_e << \tau_b$ but with the strong assumption that the contamination mechanism from a vortex to the next one is not efficient for ignition before the combustion of the preceding vortex has gone to completion. In a similar way one may define a limiting regime of the flame propagation in turbulent flows in which the flame speed is controlled by the time τ_b necessary to burn the fresh mixture which is engulfed in the burned gases by the turbulent mixing process. We will consider first this limiting case in the next subsection §6.3-a. The other extreme regime in which the speed is controlled by the fastest contamination process is presented in §6.3-b. The real turbulent flame speed must range between these two limiting values. As it will be shown in § 6.4, the order of magnitude estimates for these two limiting values merge for a flame propagating in a fully developed turbulence represented by the Kolmogorov cascade with a high Reynolds number and with all lengthscales larger than the flame thickness.

6.3-a Propagation limited by combustion of engulfed tongues

Consider the flame propagation in the Taylor flow within the above described assumption of a propagation limited by the combustion. Even if this assumption is not well justified for this simplified flow, the method is worthy to be used for following applications in more complex flows. Within this assumption, the flame speed would be :

$$U_T \equiv (1/\tau_b), \tag{15-a}$$

to give, according to (14-d) :

$$U_T \approx (u'U_L)^{1/2}, \quad \text{i.e.} \quad U_T/U_L \approx (u'/U_L)^{1/2}, \tag{15-b}$$

with a thickness of turbulent brush equal to

$$d_T = l = u' \tau_e, \tag{15-c}$$

where $\tau_e \equiv l/u'$ is the turn over time. In fact (15-b) is a lower bound for the flame speed. Assuming scale separation, a renormalisation procedure has also been used[17] to extend this rough analysis to the case of a cascade of countable vortices with different sizes. The corresponding result for the turbulent flame speed is

$$U_T/U_L \approx (u'/U_L)^m \quad \text{with } 1/2 < m < 1, \quad m = 1 - (1-2^{-n})/n \tag{15-d}$$

where the value of $m > 1/2$ reflects the effects of smaller eddies which are superimposed on the largest coherent structures. The number of different families of vortices involved in the combustion of one coherent structure is n, and one has m=1/2 for n=1 but $m \rightarrow 1$ in the limit $n \rightarrow \infty$. This is a first instructive example because it shows how the laminar burning rate U_L may disappear from the final result even when the propagation mechanism was supposed to be controlled by the combustion

A similar approach based on the combustion time τ_b of the engulfed tongues of fresh mixture may be used to determine a lower bound of the turbulent flame speed in more general flows. The result (15-a) is based on a surface increase in the absence of combustion which is too slow for usual turbulence. In the model described by Eqs.(14), Σ growths linearly in time and λ

decreases as the inverse of the time. According to Batchelor [21], the surface of a passive front in a turbulent flow has to grow exponentially with the time as:

$$<\Sigma> \approx \exp(t/\tau_e) \qquad (16\text{-a})$$

where $\tau_e^{-1} = u'/l$ is the mean rate of strain of the turbulent flow. Such a behaviour has been checked by numerical analysis in the case of a single length scale in the spectrum of turbulence[22].The corresponding evolution of the mean thickness of the engulfed tongues $<\lambda>$ may be estimated as the ratio of volume of engulfed fresh mixture ($\approx d_T S_o$) to flame area $S_o\Sigma$:

$$<\lambda> \approx d_T / <\Sigma>, \qquad (16\text{-b})$$

(see also Eq.(19-a)). Without combustion, the thickness of the turbulent brush d_T is expected to follow a diffusion law:

$$d_T \approx (D_T t)^{1/2}, \qquad (16\text{-c})$$

where, in the simplest turbulent model of "eddy mixing length", the turbulent diffusion coefficient is $D_T \approx u' l$. The corresponding time evolution of $<\lambda>$ is :

$$<\lambda> \approx (D_T t)^{1/2} \exp(-t/\tau_e). \qquad (16\text{-d})$$

When $u' >> U_L$, the time τ_b to burn the tongues of fresh mixtures is obtained in a similar way as

for the vortices (see Eq. 14 c-d), $\tau_b = <\lambda> / U_L$ yielding an implicit relation for τ_b:

$$\tau_b = (D_T \tau_b)^{1/2} \exp(-\tau_b/\tau_e) / U_L. \qquad (16\text{-e})$$

During a time equal to τ_b the mean position of the front must moves forward at least on a distance of the order of magnitude equal to d_T. This yields the following lower bound for the turbulent flame speed:

$$U_T \approx (D_T \tau_b)^{1/2}/\tau_b = (D_T/\tau_b)^{1/2}. \qquad (16\text{-f})$$

Such an expression is worthy to be compared to the laminar case: the molecular diffusion coefficient is replaced by the turbulent diffusion coefficient and the laminar reaction rate by the inverse of the mean combustion time of the engulfed tongues,τ_b^{-1}. Eqs.(16-e,f), yield:

$$U_T / U_L \approx \exp(\tau_b/\tau_e) \approx \exp (u'/U_T)^2,$$

to give:

$$(U_T/U_L)^2 \ln(U_T/U_L) \approx (u'/U_L)^2, \qquad (16\text{-g})$$

where, according to the "eddy mixing length model", the turbulent intensity is $u'^2 \approx D_T/\tau_e$. Surprisingly, the result (16-g) is exactly the same as the one obtained by Yakhot[19] who carried out an iteration procedure of (11-b) by applying the renormalisation group method to Eq.(13). In the limit of high turbulent intensity, $u'/U_L >> 1$, Eq (16-g) gives:

$$U_T / U_L \approx (u'/U_L)(\ln u'/U_L)^{-1/2} \qquad (16\text{-h})$$

where the factor $(\ln u'/U_L)^{-1/2}$ is a second order perturbative term in the limit $u'/U_L \to \infty$. At the leading order, the lower bound of the turbulent flame speed has the same value as (15-d) for $n \to \infty$:

$$u'/U_L >> 1 \; : \; U_T \approx u'. \qquad (16\text{-i})$$

In the limit of a high turbulent intensity $u'/U_L \to \infty$, this lower bound does not depend any more on the laminar flame speed U_L. This result which is, here, a direct consequence of (16-a) is similar to (15-d).

6.3-b. Propagation controlled by contamination[30]

Consider now the opposite assumption of an early contamination in a flow with a turbulent intensity u'^2. This can be defined more precisely as follows: during a turn over time $\tau_e = l/u'$ a piece of flame front is transferred on an average at a distance of the order of magnitude of the integral scale l and the smallest piece of flame front is assumed to be sufficient to ignite the fresh mixture. According to such an assumption, the propagation speed would be equal to u',

$$U_T = u' \qquad (17\text{-a})$$

which may be written as

$$U_T = (l/\tau_e) = (D_T/\tau_e)^{1/2} \qquad (17\text{-b})$$

and the turbulent brush thickness would be given by:

$$d_T = u'\, \tau_b \qquad (17\text{-c})$$

By comparison with Eqs. (15 a-c) or (16-f), it appears that τ_e replaces τ_b in the turbulent flame speed and that τ_b controlled the flame thickness instead of τ_e.

The difference between these two propagation mechanisms has something similar to the different regimes of reaction-diffusion waves propagating in an unstable medium. See reference[23] for a recent review. When the nonlinear term representing the reaction rate is a sufficiently smooth function of the field variables representing species concentration and temperature, the KPP[24] analysis holds, and the propagation speed is controlled by the upstream reaction rate in the fresh mixture. But for sufficiently stiff reaction rate, the ZFK[25] regime holds and the flame speed is controlled by an integral of the reaction rate whose the main contribution is located in the downstream part of the flame thickness, close to the burned gas side. This last case is similar to the preceding section where the flame speed was assumed to be controlled by the time necessary to burn the tongues of fresh mixture engulfed in the burned gases.

This question is worthy to be considered in more details in the framework of a model inspired from the so called "coherent flamelet model" developed by Marble and Broadwell[27] for turbulent diffusion flames and explored extensively by S. Candel and coworkers[26] in turbulent premixed combustion. The question we want to address here may be treated in the simplified one-dimensional case presented at the beginning of §4 (see fig.3) and within the simplest turbulent modelling assumption of "eddy mixing length". One introduces two field variables, the flame surface per unit volume s(x,t) and the volume fraction of engulfed fresh mixture Y(x,t). This quantities are defined here as a space average in the y and z directions perpendicular to the mean propagation along the x-axis. One has by definition:

$$\langle\Sigma\rangle = \int_{-\infty}^{+\infty} s(x,t)dx. \qquad (18)$$

Then, Y(x,t) is assumed to be governed by a transport equation similar to the one for passive scalars but with an additional reaction term $\partial Y/\partial t = - sU_L$ corresponding to the consumption of fresh mixture by the normal burning velocity of the flame front. All the lengths in the turbulent spectrum have been assumed to be much larger than the laminar flame thickness ($d < l_K \Rightarrow U_n \approx U_L$). The basic assumption of the model is that the flame surface per unit volume s(x,t) is governed also by a transport equation but with a source term taking into account the exponential growth of the area by turbulent stretch (16-a), $\partial s/\partial t = s/\tau_e$, and a sink term describing the consumption of flame area by the laminar combustion of the engulfed tongues of fresh mixture, $\partial s/\partial t = - s/\tau_b$. As in the previous section, the reaction time τ_b is still defined by $\tau_b = \lambda/U_L$ but here the x variation of the local average thickness of the tongues λ is taken into account by introducing the local ratio of volume of engulfed fresh mixture to flame area:

$$\lambda(x,t) = Y(x,t) / s(x,t). \qquad (19\text{-a})$$

An averaging procedure based on a x integration in the burned part of the flame brush $\int_0^{+\infty} Y(x,t)dx \approx d_T$ and carried out in the referential frame where the mean flame position x=0 is at rest, yields, using (18), a result which is consistent with the global relation (16-b). Thus this model includes the mechanisms considered in the preceding section. The turbulent flame problem is reduced here to find the travelling wave solutions of the following system of differential equations:

$$\partial s/\partial t = D_T \partial^2 s/\partial x^2 + \omega_s(s,Y),$$

with

$$\omega_s(s,Y) = s/\tau_e - U_L s^2/Y, \qquad (19\text{-b})$$

$$\partial Y/\partial t = D_T \partial^2 Y/\partial x^2 - U_L s, \qquad (19\text{-c})$$

and with the boundary conditions:

$$\text{fresh mixture, } x \to -\infty : s=0,\ Y=1, \qquad (19\text{-d})$$

$$\text{burned gases , } x \to +\infty : s=0,\ \partial Y/\partial x = 0. \qquad (19\text{-e})$$

and where D_T and τ_e are constant parameters characterising the prescribed turbulent flow which are related to the turbulent intensity by $u'^2 \approx D_T/\tau_e$. By definition of s and Y, the solutions of (19 b-e) which do not satisfy $s \geq 0$ and $1 \geq Y \geq 0$ have to be rejected. Moreover when the quenching mechanisms are neglected, only the solutions satisfying Y=0 downstream,

$$\text{burned gases , } x \to +\infty : s=0,\ Y=0. \qquad (19\text{-f})$$

are consistent with the physical phenomena. Using (18) with an x integration of (19-c) across the flame brush, $\int_{-\infty}^{+\infty} dx$, Eq. (9-c) is found to be verified only for steady travelling wave solutions of (19 b-c), ($s(\xi)$ and $Y(\xi)$ where $\xi = x + U_T t$), which satisfy the boundary conditions (19-d) and (19-f) ($\xi \to -\infty$: s=0, Y=1 and $\xi \to +\infty$: s=0, Y=0). Existence and unicity of such a steady solution is not yet proved, and more work remains to be done on this simplified version of the model to clarify this point before taking into account local phenomena such as (2) or local quenching. Skipping these difficulties, one may obtain interesting informations by looking only at the upstream tail of the turbulent brush in the fresh mixture side. Here, s being small and Y close to unity, the solutions are governed by the linearised version of the equations (19 b-c) which is obtained by neglecting the last term in the r.h.s. of eqs.(19-b). In this region, the equation (19-b) has a closed form for s(x,t). Such an unsteady linear diffusion equation may be solved easily. Using for example an initial condition of the type $s(x,t=0) \propto \delta(x)$ where $\delta(x)$ is a Dirac distribution, one obtains:

$$s(x,t) \propto (4\pi D_T t)^{-1/2} \exp(-x^2/4D_T t + t/\tau_e). \qquad (19\text{-g})$$

Eq. (19 g) may be used in the limit $t \to +\infty$ to represent the upstream tail of the s-profile. It appears that no travelling waves with a speed smaller than the following value

$$U^*_T = 2(D_T/\tau_e)^{1/2} \approx u' \qquad (19\text{-h})$$

can exist because such solutions would be absolutely unstable, i.e. s(x,t) will diverge with t for $\xi = x + U_T t = Ct$ as $(4\pi D_T t)^{-1/2} \exp\{ (U^{*2}_T - U_T^2)\, t/4D_T\}$. In the K.P.P. case, the marginal solution (19-h) is the one which is selected in the long time limit for initial conditions localised in space[29]. Such a statement is not proved for (19 b-f). Nevertheless one may conclude that, without ignition threshold, a flame area increase by the turbulence represented by $\partial s/\partial t = s/\tau_e$ and associated with a crude turbulent mixing model such as the "eddy mixing length" $D_T \partial^2 s/\partial x^2$, leads to a lower bound for the turbulent flame speed controlled only by a contamination mechanism of the turbulent flow independently of the laminar combustion represented by the last term in the r.h.s. of (19-b) and (19-c). This is in agreement with (17 a-b) and shows that modifications to (19-h) such as (16-h) due to the combustion time τ_b, are unlikely in a pure flamelet regime. Modifying the nonlinear part of $\omega_s(s,Y)$ by introducing stretch and curvature effects such as (2) or nonlinear quenching phenomena, will not modify this lower bound U^*_T. The only way to change the result (19-h) is to modify the linear part of $\omega_s(s,Y)$ for example by introducing an ignition threshold. But as shown in the next section, this is unlikely for the basic case which has to be considered first: flames, far from the flammability limits, and propagating in a fully developed turbulence with a high turbulent intensity, $u' >> U_L$, and with all length scales larger than the laminar flame thickness, $d < l_K$.

Moreover, according to (16-i), the leading order in the limit $u'/U_L \to \infty$ of the flame speed lower bound controlled by combustion leads to a result similar to (19-h). This is because τ_e and τ_b become of the same order of magnitude for strongly turbulent flows. This question is considered in more details in the next section.

6.4. Flame propagation in a fully developed turbulence

The main purpose of this section is to take into account the multiscale aspect of a fully developed turbulence and to investigate its consequences upon the flame propagation. Consider as an example, the self similar scaling laws of the Kolmogorov cascade, valid in the inertial range between the integral length scale l_Λ and the Kolmogorov length scale l_K,

$$u' \propto \varepsilon^{1/3} l^{1/3} \ , \quad \tau' \equiv l/u' \propto \varepsilon^{-1/3} l^{2/3} \qquad \text{(20-a)}$$

with

$$l_K/l_\Lambda = Re^{-3/4}, \quad u'_K/u'_\Lambda = Re^{-1/4} \ , \quad \tau_K/\tau_\Lambda = Re^{-1/2}, \qquad \text{(20-b)}$$

where the Reynolds number is based on the integral scale $Re = u'_\Lambda l_\Lambda/\upsilon$ and where ε is the energy dissipation rate. By definition of the Kolmogorov scale and by definition of the laminar flame structure one has

$$l_K u'_K/\upsilon \approx d\, U_L/\upsilon \approx 1. \qquad \text{(20-c)}$$

In the flamelet regime one has $d < l_K$ and according to (20-c), $u'_K < U_L$. This insures that the transit time of the laminar flame $\tau_t \equiv d/U_L$ is smaller than the turn over time $\tau' \equiv l/u'$ of every eddy of the turbulent flow. As a consequence according to (2), the laminar flame structure cannot be substantially modified by the flow, $U_n \approx U_L$ and the quenching effects can be neglected. The Gibson scale[28] l_G defined by $u'_G = U_L$,

$$U_L = \varepsilon^{1/3} l_G^{1/3}, \qquad \text{(20-e)}$$

corresponds to the smaller eddies which can influence the geometry of the flame front because the interaction time l_G/U_L between the laminar flame front and those eddies is as small as the turnover time $\tau'_G = l_G/u'_G$. In the spirit of the assumption of scales separation, consider a geometrical series of lengthscales defined by:

$$l_i = l_\Lambda / a^i \quad \text{with } i = 1, 2, ...n \ \text{ and } a \geq 2, \qquad \text{(20–f)}$$

and the two companion series, u'_i and τ'_i given by (20-a). The sum $\Sigma_{i=1}^{i=n} \tau'_i$ converges to a value of the same order of magnitude as the integral time τ_Λ and, as it is well known, the sum $\Sigma_{i=1}^{i=n} (1/\tau'_i)$ diverges in the limit $Re \to \infty$ (see Eqs. 20 a-b). The different families of vortices having a multiplicative effect on the area of a passive surface one must have,

$$<\Sigma> \approx \exp \{ t [\Sigma_{i=1}^{i=n} (1/\tau'_i)] \} \qquad (21\text{-a})$$

instead of (16-a), with a growth rate which diverges in the limit $Re \to \infty$. The mean thickness of the engulfed tongues decreases to zero at a similar rate. This is the basic reason for a passive surface in a fully turbulent flow to become fractal after a time of the order of magnitude of the integral time τ_Λ. As a consequence, the molecular diffusion disappears from the turbulent mixing process in the limit $Re \to \infty$. For flame fronts, the Gibson cut off l_G yields:

$$\Sigma_i (1/\tau'_i) \approx \int_{lG}^{l\Lambda} (1/\tau')\, l^{-1}\, dl \approx (\tau'_G{}^{-1} - \tau'_\Lambda{}^{-1}) \to \tau'_G{}^{-1} \text{ when } Re \to \infty, \qquad (21\text{-b})$$

$$\Sigma_i\, \tau'_i \approx \int_{lG}^{l\Lambda} \tau'\, l^{-1}\, dl \approx (\tau'_\Lambda - \tau'_G) \to \tau'_\Lambda \text{ when } Re \to \infty, \qquad (21\text{-c})$$

and a direct consequence of the multiscale aspect of the turbulence is, according to (21 a-c), that the characteristic time for the area growth rate is the Gibso scale τ'_G (instead of τ_e in (16 a)), which is much shorter than the integral time, $\tau'_G << \tau'_\Lambda$. As the thickness of the corresponding tongues of engulfed fresh mixture decrease toward zero at a similar rate, the combustion of a volume of fresh mixture with a characteristic length l_Λ will be limited by the integral turn over time τ'_Λ. Thus, as already suggested by (15-d) and (16-i), the combustion time is no more influence by the laminar burning and is controlled by the integral time scale:

$$\tau_b \approx \tau'_\Lambda . \qquad (21\text{–d})$$

On another hand, the fastest contamination process is, as always, insured by the integral scales

$$\tau_e = \tau'_\Lambda . \qquad (21\text{–e})$$

As a consequence, the contamination and the combustion times have the same order of magnitude in the fully developed turbulence described above. Following previous analysis[23], this provides a strong argument in favour of the KPP scenario[24], which has been already mentioned for the model equations (19 b-c), and in which the propagation is fully controlled by the contamination mechanism. Thus, in the singular limit $U_L/u'_\Lambda \to 0$, the characteristics of the laminar combustion rate do not influence the propagation and the turbulent flame speed as well as the flame thickness are controlled by the integral scales of the turbulent flow:

$$U_T \approx u'_\Lambda \approx l_\Lambda / \tau'_\Lambda \qquad (21\text{-f})$$

with $$d_T \approx l_\Lambda. \qquad (21\text{-g})$$

Eq.(21-f) can be written in a more precise form as $\ln (U_T/U_L) = \ln(u'_\Lambda/U_L)\ \{1 + o(1)\}$ in the limit $Re \to \infty$ with $u'_\Lambda / U_L \to \infty$ in such a manner that the flamelet regime remains valid, $u'_K << U_L << u'_\Lambda$. This result is in agreement with the leading order of the Yakhot analysis[19]. But, by analogy with the KPP scenario, one may expect that the propagation must remain controlled by the turbulent contamination in all a wide range $u'_K << U_L << u'_\Lambda$. In this domain, U_T must be proportional to u'_Λ with a universal constant coefficient which could well be unity when the intermittence effects are neglected. Thus, corrective terms such as (16-h) are unlikely. The multiscale aspect becomes less pronounced when the ratio u'_Λ/U_L decreases, but this will not change the final results up to a critical value close to $u'_\Lambda/U_L = O(1)$ at which the KPP scenario fails in a manner similar to ordinary reaction diffusion waves[24]. Beyond this critical value, the contamination don't control any more the flame speed and (21-f) is modified to match (11-b) as described by (11-c). But hydrodynamical phenomena can surely not be neglected in this region and one has to deal with the cusped flame regime described in §IV. On the other limit, $u'_\Lambda/U_L \to \infty$, the results can be modified only close to the boundary of the flamelet regime, when the thick flame regime defined by $U_L<u'_K$ (or $\tau'_K<\tau_t$) is approached.

On another hand, results (21 e-g) will be modified drastically close to the flammability limits where ignition thresholds become important and perturb the contamination process. This is more specifically the case in lean hydrocarbon mixtures with air or in rich hydrogen mixtures with oxygen for which the overall Lewis number being larger than unity, the critical radius for ignition may become very large. Flame propagation is no more possible when this critical radius becomes larger than l_Λ. In such cases, lengthscales are expected to play an important role in the flame propagation.

The geometry of the turbulent flam is known to be a difficult task. Assuming self similar properties of the front, one obtains [28] according to (9-c) and (20 a-e):

$$U_T/U_L \approx (l_\Lambda / l_G)^{\delta_F-2} \approx (u'_\Lambda/U_L)^{3(\delta_F-2)} \qquad (21\text{-h})$$

where δ_F is the fractal dimension of the flame front. Moreover a direct comparison with (21-f) yields:

$$\delta_F = 2.333... \qquad (21\text{-i})$$

It is important is to notice that Eq.(21-d) beeing in fact valid for each lengthscale, the combustion time is equal to the corresponding turnover time, $(\tau_b)_i \approx \tau'_i$ everywhere in the length spectrum. Thus, the interaction time between vortices of size l_i and the effective flame front caracterised by a relative velocity renormalised by all the smaller lengthscales, is the turnover time τ'_i. Then, the amplitudes $u'_i\tau'_i$ of the flame corrugations produced by the ith family of vortices are of the same order of magnitude as the wavelength l_i and the effect of each family of vortices are well separated in the length spectrum of the flame. This justifies the assumption of a self similar property of the front geometry underlying (21-h) and (21-i). Another consequence concerns dynamical properties of the flame: the power spectra of flame area will reflect closely the dynamics of the turbulent flow. Moreover, because of the multiplicative process leading to (21-a), the distribution of fluctuations in the total flame area has to be of a log-normal type. Effect of intermittence is an open question. Spectral analysis of flame cross sections by laser tomography[31] are appropriate tools for such studies and more specially in the intermediate asymptotic cases $u'_K<<U_L<<u'_\Lambda$ for which the flame front geometry is a a mirror of the properties of the turbulent flow in the inertial range. All these results are also useful for the study of sound generation by turbulent flame[32].

ACKNOWLEDGEMENTS

Prof. Eric Siggia and Prof. Yves Pomeau are acknowledged for very stimulating discussions. Financial support was partly provided by a CCE twinning contract ST25-0029-1-F and the French ARC programme of CNRS “Combustion dans les moteurs alternatifs”.

REFERENCES

1. P. Clavin, F.A. Williams. J. Fluid Mech. **116**, 251-282 (1982)
2. P. Pelcé, P. Clavin, J. Fluid Mech. **124**, 219-237 (1982)
3. P. Clavin, G. Joulin, J. Phys.Lettre, **44**, L 1-12 (1983)
4. G.H. Markstein "*Nonsteady Flame Propagation*" Pergamon Press (1964)
5. B. Karlovitz et al. IV Combustion Symposium, 613-620. The Williams & Wilkins Company (1953)
6. L. Landau. Acta Physicochimica URSS, **19**, 77 (1944)

7. G.I. Sivashinsky, Acata Astronautica, **4**, 1177-1206 (1977)
8. Y.B. Zeldovich, A.G. Istratov, N.I.Librovich, Combust. Sci.Technol. **24**, 1 (1980)
9. G. Searby, P. Clavin, Combust. Sci. Tech.**46**, 167-193 (1986)
10. B. Denet. *"Simulation numérique d'instabilités de fronts de flammes"* Thèse de Doctorat Université Aix-Marseille I (1988)
11. N. Peters, F.A. Williams. XXII Combustion Symposium, 495-502, (1988)
12. G.I. Taylor. *The Scientific papers of G.I. Taylor*, **II**, 190-192 Cambridge (1960)
13. P. Clavin, F.A. Williams. J. Fluid Mech. **90**, 589-604 (1979)
14. P. Pelcé, *"Dynamics of curved fronts"*, Academic Press, (1988)
15. P. Pelcé, P. Clavin, Europhysics Letters, **3** (8) 907-913 (1987)
16. G.Joulin, "On the hydrodynamic stability of curved premixed flame" Journal de Physique, Mai (1989)
17. P.Clavin, in *"Disorder and Mixing"* NATO ASI Series E Vol **152**, 293-315, (1988)
18. A.R. Kerstein, W.T. Ashurst, F.A. Williams, Phys. Rev. **A** **37**, 2728-31, (1988)
19. V. Yakhot, Combust. Sci. Tech.**60**, 191, (1988)
    Combust. Sci. Tech.**62**, 127, (1988)
20. G. Sivashinsky , Combust.Sci. Tech. **62**, 77, (1988)
21. G.K. Batchelor, Proc. Roc. Soc. **A** **213**, 349 (1952)
22. W.H.P. Münch *"Turbulent distorsion of line and surface elements"* Ph.D. Thesis Cambridge, (1989)
23. P. Clavin, A. Liñan, NATO ASI Series B, Physics,**116**, 291-338 (1984)
24. A. Kolmogorv, I. Petrovsky, N.Piscounov, Bull.Univ.Moscou, vol.**N°6** (1937)
25. Y.B. Zeldovich, D.A. Frank-Kamenetskii. Acta Phys. URSS; **IX** 341-350 (1938)
26. E. Maistret et al. in Lecture Notes in Physics, **351**, 98-117 Springer (1989)
27. F.E. Marble, J.E. Broadwell, "The coherent flame model for turbulent chemical reaction" Tech. Report. Purdue Univ. (1977)
28. N. Peters, XXI Combustion Symposium , 1231-1250 (1986)
29. D.G. Aronson, F.H. Weinberger, Advances in Mathematics Vol. **30**, **1**, 33 (1975)
30. P. Clavin, E. Siggia, Combust. Sci. Tech. *to appear* (1991)
31. L. Boyer, Combust. Flame, 39, 321 (1980)

Paul CLAVIN Laboratoire de Recherche en Combustion. CNRS / Univ. Aix-Marseille I URA 1117 . S 252. Centre Universitaire de St Jérôme. 13397 Marseillle cedex 13 France.

C M BRAUNER AND C SCHMIDT-LAINE

Nonlinear mathematical analysis of some combustion problems

1. INTRODUCTION

Combustion modeling leads to interesting mathematical problems involving systems of nonlinear ODEs or PDEs of various types. It would be a very difficult task to present a unified approach to these problems. We shall however present a concise summary of some common difficulties appearing in their mathematical treatment.

Firstly, most of these models are derived via an asymptotic analysis, e.g. activation-energy asymptotics (see the pioneer results from the Russian school [26], [27]). The stretched reaction zone gives rise to an unbounded domain, most frequently the whole real line. One cannot therefore except compactness, although exponential decay of the solutions at infinity may compensate for this inconvenience. Many models lead to nonlinear eigenvalue problems which usually possess a trivial solution. The mathematical proof must oversome this obstacle. Obviously several physical constants are involved (Damköhler number, Lewis number, etc.), leading to multiparameter bifurcation.

Within the realm of the mathematical aspects of combustion, unsteady problems are of special interest. We mention the famous Kuramoto-Sivashinsky equation and the study of attractors, where several inroads have recently been made [17]. Among these evolution problems we find travelling waves, which are a part of the description of premixed flame. The location of the flame front is one of the unknowns, therefore free boundary/moving boundary problems arise. Geometrical tools (dynamical system theory) provide an adequate machinery for the qualitative behavior of the solutions.

Many other exciting problems also stem from combustion theory. References include the monograph of J. Buckmaster and G.S.S. Lufford [9], and P.C. Fife [12].

The following paper presents two examples of such nonlinear analysis in two different physical frameworks.

The first part is devoted to the study of a nonlinear differential system modelling a two-step reaction, considered for a certain distinguished limit of parameter values corresponding to fast recombination of the radicals. These kinetics are the most attractive of the two-step schemes that have been proposed for explaining interesting phenomena not covered by one-step kinetics [2]. The purpose is to find the "laminar-flame-eigenvalue" representing the burning rate as a function of essentially the ratio of the two reaction rates.

The second part is concerned with the nonlinear stability analysis of the travelling wave solution of a canonical problem from "Near Equidiffusional Flames" (NEF) theory. Two basic phenomena are usually found in the stability of plane deflagration waves : the hydrodynamic and Lewis number effects. The treatment of flame stability from a hydrodynamic point of view is due to Darrieus and Landau. However, if the constant-density approximation is adopted, the equations for temperature T and mass fraction Y have constant coefficients. This eliminates the hydrodynamic effect, leaving the Lewis-number effect.

The linear stability of "Slowly Varying Flames" (SVFs) and "Near Equidiffusional Flames" (NEFs) is investigated in [10, Lect. 5]. It is shown there that SVFs are unstable to plane disturbances for Lewis number > 1. Nonplanar disturbances are also unstable now for Lewis < 1 as well. Therefore the SVFs are unstable for all values of Lewis, which decreases their value as a class of solutions. In pursuit of stable flames, the more usual phenomenon, NEFs should be considered. We present here a nonlinear stability analysis of NEFs in the special case of equidiffusion of heat and species, namely Lewis $= 1$ (see also [6], [7], [18]), referring to [1] for a general study. In this case the free boundary problem simplifies to a one-phase problem, i.e. on one side of the flame sheet. We are going to study the stability of this one-phase problem in detail since it can be considered as a paradigm reflecting many situations.

Recently this class of problems has been reformulated as *fully nonlinear* parabolic equations [4, 5]. Therefore existence and stability follow from the general theory of abstract evolution equation. This new approach appears to be very promising, and is essential for the study of instabilities.

2. A PLANE PREMIXED FLAME PROBLEM WITH TWO-STEP KINETICS.

2.1 The set of equations

In [16], we introduced a two-step irreversible reaction for a steady plane flame, with chain-branching/chain-breaking kinetics :

$$A + X \to 2X \ , \quad 2X + M \to 2P + M \tag{2.1}$$

Radical X is obtained in the production step, which has a very large activation energy θ, and provides product P in the recombination step for which the activation energy is taken to be zero ; A is the reactant and M a third body.

This two-step scheme is presented as an alternative to classical single-step kinetics and allows the description of a wider range of phenomena.

The equations are derived in the stetched flame zero, described by the one dimensional space variable $\eta, -\infty < \eta < +\infty$, by assuming a fast recombination, i.e. that both production and recombination of radicals take place in the same thin zone ; the system reads (see [16, p. 423])

$$(2.2) \qquad \begin{cases} u'' = q_1\ r\ \delta(u-v)ve^{-u} + \delta(u-v)^2 \\ v'' = r\ \delta(u-v)ve^{-u} \end{cases}$$

and the associated boundary conditions are :

$$(2.3) \qquad \begin{cases} u = -\eta + o(1), v = -\eta + o(1) \text{ as } \eta \to -\infty \\ u = o(1), v = o(1) \text{ as } \eta \to +\infty\ . \end{cases}$$

The unknowns are positive functions u and v, representing temperature and mass fraction of the reactant and a positive constant δ, representing the burning rate. The parameters q_1 and q_2 are the proportions of the total heat released in the first and second steps of the reaction, so that $q_1 + q_2 = 1$. Physical considerations require the recombination step to be exothermic, so that $q_2 > 0$.

Finally r is a positive parameter, corresponding to the ratio of the two reaction rates. The boundary conditions (2.3) are obtained by matching with expansions on either side of the flame sheet.

The mathematical problem is the following : q_1, q_2 and $r > 0$ being given, find two functions $u > 0, v > 0$ and the constant $\delta > 0$ satisfying (2.2), (2.3).

It is particularly convenient for our study to deal with an equivalent formulation of the system involving the radical mass fraction $w = u - v > 0$. It consists of both systems :

$$(2.4) \qquad \begin{cases} v'' = r\ \delta\ v\ we^{-v}e^{-w} \\ w'' = -q_2\ r\ \delta v\ w\ e^{-v}e^{-w} + \delta w^2 \end{cases}$$

and

$$(2.5) \qquad \begin{cases} u'' = q_1\ r\ \delta(u-w)w\ e^{-u} + \delta w^2 \\ w'' = -q_2\ r\ \delta(u-w)we^{-u} + \delta\ w^2 \end{cases}$$

together with the boundary conditions :

$$(2.6) \qquad \begin{cases} v = -\eta + o(1),\ w = o(1) \text{ as } \eta \to -\infty \\ v = o(1),\ w = o(1) \text{ as } \eta \to +\infty \end{cases}$$

and

$$(2.7) \qquad \begin{cases} u = -\eta + o(1),\ w = o(1) \text{ as } \eta \to -\infty \\ u = o(1),\ w = o(1) \text{ as } \eta \to +\infty\ . \end{cases}$$

2.2 The existence question[4], [5], [23]

Let us introduce the following problem in the x-variable, which is obtained from (2.4) by taking δ equal to 1 :

(2.8)
$$\begin{cases} \dfrac{d^2v}{dx^2} = rv\ w\ e^{-v}\ e^{-w} \\ \dfrac{d^2w}{dx^2} = -q_2\ rv\ w\ e^{-v}\ e^{-w} + w^2 \end{cases}$$

We consider this nonlinear system subject the following boundary conditions :

(2.9)
$$\begin{cases} v' \to 0 \text{ as } x \to +\infty \\ w\ \to 0 \text{ as } x \to \pm\infty \end{cases}$$

together with the normalization condition

(2.10)
$$v(0) = 1.$$

We prove the following theorem :

THEOREM 2.1 For any $r > 0$ fixed, there exists a solution $(v,\ w) \in (C^\infty(\mathbb{R}))^2$ to problem (2.8)-(2-10), such that

(2.11)
$$0 < w < M_0 = q_2\ r/e$$

(2.12)
$$v > 0\ ;\ v' < 0\ ;\ v \to +\infty \text{ as } x \to -\infty,\ v \to 0 \text{ as } x \to +\infty.$$

Moreover, $w \in H^2(\mathbb{R})$ and, as $x \to -\infty$, $v(x) = -\ell(x - x_0)$+E. S. T., where $x_0 \in \mathbb{R}$ and

(2.13)
$$\ell = \frac{1}{q_2} \int_{-\infty}^{+\infty} w^2(x)\ dx = \frac{1}{q_2}\ \|w\|^2_{L^2(\mathbb{R})}.\square$$

To return to the initial problem (2.4) (2.6), we make the transformation

(2.14)
$$\eta = \ell(x - x_0)$$

Then (2.8) becomes

(2.15)
$$\begin{cases} \dfrac{d^2v}{d\eta^2} = \dfrac{1}{\ell^2}\ rvwe^{-v}\ e^{-w} \\ \dfrac{d^2w}{d\eta^2} = -\dfrac{1}{\ell^2}\ q_2rvwe^{-v}\ e^{-w} + \dfrac{1}{\ell^2}\ w^2 \end{cases}$$

and, as $\eta \to -\infty$, $v(\eta) = -\eta +$ E. S. T. Thus it is clear that (2.4) (2.6) is solved by the pair $(v(\eta),\ w(\eta))$ with

$$\delta = 1/\ell^2 \tag{2.16}$$

Therefore, we have determined

$$\delta = q_2 / \int_{-\infty}^{+\infty} w^2(\eta\ d\eta. \tag{2.17}$$

In [5], we give a detailed demonstration of Theorem 2.1. A sketch of the proof is the following : First, we exhibit positive upper and lower solutions of the problem in a formal way. Next, we consider the system (2.8) on a bounded interval $[-a, b]$ with suitable boundary conditions and we prove the existence of a solution (v, w) in a closed convex set K, by a fixed point argument (the convex K involves the upper and lower solutions). We derive a priori estimates in the C^1-norm as well as in the H^1-norm. Finally, we let a and b tend to $+\infty$, and prove the existence of a limit which solves (2.8) (2.10). Properties (2.12)(2.13) appear as a "spin-off" of the existence proof. An alternative proof by a topological shooting method has been presented by S. P. Hastings, C. Lu and Y. H. Wan [14].

In [16], a numerical shooting method is presented leading to the description of the solution and the response curve $\delta(r)$. The mathematical proof leads to a different iterative scheme whose convergence to a non trivial solution is ensured by the knowledge of explicit lower solutions. Figure 1 shows typical u-and w-profiles. Figure 2 gives the r, δ-curves for various values of q_1 less than 1, i.e. $q_2 > 0$. In particular, two negative values of q_1 are included, corresponding to endothermic production steps.

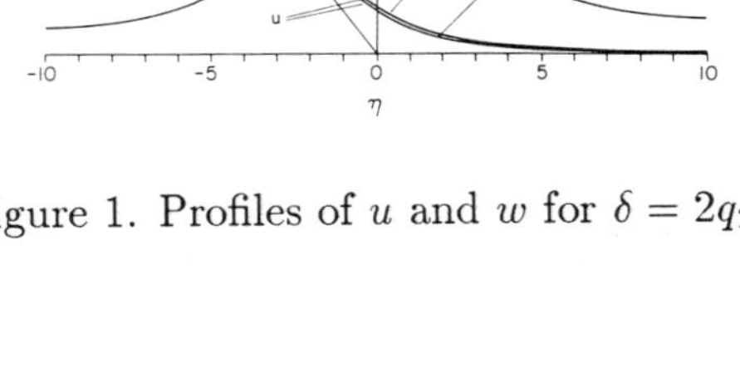

Figure 1. Profiles of u and w for $\delta = 2q_2$.

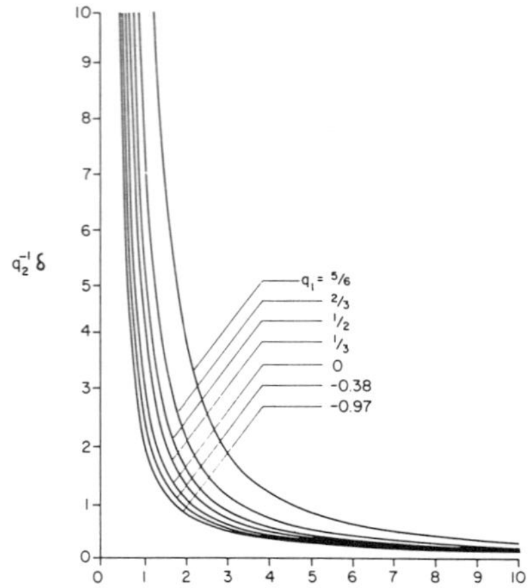

Figure 2. The response curves $\delta(r)$

2.3. The stability question[23], [24]

System (1.5) can be rewritten in the canonical form :

$$(2.18)\qquad \begin{cases} u' = p \\ p' = q_1\ r\delta(u-w)we^{-u} + \delta w^2 \\ w' = q \\ q' = -q_2\ r\delta(u-w)we^{-u} + \delta w^2 \end{cases}$$

for which $0 \in \mathbb{R}^4$ appears to be a fixed point, derived from the boundary conditions (2.7). By linearizing (2.18) about this point, it comes 4 zero eigenvalues ; such a degenerate fixed point requires more sophisticated treatment. Let us then consider the second order approximation of (2.18) near this point :

$$(2.19)\qquad \begin{cases} u' = p \\ p' = q_1\ r\delta(u-w)w + \delta\ w^2 \\ w' = q \\ q' = -q_2\ r\delta(u-w)w + \delta w^2\ . \end{cases}$$

This system only contains quadratic terms and can be considered in the general form of second order homogeneous problem. So, we consider the problem

$$(2.20)\qquad \frac{d^2X}{d\eta^2} = X'' = F(X)$$

where $X \in \mathbb{R}^n$ and F is a second order homogeneous function. More precisely, we study the stability of the degenerate fixed point $0 \in \mathbb{R}^{2n}$ of the autonomous dynamical system :

$$(2.21)\qquad \begin{cases} X' = Y \\ Y' = F(X) \end{cases}$$

The homogeneity property allows to reduce the dimension of problem (2.21). By means of the change of function :

$$x = \frac{X}{\|X\|}; y = \frac{Y}{\|X\|^{3/2}} \tag{2.22}$$

such that $(X,Y) \in \mathbb{R}^n \times \mathbb{R}^n \longmapsto (x,y) \in S^{n-1} \times \mathbb{R}^n$ where S^{n-1} is the unit sphere of $\mathbb{R}^n$, together with the change of variable defined by

$$\frac{ds}{d\eta} = \|X\|^{1/2}. \tag{2.23}$$

The reduced autonomous dynamical system is the following :

$$\begin{cases} \dfrac{dx}{ds} = y - x(x.y) \\ \dfrac{dy}{ds} = F(x) - \dfrac{3}{2}\ y(x.y) \end{cases} \tag{2.24}$$

where (.) denotes the euclidian scalar product in $\mathbb{R}^n$, and $\| \, . \, \|$ the associated norm.

Then we prove existence of at least a couple of symmetric points of (2.24) : $P_i = (x_i, y_i) \in S^{n-1} \times \mathbb{R}^n$ and exhibit the mapping between trajectories of (2.21) and trajectories of (2.24) ; an essential remark is that a trajectory of (2.24) provides by lifting a *family* of trajectories of (2.21), due to an arbitrary integration parameter ; the stability result is the application of the former property to invariant manifolds :

THEOREM 2.2. *Let $P_0 = (x,y) \in S^{n-1} \times \mathbb{R}^n$, $n \geq 1$, be a fixed point of (2.24), such that $(x_0.y_0) < 0$. Then*

(i) *The stable manifold $W^S(P_0)$ of problem (2.24) lifts to the stable manifold $W^S(0)$ of homogeneous problem (2.21).*

(ii) *dim $W^S(0) =$ dim $W^S(P_0) + 1$.*□

The main interest of this result is that investigation of the stable manifold $W^S(P_0)$ of problem (2.24) is often accessible by classical tool, namely by linearization, because the fixed points of (2.24) are usually hyperbolic. A complete stability analysis of fixed points P_0 of (2.24) is presented in [24], leading to results on the dimension of the stable manifold $W^S(0)$ of (2.21). These results are summarized as follows :

THEOREM 2.3. *Let us suppose that $F'(x_0)$ is similar to a diagonal matrix and that its eigenvalues λ_j are such $\lambda_j = a_j + ib_j$:*

$$\begin{cases} 6b_j^2 \neq 25\ \rho(\rho - a_j) \\ \rho = \dfrac{3}{2}(x_0.y_0)^2\ . \end{cases} \tag{2.25}$$

Then by lifting of $W^S(P_0)$ to $W^S(0)$

(2.26) $$\dim W^S(0) = \sum_{j \in J} m(\lambda_j)$$

where

$$J = \{j \in \mathbb{N} / \ 6b_j^2 > 25\ \rho(\rho - a_j)\}$$

and $m(\lambda_j)$ is the geometrical multiplicity of the eigenvalue λ_j.□

Applying this general method to (2.19) leads to a reduced system for which two fixed points P_0 and P_1 are obtained, verifying $(x_0.y_0) < 0$, $(x_1.y_1) < 0$. As $r = 1$, the two fixed points are no more distinct and this unique fixed point is degenerate. For $r \neq 1$, we obtain by linearization and lifting to (2.19) :

(2.27) $\quad \dim W^S(0) = 2$

The transcritical bifurcation case $r = 1$ is solved by means of the center manifold theorem.

3. NONLINEAR STABILITY ANALYSIS OF SINGULAR TRAVELLING WAVES IN COMBUSTION : A ONE-PHASE PROBLEM.

3.1. The NEF problem

The NEF (Near Equidiffusional Flames) theory is characterized by the requirements [10]:

(i) $$\mathcal{L}^{-1} = 1 - \ell/\theta$$

where $\ell = 0(1)$ is the reduced Lewis number and θ the activation energy and

(ii) $$H = H_f + 0(\theta^{-1})$$

where H is the enthalpy $Y + T$. The subscrip f (for fresh) stands for the value at $-\infty$ while b (for burnt) corresponds to the value at $+\infty$. The second requirement corresponds to using the following expansions for the temperature T and the reactant mass fraction Y:

(3.1) $$\begin{cases} T = T_0 + \theta^{-1}\ T_1 + \cdots \\ Y = (H_f - T_0) + \theta^{-1}(H_1 - T_1) + \cdots \end{cases}$$

Within this framework, the basic equations for a planar flame become on either side of the flame sheet $\xi = \xi(t)$ [10] :

(3.2) $$\begin{cases} \dfrac{\partial T_0}{\partial t} = \dfrac{\partial^2 T_0}{\partial x^2} \\ \dfrac{\partial H_1}{\partial t} = \dfrac{\partial^2 H_1}{\partial x^2} + \ell\ \dfrac{\partial^2 T_0}{\partial x^2}\ . \end{cases}$$

In the burnt gas the assumption of equilibrium leads to

$$T_0 = T_b, \ \frac{\partial H_1}{\partial t} = \frac{\partial^2 H_1}{\partial x^2} \ . \tag{3.3}$$

At the flame sheet, $\xi(t)$, the jump conditions are

$$[T_0] = 0 \ , \ [H_1] = 0 \tag{3.4}$$

$$\begin{cases} [\frac{\partial T_0}{\partial x}] = -Y_f \exp\,(H_1/2T_b^2) \\ [\frac{\partial H_1}{\partial x}] = -\ell\,[\frac{\partial T_0}{\partial x}] \end{cases} \tag{3.5}$$

where $[f] = f(\xi(t)^+) - f(\xi(t)^+)$. The boundary conditions are

$$T_0(-\infty, t) = T_f = 0, \ H_1(\pm\infty, t) = 0. \tag{3.6}$$

System (3.2)-(3.6) defines a two-phase Free Boundary Problem with the flame sheet as the moving boundary.

We point out that NEFs are a restricted class of solutions, therefore initial conditions must be consistent with the assumption that H is constant to leading order. As far as the stability question is concerned, the initial conditions are chosen as small disturbances of the traveling wave solutions of the above problem, which may be easily derived. These disturbances should not affect the conditions at the front, otherwise the equilibrium assuption would be violated.

Here we are interested in the nonlinear stability analysis of the traveling waves in the *equidiffusional case* $\ell = 0$, under a special class of disturbances, namely which pertubs the temperature only, so the (perturbation of) enthalpy H_1 is identically zero. Therefore we consider the following *one-phase problem* , subject to an adequate initial condition : Find a temperature T_0 and a front ξ obeying

$$\begin{cases} \frac{\partial T_0}{\partial t} = \frac{\partial^2 T_0}{\partial x^2}, \ x < \xi(t) \\ T_0 = T_b, \ x \geq \xi(t) \\ \frac{\partial T_0}{\partial x}_{|x=\xi^-(t)} = Y_f \ . \end{cases} \tag{3.7}$$

We are going to extend to this *singular* case a method developed by Sattinger [21, 22] for *smooth traveling waves* (see also Fife and Mc Leod [13]), Here the framework is rather different since the speed of the front is involved in the formulation, but is eliminated thanks to a Lyapunov - Schmidt-like procedure. For details the reader is referred to [18] and [7].

3.2. Mathematical formulation.

We are concerned with the nonlinear stability of the following one-phase problem, in whch a front $\xi = \xi(t)$ and a function (temperature) $u = u(\eta, t)$ are to be determined :

$$(3.8) \qquad \begin{cases} u_t = u_{\eta\eta}, & -\infty < \eta < \xi(t) \\ u(\xi(t), t) = u_*, u_\eta(\xi(t)^-, t) = 1 \\ u(-\infty, t) = 0 \end{cases}$$

where u_* is a fixed positive number.

It is straightforward to check that (3.8) admits a Travelling Wave (TW) solution $U(x,t) = U(x+ct)$ with velocity $-c, c > 0$:

$$(3.9) \qquad \begin{cases} y = x + ct, c = 1/u_* \\ U(y) = u_* e^{cy}. \end{cases}$$

Because of the translation invariance, $U^h(y) = U(y-h)$ is also a TW solution whenever $h \in \mathbb{R}$. In the moving coordinate frame $y = x + ct$, problem (1.1) now reads $(h(t) = \xi(t) - ct)$:

$$(3.10) \qquad \begin{cases} u_t + cu_y = u_{yy}, -\infty < y < h(t) \\ u(h(t), t) = u_*, u_y(h(t)^-, t) = 1 \\ u(-\infty, t) = 0. \end{cases}$$

The stability question consists in assigning an initial data

$$(3.11) \qquad u(y, 0) = U(y) + \varepsilon u_0(y), h(0) = 0.$$

where $\varepsilon > 0$ and u_0 is a given smooth function with compact support in $(-\infty, 0)$. The latter assumption on u_0 is to meet the requirements of NEFs and for the sake of simplicity.

As usual in autonomous TW problems, we do not think of classical stability, but rather of "orbital stability" ([21]) : As $t \to +\infty$, $h(t)$ converges to a phase shift h_∞ which can be calculated by formal integration of (3.10) a :

$$(3.12) \qquad h_\infty = -\varepsilon c \int_{-\infty}^{0} u_0 \, dy.$$

Therefore we do not expect u to converge to U but rather to approach U^{h_∞}, in an exponential way.

Finally it is convenient to reformulate (3.10) in a fixed coordinate frame $x = y - h(t)$:

$$(3.13) \qquad \begin{cases} u_t + (c - h_t)u_x = u_{xx}, -\infty < x < 0 \\ u(0, t) = u_*, u_x(0^-, t) = 1, u(-\infty, t) = 0 \\ u(0, x) = U(x) + \varepsilon u_0(x). \end{cases}$$

Our goal is to establish the existence of a function $u = u^\varepsilon(x,t)$ and a front $h = h^\varepsilon(t)$, solutions of (3.13), such that $u^\varepsilon \to U$ and $h^\varepsilon \to h_\infty$ exponentially when $t \to +\infty$. More precisely we will look for u^ε and h^ε in the form :

$$u^\varepsilon(x,t) = U(x) + \varepsilon v^\varepsilon(x,t), h^\varepsilon(t) = \varepsilon z^\varepsilon(t). \tag{3.14}$$

Clearly it is equivalent to prove the existence and exponential decay of v^ε to 0 and of z^ε to some value z_∞. Substituting (3.14) into (3.13) leads to :

$$\begin{cases} v_t^\varepsilon + cv_x^\varepsilon - v_{xx}^\varepsilon - z_t^\varepsilon\, U_x = \varepsilon z_t^\varepsilon v_x^\varepsilon \\ v^\varepsilon(0,t) = v_x^\varepsilon(0,t) = 0 \\ v^\varepsilon(x,0) = u_0(x). \end{cases} \tag{3.15}$$

3.3. Linear stability analysis.

The linear stability analysis corresponds to the resolution of (1.8) with $\varepsilon = 0$:

$$\begin{cases} v_t^0 + cv_x^0 - v_{xx}^0 - z_t^0\, U_x = 0 \\ v^0(0,t) = v_x^0(0,t) = 0 \\ v^0(x,0) = u_0(x). \end{cases} \tag{3.16}$$

For this purpose we start by defining a class of weighted Banach spaces introduced by Sattinger in [21, 22]. The space of real continuous, bounded, functions with the *sup norm* $\|\ .\ \|_0$ will be denoted by C_B^0. As for as the space variable is concerned, x belongs to $(-\infty, 0)$ and *uniform continuity* is implicit.

DEFINITION 3.1. *For $q(x) = e^{-cx/2}, C_{q,0} = \{v, qv \in C_B^0\}$ endowed with the norm $\|v\|_{q,0} = \|qv\|_0$. For j an integer $\geq 1, C_{q,j} = \{v, v^{(p)} \in C_{q,0}, p = 0, \ldots, j\}$.*

In the sequel we will denote $C_{q,0}$ by X as the reference space. On the space X, we define the unbounded operator L by :

$$\begin{cases} L\varphi = \varphi_{xx} - c\varphi_x \\ D(L) = \{\varphi \in C_{q,2}, \varphi_x(0) = c\varphi(0)\}. \end{cases} \tag{3.17}$$

THEOREM 3.1. *L is a sectorial operator in the space X. It has an isolated simple eigenvalue at the origin, while the essential spectrum is the half-line $(-\infty, -c^2/4]$.*

REMARK 3.1. A straightforward computation shows that, in the space C_B^0, 0 is not an isolated eigenvalue, which compromises the stability. The introduction of the weight q shifts the essential spectrum to the left and leaves the eigenvalue 0 (due to the translation invariance) isolated.

Sketch of the proof : To determine the spectrum $\sigma(L)$and the resolvent set, it is convenient to introduce the unbounded operator M in C_B^0 :

$$\begin{cases} Mv = v_{xx} - \dfrac{c^2}{4}v \\ D(v) = \{v \in C_B^2, v_x(0) = \dfrac{c}{2}v(0)\} \end{cases} \tag{3.18}$$

Clearly M is derived from operator L by the straightforward transformation $L = q^{-1}Mq$, and we see that $\rho(M) = \rho(L)$. The estimation of the resolvent is only technical.

For 0 is an isolated eigenvalue, $\sigma_1 = \{0\}$ and $\sigma_2 = (-\infty, -c^2/4] \cup \{-\infty\}$ are spectral sets. Let E_j be the projections associated with these spectral sets and $X_j = E_j(X), j = 1, 2$. Then $X = X_1 \oplus X_2$, the X_j are invariant under L, and, if L_j is the restriction of L to X_j, then

$$L : X_1 \to X_1 \text{ is bounded, } \sigma(L_1) = \sigma_1$$
$$D(L_2) = D(L) \cap X_2 \text{ and } \sigma(L_2) = \sigma_2.$$

X_1 is spanned by $U_x = e^{cx}$ and E_1 is defined on X by $E_1(\varphi) = (c \int_{-\infty}^{0} \varphi \, dx) \, U_x$.

The following results about L_2 follow from standard analytic semigroup theory (cf. e.g. [11, 15]) :

COROLLARY 3.1. *(i) L_2 is also a sectorial operator, which generates an analytic semi-group e^{tL_2} ;*

(ii) $\forall \delta \in (0, c^2/4)$, $\|e^{tL_2}\|_{\mathcal{L}(X_2)} \leq cst \; e^{-\delta t}, \forall t > 0$;

(iii) e^{tL_2} has the representation :

$$e^{tL_2} = \frac{1}{2\pi i} \int_{\Gamma} (\lambda I - L_2)^{-1} e^{\lambda t} d\lambda \tag{3.19}$$

where Γ is any contour which lies strictly in $\rho(L_2)$, $\arg\lambda \to \pm\theta$, $\theta \in (\pi/2, \pi)$, as $|\lambda| \to \infty$.

We now return to problem (3.16), to which we apply the operators E_1 and E_2. The splitting :

$$v^0 = E_1 \, v^0 + E_2 \, v^0 = p^0 \, U_x + w^0 \tag{3.20}$$

yields, since $E_1 v_x = E_1 v_{xx} = 0$:

$$\begin{cases} p_t^0 = z_t^0 \\ w_t^0 + c w_t^0 - w_{xx}^0 = 0. \end{cases} \tag{3.21}$$

In order to derive the boundary conditions associated to w_0, we calculate :

$$\begin{cases} w^0(0,t) = v^0(0,t) - p^0(t) U_x(0) = -p^0(t) \\ w_x^0(0,t) = v_x^0(0,t) - p^0(t) U_{xx}(0) = -c p^0(t) \end{cases} \tag{3.22}$$

hence p^0 may be eliminated through (3.22) a and b :

$$w_x^0(0,t) = c w^0(0,t). \tag{3.23}$$

Summing up, we find that w^0 verifies the following problem in the space X_2 :

(3.24) $$w^0_t = L_2 w^0, w^0(0) = E_2 u_0 \in X_2$$

whose solution is $w^0(t) = e^{tL_2} E_2 u_0$ from Corollary 3.1. Next, p^0 is explicitely given by (3.22)a, viz.

(3.25) $$p^0(t) = -w^0(0,t).$$

After (3.21)a, p^0 and z^0 differ up to a constant only. For $p^0(0)U_x = E_1 u_0$ and $z^0(0) = 0$

(3.26) $$z^0(t) = p(t) - c \int_{-\infty}^{0} u_0 \, dx.$$

REMARK 3.2. The trick (3.20) enables us to fully decouple (3.16), and reduces the problem to a PDE (3.24) in space of codimension 1, plus a trivial relation (3.25) in dimension 1. Therefore we refer to it as a Lyapunov-Schmidt like procedure.

The next step is to examine the asymptotic behaviour of w^0 for t large.

LEMMA 3.1. *Whenever* $0 < \omega < c^2/4$,
$$\|w^0(t)\|_{q,1} \leq cst \; e^{-\omega t}, \forall t > 0.$$

It follows from this Lemma and (2.9) that $p^0(t) \to 0$ as $e^{-\omega t}$ when $t \to \infty$, and so does $\|v^0(t)\|_{q,1}$. From (3.26),

(3.27) $$z^0(t) \to z_\infty = -c \int_{-\infty}^{0} u_0 \, dx.$$

To take into account this exponential decay as time approaches infinity it is convenient to introduce a class of suitable Banach spaces (Sattinger [21]) :

DEFINITION 3.2. *Let* Y *be a Banach space,* $\omega > 0$:

$$\mathcal{E}^\omega([0,+\infty);Y) = \{u \in C^0_B([0,+\infty);Y), \sup_{t>0} e^{\omega t} \|\varphi(t)\|_Y < \infty\}.$$

With this definition, Lemma 3.1 simply reads : $w^0 \in \mathcal{E}^\omega([0,+\infty);C_{\varsigma,1}), 0 < \omega < c^2/4$.

3.4. Nonlinear Analysis

We now turn to the full nonlinear problem (3.15) with $\varepsilon > 0$. We apply the same Lyapunov-Schmidt-like procedure as in Section 3.3. However the speed of the front may not be completely eliminated.

(3.28) $$v^\varepsilon = E_1 v^\varepsilon + E_2 v^\varepsilon = p^\varepsilon U_x + w^\varepsilon$$

(3.29) $$p_t^\varepsilon = z_t^\varepsilon, \; p^\varepsilon(0) = -c \int_{-\infty}^{0} u_0 \, dx$$

(3.30) $$\begin{cases} w_t^\varepsilon = L_2 w^\varepsilon + \varepsilon p_t^\varepsilon v_x^\varepsilon, \; p^\varepsilon(t) = -w^\varepsilon(0,t) \\ w^\varepsilon(0) = E_2 u_0. \end{cases}$$

We need some insight on abstract Cauchy problems in Banach spaces and Hölder regularity of solutions. First we recall the following definition :

DEFINITION 3.3. *Let Y be a Banach space, $k \in \mathbb{N}$, $0 < \theta < 1$.*

$$C^{k,\theta}([0,+\infty); Y) = \{\varphi \in C_B^k([0,+\infty); Y, < \varphi^{(k)} >^{(\theta)} > \infty\}$$

where $$< \varphi >^{(\theta)} = \sup_{\substack{t,t+h>0 \\ |h| \le h_0}} \frac{\|\varphi(t+h) - \varphi(t)\|_Y}{|h|^\theta}$$

We also need the space $C_0^{0,\theta}([0,+\infty); Y)$ of functions of $C^{0,\theta}([0,+\infty); Y)$ such that $\varphi(0) = 0$.

Next, we define the operator $K : f \longmapsto u$ where u is the solution of the auxiliary Cauchy problem :

(3.31) $$u_t = L_2 u + f, \; u(0) = 0.$$

Operator K enjoys two kinds of important properties which are collected into two Lemmas :

LEMMA 3.2. *For any $\omega, 0 < \omega < c^2/4$, the operator K is a bounded transformation from $\mathcal{E}^\omega([0,+\infty); X_2)$ into $\mathcal{E}^\omega([0,+\infty); E_2\, C_{q,1})$.*

LEMMA 3.3. *The operator K is bounded transformation from $(C_0^{0,\theta}([0,+\infty); X_2)$ into $C^{1,\theta}([0,+\infty); X_2) \cap C^{0,\theta}([0,+\infty); D(L_2))$. Moreover $u_t(0) = 0$.*

This lemma is based on Corollary 3.1 and the following abstract result :

THEOREM 3.2. *Let Y be a Banach space, A the infinitesimal generator of an analytic semi-group e^{tA} such that*

(3.32) $$\|e^{tA}\|_{\mathcal{L}(Y)} \le M e^{-\delta t}, \; M \text{ and } \delta > 0.$$

Let u be the solution of the Cauchy problem

(3.33) $$u_t = Au + f, \; u(0) = 0.$$

Then the mapping $f \longmapsto u$ *is bounded transformation from* $C_0^{0,\theta}([0,+\infty);Y)$ *into* $C^{1,\theta}([0,+\infty);Y)$.

This theorem is a by-product of abstract results for evolution equations in Banach spaces (cf. Sinestrari [25, p. 53], Pazy [19, p. 478]), extended to the case $T = +\infty$ under the additional hypothesis (3.32) - see [18].

We are now in position to reformulate (3.29) (3.30) with the help of the operator K :

$$\begin{cases} \mathcal{F}(w^\varepsilon;\varepsilon) = w^\varepsilon + \varepsilon\ K[p_t^\varepsilon(cp^\varepsilon\ U_x + w_x^\varepsilon)] - w^0 = 0 \\ p^\varepsilon(t) = -w^\varepsilon(0,t). \end{cases} \tag{3.34}$$

For $0 < \omega < c^2/4$ and $0 < \theta < 1$, we define the functional space :

$$\begin{aligned} W = \{w \in C^{1,\theta}([0,+\infty);X_2) \cap C^{0,\theta}([0,+\infty);D(L_2)), \\ w_t(0,0) = 0\} \cap \mathcal{E}^\omega([0,+\infty);E_2C_{q,1}) \end{aligned} \tag{3.35}$$

equipped with the natural norm.

Thanks to Lemmas 3.2 and 3.3 we are able to prove the

THEOREM 3.3. $w^0 \in W$ *and the mapping* $\mathcal{F}$ *defined by (3.34) is a Fréchet differentiable mapping from* $W \times \mathbb{R}^+$ *into* W.

We may thus apply the implicit function theorem to the mapping $\mathcal{F}$. The Fréchet derivative of $\mathcal{F}$ at $(w^0,0)$, is clearly the identity. Therefore we have the following result, *where* ω *is any fixed number in the interval* $(0,c^2/4)$:

THEOREM 3.4. *There exists a* C^1 *mapping* $\varepsilon \to w^\varepsilon$ *from some interval* $[0,\varepsilon_0)$ *into* W *such that* $\mathcal{F}(w^\varepsilon;\varepsilon) = 0$. *Therefore problem (3.3) admits a unique solution* $(w^\varepsilon,p^\varepsilon)$ *for* $\varepsilon \in [0,\varepsilon_0)$ *such that, as* $t \to +\infty$:

$$\|w^\varepsilon(t)\|_{q,1} \le cst\ e^{-\omega t}. \tag{3.36}$$

For $p^\varepsilon(t) = -w^\varepsilon(0,t)$, $|p^\varepsilon(t)| \le cst\ e^{-\omega t}$ *also.*

Finally we have the following Theorem which provides a complete answer to the stability question.

THEOREM 3.5. *Define* $h_\infty = h_\infty^\varepsilon = -\varepsilon c \int_{-\infty}^{0} u_0\, dx$. *Whenever* $\varepsilon \in [0,\varepsilon_0)$, *problem (1.6) admits a unique solution* $(u,h) = (u^\varepsilon,h^\varepsilon)$ *such that, as* $t \to +\infty$:

$$\begin{cases} \|u^\varepsilon(t) - U\|_{q,1} \le cst\ e^{-\omega t} \\ |h^\varepsilon(t) - h_\infty^\varepsilon| \le cst\ e^{-\omega t}. \end{cases} \tag{3.37}$$

REFERENCES

[1] F. ALABAU, A. LUNARDI, Behaviour near the travelling wave solutions of a free boundary system in NEF theory, in preparation.

[2] H.BERESTYCKI, B. NICOLAENKO, B. SCHEURER, Travelling wave solutions to combustion models and their singular limits, SIAM J. Math. Anal., 16 (1985), pp. 1207-1242.

[3] M. BERTSCH, H. HILHORST, Cl. SCHMIDT-LAINÉ, The well-posedness of a free-boundary problem arising in combustion theory, to appear.

[4] C.-M. BRAUNER, A. LUNARDI, Cl. SCHMIDT-LAINÉ, Une nouvelle formulation de modèles de fronts en problèmes totalement non linéaires, C. r. Acad. Sci. Paris, 311(1990), pp. 597-602.

[5] C.-M. BRAUNER, A. LUNARDI, Cl. SCHMIDT-LAINÉ, Stabilité et instabilité des ondes stationnaires d'un problème de détonation à deux phases, C. r. Acad. Sci. Paris, 311(1990), pp. 697-700.

[6] C.-M. BRAUNER, S. NOOR EBAD, Cl. SCHMIDT-LAINÉ, Sur la stabilité d'ondes singulières en combustion, C. r. Acad. Sci. Paris, 308 (1989), pp. 159-162.

[7] C.-M. BRAUNER, S. NOOR EBAD, Cl. SCHMIDT-LAINÉ, Nonlinear stability analysis of singular travelling wave in combustion : a one-phase problem, Nonlinear Analysis, 16 (1991), pp. 881-892.

[8] C.-M. BRAUNER, Cl. SCHMIDT-LAINÉ, Existence of a solution to a certain plane premixed flame problem with two-step kinetics. SIAM J. Math. Anal., 18 (1987), pp. 1406-1415.

[9] J. BUCKMASTER, G.S.S LUDFORD, Theory of laminar flames, Cambridge Univ. Press, New-York, (1982).

[10] J. BUCKMASTER, G.S.S LUDFORD, Lectures on mathematical combustion, CBMS NSF Conference series in Appl. Math. 43 (1983), SIAM.

[11] N. DUNFORD, J. T. SCHWARTZ, Linear operators I, Interscience, New-York, (1957).

[12] P.C. FIFE, Dynamics of internal layers and diffusive interfaces, CBMS-NSF Conference series in Appl. Math., 53 (1988), SIAM.

[13] P.C. FIFE, J.B.Mc LEOD, The approach of solutions of nonlinear diffusion equations to travelling front solutions, Arch. Rat. Mech. Anal. 65 (1977), pp. 335-361.

[14] S.P. HASTINGS, C. LU, Y.H. WAN, A three dimensional shooting method as applied to a problem in combustion theory, Phys. 19D (1986), pp. 301-306.

[15] D. HENRY, Geometric theory of semilinear parabolic equations, Lecture Notes in Math., 840, Springer, New-York (1981).

[16] G. JOULIN, A. LINAN, G.S.S. LUDFORD, N. PETERS, Cl. SCHMIDT-LAINÉ, Flames with chain-branching/chain-breaking kinetics, SIAM J. Appl. Math. (1985), pp. 420-434.

[17] B. NICOLAENKO, B. SCHEURER, R. TEMAM, Some global dynamical properties of the Kuramoto-Sivashinsky equation : Nonlinear stability and attractors, Phys. 16D (1985), pp. 155-183.

[18] S. NOOR EBAD, Aspects théoriques et numériques de stabilité d'ondes dans des modèles de combustion, Ph.D. Thesis, Lyon (1989).

[19] A. PAZY, Semi-groups of linear operators and applications to partial differential equations, Appl. Math. Sci. 44, Springer-Verlag, New-York.

[20] V. ROYTBURD, A Hopf bifurcation for a reaction-diffusion equation with concentrated chemical kinetics, J.D.E. 56 (1985), pp. 40-62.

[21] D.H. SATTINGER, On the stability of waves of linear parabolic systems, Adv. in Math. 22 (1976), pp. 312-355.

[22] D.H. SATTINGER, Weighted norms for the stability of travelling waves, J.D.E., 25 (1977), pp. 130-144.

[23] Cl. SCHMIDT-LAINÉ, Sur quelques problèmes non linéaires en Mécanique des Fluides, Chimie et Combustion, Thèse d'Etat, Univ. Lyon I, (1985).

[24] Cl. SCHMIDT-LAINÉ, D. SERRE, Etude de stabilité d'un système non linéaire de dimension 4 en combustion et généralisation à une classe de problèmes homogènes de degré 2, Phys. 12D (1986), pp. 42-60.

[25] E. SINESTRARI, On the abstract Cauchy problem of parabolic type in spaces of continuous functions, J. Math. Anal. Appl. 107 (1985), pp. 16-66.

[26] Ya. B. ZELDOVICH, D.A. FRANCK-KAMENETSKII, Theory of uniform flame propagation, Dolk. Akad. Nauk. SSSR, 19 (1938).

[27] Ya. B. ZELDOVICH, On the theory of flame propagation, Zh. Fiz. Khim., 22 (1948).

C.-M. BRAUNER
Centre de Recherche en Mathématiques de Bordeaux
Université Bordeaux I
351, Cours de la Libération
33405 Talence Cedex - France.

Cl. SCHMIDT-LAINÉ
Laboratoire de Mathématiques, CNRS UMR 128,
Ecole Normale Supérieure de Lyon
46, Allée d'Italie
69364 Lyon Cedex 07 - France.

2 Turbulent combustion

F A WILLIAMS

Premixed turbulent combustion fundamentals

1. INTRODUCTION

These lectures attempt to describe the wide variety of approaches that have been used in addressing premixed turbulent flame propagation. The presentation here focuses mainly on summaries of the current status and on literature citations for entry into recent work. In addition, in a final section a summary of the course, and an appraisal of future prospects is attempted.

2. REGIMES OF PREMIXED TURBULENT COMBUSTION

It is well known that at high turbulence Reynolds numbers there is a range of length scales of turbulent eddies, from the Kolmogorov scale to the integral scale, with the Kolmogorov subrange between. There are corresponding Kolmogorov and integral time scales, and Damköhler numbers may be defined as the ratio of either of these to a chemical time. The simplest way to view regimes of turbulent combustion is in a plane of Damköhler number and Reynolds number [8]. In this plane there is a tendency at large Damköhler numbers to have reaction-sheet regimes and at small Damköhler numbers to have distributed-reaction regimes. Other parameters can be important in identifying subregimes. For premixed combustion the interpretations are especially rich, with single-sheet and multiple-sheet subregimes of the reaction-sheet regime being defined [7,8].

Numerical estimates have shown that premixed turbulent combustion often occurs in the reaction-sheet regime. In particular, spark-ignition engines have generally been calculated to operate in the reaction-sheet regime. Therefore much of the attention is focused on this regime and on its subregimes.

3. CLASSIFICATION OF APPROACHES TO THE THEORY OF PREMIXED TURBULENT COMBUSTION

It is possible to place the various theoretical approaches that have been employed into eight categories:

a. Direct numerical integration.

   This simulation method is becoming more and more important as computer capabilities increase.

b. Evolution of probability-density functions.

This approach has a firm statistical foundation but requires modeling, mainly of two-point quantities. Since we still have insufficient experience in this modeling, there are uncertainties in the predictions, despite the apparent ease with which chemistry can be included.

c. Approximation of probability-density functions.

This "presumed-PDF" approach introduces parameterized probability-density functions, with moment methods employed to calculate the evolution of the parameters. It is useful in reaction-sheet regimes and is now overcoming difficulties with countergradient diffusion for premixed flames by continuing to second-order moment-method closures.

d. Moment methods.

These include mixing-length theories, k-ε-g approaches ("one-and-a-half" order closures) and full second-order closures that are finding increasing application. The methods are among the oldest, yet remain the workhorses for complex flows.

e. Age theories.

These approaches, such as ESCIMO, are phenomenologically based rather than being derived from Navier-Stokes equations with chemical reactions. They play important roles in certain chemical-engineering situations, for example.

f. Perturbation methods.

Restricted mainly to low-intensity turbulence, these methods give well-justified predictions over their limited range of applicability.

g. Renormalization methods.

This approach, from theoretical physics, has recently been applied to extend the low-intensity reaction-sheet results to high-intensity turbulence [1,6,9]. The results are attractive, although their justification remains unclear.

h. Fractal and other approaches.

There are important recent ideas on fractal descriptions of premixed turbulent combustion [2,4]. Fractals tend to be descriptive and to lack dynamical content enabling burning velocities to be obtained, but the ideas are convenient for data reduction.

4. TURBULENT FLAME PROPAGATION IN THE REACTION-SHEET REGIME

It is possible to write a field equation describing premixed turbulent flame propagation in the reaction-sheet regime [3]. Front-tracking methods offer an alternative to solving the field equation. The most efficient approach for numerical studies is not entirely clear, but it likely depends not only on whether the field or front description is adopted but also on the numerical procedures employed. The challenge is to have an efficient procedure that includes front-shortening in the fractal subregime of multiple reaction sheets.

Turbulent burning velocities in the reaction-sheet regime can be related to laminar burning velocities and to average wrinkled-flame areas, sometimes with strain-curvature corrections to laminar burning velocities included. Comparisons of different burning-velocity formulas reveal substantial differences, depending on the approach adopted. A general conclusion is that the ratio of turbulent to laminar burning velocity is independent of turbulence scale in the limit of large-scale turbulence. Experiments on turbulent burning velocities show much scatter; better experiments are needed.

5. PREMIXED FLAMES IN VORTICES AND IN OTHER TURBULENCE STRUCTURES

Descriptions of flames in turbulent flows in reaction-sheet regimes are needed in analyzing premixed turbulent flame propagation. In the simplest approximation these flamelets propagate normal to themselves, with respect to the fresh mixture, at the normal steady laminar burning velocity. In more accurate approximations flamelet-structure modifications by turbulence are included. The structure of a planar sheet can be analyzed most conveniently by considering a simple stagnation flow. Many corresponding laminar-flame experiments have been performed, and thorough theoretical analyses now exist. These are deemed relevant because a general flow may locally be decomposed into a uniform translation plus a rigid-body rotation plus a pure strain in the constant-density fresh mixture, and the first two do not affect the structure of a planar flame sheet. The stagnation flow describes the effect of the pure strain on the planar flame sheet.

It is possible to define weak, moderate, and strong strain rates (or flame stretch) on the basis of an appropriate Damköhler number employing a fluid time that is the reciprocal of a local strain rate. Extinction occurs when this Damköhler number is small enough, and depending on the Lewis number, the extinction may be gradual or abrupt. Extinction removes the turbulent flame from the reaction-sheet regime and therefore needs to be considered. A great deal of information is now available on these extinctions and on structures of strained flames.

Another configuration of relevance to turbulent combustion is the premixed flame in a vortex [5]. Analysis shows that rollup of the flame can lead to a burnt-gas core growing as the 2/3 power of time, and extinction can occur in the near wings where the strain rate is highest. More studies are needed of flamelets in other configurations to see how general the canonical stagnation-flow configuration really is.

6. NEEDS IN PREMIXED TURBULENT COMBUSTION

Current needs can be listed as follows:

a. Careful evaluation of data on turbulent burning velocities.
b. Clarification of the basis of renormalization procedures.
c. Analysis of alternative flamelets, such as flame propagation along vortex cores.
d. Study of effects of laminar-flame instabilities.
e. Analysis of flamelet extinction dynamics.
f. Determination of the extent of multiple-sheet occurrence.

g. Analysis of the dynamics of sheets with holes in them.

h. Development of theories for turbulent burning velocities in the regime of broken flamelets.

i. Analysis of further real-chemistry effects such as afterburning.

7. SUMMARY AND FUTURE PROSPECTS

These lectures have uncovered a number of general conclusions, which may be listed as follows:

a. Under many conditions combustion occurs in thin reaction zones. This has been seen in laminar flames, in detonations, and in turbulent flames. It may be concluded that understanding of thin reaction-zone structure is a central aspect of understanding combustion.

b. Methods of matched asymptotic expansions for combustion problems have been well developed in the past twenty years. They are extremely important in combustion – even the idea of the existence of a laminar burning velocity is basically an asymptotic concept. There have been many developments in activation-energy asymptotics for one-step chemistry approximations, such as theories of slowly varying flames and of nearly equidiffusional flames. More recently, Damköhler-number asymptotics and reaction-rate-ratio asymptotics are playing prominent roles in developing understanding of flames with real chemical kinetics. Reduced chemical-kinetic mechanisms are key to the comprehension of influences of chemistry on flames.

c. Instabilities are prominent throughout combustion and contribute to the fascination of the subject. There are acceleration, hydrodynamic, and diffusive-thermal instabilities of laminar flames, for example, as well as cellular instabilities of detonations. Diffusion flames exhibit fewer varieties of instabilities (lacking the hydrodynamic instability), but cases of diffusive-thermal instabilities in these flames have been encountered, for example. Studies of instabilities thus are significant in combustion.

d. Computation has become a major third of laminar and turbulent combustion. For laminar flames, full detailed chemistry with more than 300 elementary steps has now been handled. For turbulent premixed flames and turbulent diffusion flames there have been great advances, but there still are many significant limitations for computational methods. A mark of the emerged role of computation is that nowadays theory strives as much to explain computational results as to explain experimental results. The three thirds of combustion are experiment, computation, and theory.

e. Turbulent combustion is a multifaceted subject, spanning from practical engineering, through mathematics and chemistry, into theoretical physics. There has been remarkable growth in activity and in understanding in turbulent combustion in the past twenty years. Yet, many challenging problems remain. There are important roles for theory, for experiment, and for computation in meeting these challenges.

Where do we go from here? Although it is always hazardous to attempt to predict the future, some general guesses can be made:

a. Studies of reduced kinetic mechanisms will bring vast improvement in our knowledge of laminar combustion, stability, and ignition and extinction. Systematic methods for pursuing these studies are now in place, and exploitation of these methods may be expected to provide great advances.

b. Although the heyday of activation-energy asymptotics is over, continuing advancements will occur through these methods, for example in analyzing ignition and extinction dynamics and the behaviors of holes in reaction sheets. The approach has given us considerable understanding and will continue to be of service in the future.

c. Theoretical concepts of turbulent combustion may experience unbridled growth. There are now such a variety of ideas here, and so many more coming about, that explosive development may be anticipated.

d. Experiments may be anticipated to provide continued improvement in combustion knowledge. Laser techniques and ingenuity rank equally in experimental impetus. Sometimes with the latter the former can be avoided, but seldom vice versa. Good experiments are never really easy – care and thought always is essential.

e. Computation may be anticipated to provide increasing contributions to fluid dynamical aspects of combustion theory. Gradual improvements in computer capabilities and in computational methods will lead to a gradually increasing role of computation.

8. CONCLUDING COMMENTS

It is hoped that the preceding summaries will be of use in connection with the more detailed presentations in this volume. The fluid dynamics of combustion theory has become an active and multi-faceted field of research that many practitioners find fascinating. In view of these circumstances, impressive advances may be anticipated in the future.

F. A. Williams is Professor of Engineering Physics and Combustion in the Department of Applied Mechanics and Engineering Sciences and Director, Center for Energy and Combustion Research at the University of California, San Diego.

REFERENCES

1. A.R. Kerstein, CST 60, 163-165 (1988).
2. A.R. Kerstein, CST 60, 441-455 (1988).
3. A.R. Kerstein, W.T. Ashurst, and F.A. Williams, Phys. Rev. A. Rapid Communications, 37, 2728-2731 (1988).
4. N. Peters, "Laminar Flamelet Concepts in Turbulent Combustion," Twenty-First Symposium (International) on Combustion, The Combustion Institute, Pittsburgh, 1988, pp. 1231-1250.
5. N. Peters and F.A. Williams, "Premixed Combustion in a Vortex," Twenty-Second Symposium (International) on Combustion, The Combustion Institute, Pittsburgh, 1989, pp. 495-503.
6. G.I. Sivashinsky, CST 62, 77-96 (1988).
7. F.A. Williams, Combustion Theory, Second Edition, Addison-Wesley, Reading, Mass., 1985.
8. F.A. Williams, "Turbulent Combustion," Chapter III of The Mathematics of Combustion, John D. Buckmaster, Editor, SIAM, Philadelphia, 1985, pp. 97-128.
9. V. Yakhot, CST 60, 191-214 (1988).

R BORGHI

Turbulent combustion modelling: an overview of current methods related to the structure of nonpremixed flames

1.INTRODUCTION

The design of combustion chambers uses more and more Computational Fluid Dynamics techniques. This is allowed by the recent progress of computer, in terms of computing time and memories. But the possibility to use big computers does not solve all the problems; the physical (including chemical) phenomena involved in combustion are so complex, and comprise so different temporal and spatial scales, that the equations themselves, to be feeded into the computer, are not well established.

Among the various problems (turbulence, chemistry of the combustion and of pollutants formation, radiation...) the problem of turbulent combustion is particularly important. The turbulent combustion occurs when the chemical and thermal processes of the combustion interact with turbulent flows; the turbulent flows here include not only random fluctuations of velocity, but also fluctuations of temperature and concentrations; these fluctuations can be present in the gases prior to the combustion, or not in any cases the combustion will modify deeply the fluctuations.

With the point of view of the CFD, the turbulent combustion consequences are simply that it modify the modelling of the turbulent fluxes and, mainly, of the volumetric combustion rate for any interesting species. Indeed, with models for these quantities, all the properties and consequences of turbulent combustion could be computed. But this simple formulation leads to complicated models and equations...

Before describing the state of the art in the numerical modelling of turbulent combustion, it is necessary to try to describe in more details the physical phenomena that are interacting in turbulent flames and turbulent combustion chambers. In the second chapter, we will show that there are different types of turbulent flames and we shall explain physically their differences; turbulent premixed flames and turbulent diffusion flames will be discussed separately. We will show also how different are the turbulent flames from the laminar ones, just with few examples.

We will review then the modelling of these turbulent flames, keeping in mind the physics. We will concentrate, for this survey, on non premixed flames. Specific developments concerning premixed flames can be found in [2]. We will consider first the case of flames where the chemical processes are fast will respect the turbulent processes; in that limit the models for premixed or non premixed flames are quite different, and the chemistry plays a very little role. We will describe then the models proposed for non infinitely fast chemistry; the details of chemical processes now has to be included in the model, but can be included only to a moderate extend, and the differences between premixed and diffusion flames models are lowered.

Turbulent combustion modelling is a large domain, even when restricted to the case of turbojet combustion chambers. So, this review will be incomplete and we shall have no possibility to explain in deep details any models. The review is incomplete because we consider only combustion in gases, without taking into account sprays or solid particles (soot); almost nothing will be said about the prediction of pollutants, and flame radiation will not be taken into account. Instead of describing with many details few theories or models, we have chosen to present a picture of almost all existing theories, with few details about each of them. Our purpose has been to compare the basic assumptions of the theories between them, in order to give their respective application domains, their respective capabilities and weak points. We hope that the reader will have a good view of the situation, and will be able then to find more details concerning the particular method, in which he will be eventually interested, in the cited references.

2.THE DIFFERENT TYPES OF TURBULENT FLAMES; THEIR PROPERTIES

2.1 The different types of premixed turbulent flames

Let us consider a laminar flame in the propagation regime, with a fresh mixture far from self ignition. What happens when this flame is subjected to a turbulent upstream flow? We have now to distinguish different cases, depending on the length and velocity scales of both the turbulence and the flame.

We will follow here the discussions of ref. [1], and [2] which presents a classification of several types of flames depending on two non dimensional parameters: $k^{1/2}/u_L$ and l_t/e_L (where k and l_t are the turbulence kinetic energy and integral scale, u_L and e_L the velocity and thickness of the laminar flame). Other parameters could be chosen (for instance $Re_T = k^{1/2}\ l_t/\nu$ and $D_a = l_t/k^{1/2}\ \tau_c$ in [3], but the physical discussions are clearer with our chosen parameters.

a) A small turbulence intensity where all length scales are larger than e_L only has the effect of wrinkling the laminar flame front. That occurs when $e_L < \eta$, η being the Kolmogorov microscale, the smallest length scale of the turbulent velocity field. This first regime has been identified by Damköhler [4] as the "wrinkled flame" regime. The turbulent flame here, when seen instantaneously, displays a thin quasi-laminar front wrinkled and lengthened by the turbulent movements. As long as $k^{1/2}$ is small with respect to u_L, this instantaneous flame front does not follow the fluid particles; as a consequence, the flame front is much less wrinkled than a filament of fluid particles, and shows probably some cusps towards the burnt gases. Of course, this thin front is fluctuating, and the mean turbulent flame appears thick.

b) When the fluctuations of velocity, $k^{1/2}$, become larger with respect to u_L, one can expect that the flame front resembles more a filament of fluid particles. The occurrence of pockets of fresh gases or burnt gases in then possible, as emphasized by Schelkin [5], due to large scale interactions of two elements of the instantaneous flame front.

One can obtain a rough estimation of the ratio $k^{1/2}/u_L$, above which these pockets could be seen Let us consider for example a pocket of fresh gases; it is created by these interactions due to the large turbulent movements, and consequently its length scale is l_t and the time between the birth of two such pockets is $\tau_t = l_t/k^{1/2}$. But this pocket is consumed by the laminar flame surrounding it in a time of the order of l_t/u_L. If l_t/u_L is larger than τ_t, we can consider that these pockets will be steadily present: that gives $k^{1/2}/u_L > 1$.

If we are interested next in pockets of burned gases, we have to consider not the size of this pocket but its distance from the continuous flame front, which can be decreased further by the laminar propagation velocity; the rest of the reasoning is identical, and shows again that these pockets can remain separated from the burned gases domain if $k^{1/2}/u_L > 1$.

We have then identified another regime, called the "wrinkled flame with pockets" regime or the "corrugated flame" regime (see N. Peters [6]).

c) Considering again a turbulence with $l_t > e_L$, but increasing k, the influence of the turbulence on the flame fronts becomes more complicated. First, the interactions occur more and more frequently; in a given volume small domains where the flame fronts are very close together are more and more numerous. Second, the instantaneous flame front is affected more by the velocity gradient imposed on it by the turbulence; it is known that too intense a stretching can extinguish laminar flames. Third, if l_t is given and the turbulence spectrum is of the Kolmogorov type, η becomes smaller and smaller and there will be a point where $\eta = e_L$ even if $l_t \gg e_L$.

Fourth, the frequencies involved in the turbulence become higher and higher and the quasi-laminar fronts find it more and more difficult to follow, as their own characteristic frequency is nothing but τ_c^{-1}.

The equality $\eta = e_L$ can give us an upper limit for the "wrinkled flame with pockets" regime. It is very easy, from dimensional arguments, to see that it corresponds also to $\tau_c \alpha \tau_K$, where τ_K is the Kolmogorov time defined as $\tau_K = (k^{3/2}/l_t\nu)^{-1/2}$; then $e_L = \eta$ could also be seen as a limit when the highest turbulence frequency $\tau_{K^{-1}}$ higher than τ_c^{-1}. In addition, detailed studies of stretched laminar flames reveal that they can be extinguished when the stretching rate γ (a measure of the velocity gradient to which the flame is submitted) becomes larger than the reciprocal of an extinction time (which is proportional to τ_e); then $\eta = e_L$ could also represent, with a multiplicative constant, the limit where local and instantaneous extinctions of flamelets could occur in a turbulent flame, whilst a measure of the highest velocity gradient is just $\tau_{K^{-1}}$.

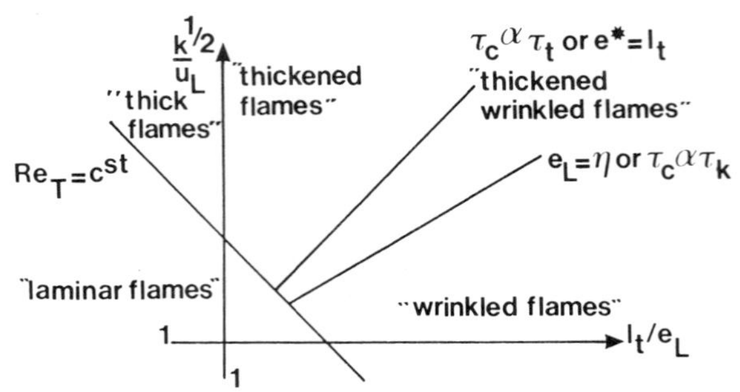

Figure 1. The different regimes of turbulent premixed flames. (Log-Log plot)

The curves $\eta = e_L$ and $k^{1/2} = u_L$ can be easily plotted in the $(k^{1/2}/u_L, l_t/e_L)$ plane. The first one is readily plotted; using the formulae of u_L, e_L and the definitions of Re_T and η, the second one corresponds to $k^{1/2}/u_L = K'''\ (l_t/e_L)^{1/3}$; as the Kolmogorov length cannot be conveniently defined if Re_T is too low, we have plotted the curves only on the right hand side of a curve $Re_T = c^{st}$, which is nothing but $(k^{1/2}/u_L)\cdot(l_t/e_L) = \underline{c^{st}}$. Figure 1 shows these curves in a log-log plot.

d) Increasing $k^{1/2}$ over the $e_L = \eta$ curve, we enter the domain where the flamelets are modified by the turbulence in a complicated way. Following [7], one can get an estimation of the thickening of the flamelets by the fine scale turbulence; as described in [1], it becomes:

$$e^* = K'(k^{3/2}\tau_c^3/l_t)^{1/2} \tag{1}$$

where e^* is the thickness of these modified flamelets, and K' a constant. The flamelets are thicker than laminar flames, because the turbulence whose scale is smaller than e_L enhances an "effective diffusion coefficient"; they are always wrinkled, as long as $e^* < l_t$.

The regime of "thickened-wrinkled flame" is shown in Fig. 1, between the curves $e_l = \eta$ and $e^* = l_t$, which turns out to be equivalent, except for a multiplicative constant, to $\tau_c = \tau_t$. The formula (1), based on the estimation of Zimont, neglects the possible extinctions and the unsteady character of the thickened flamelets: the picture of a wrinkled but thickened flamelet with thickness e^* is probably oversimplified.

For the same reason, the limit $e^* = l_t$ is not quite definite. But, on th other hand, τ_t is a dissipation time scale for the turbulent fluctuations and it can be easily conceived that τ_t/τ_c, the Damköhler number, is a relevant parameter of the turbulence-chemistry interactions. So a given value of τ_t/τ_c, if not the one corresponding to $e^* = l_t$, is probably a good limit for this regime.

e) When $k^{1/2}$ is so high that this limit is crossed over, flamelets can no longer be distinguished; the reacting turbulent mixture does not possess anymore a "flamelet" type small scale structure; all the length scales of the fluctuations are smaller than the turbulent flame thickness.

We have called in Fig. 1 "thickened flame" the regime where l_t is again larger than e_L, and "thick flame" the regime where l_t irrespective to a possible thickening of the flame, is smaller than e_L. These regimes have been called also "distributed combustion" regimes, by Damköhler [4].

2.2 Typical properties of turbulent premixed flames

A good example of turbulent premixed flame, useful to emphasize the global properties of such flames, is found in the following experiment, performed by P. Moreau in 1975 [8].

This flame is similar to the flame in a turbojet after burner, but simpler.

The turbulent flame is stabilized in a premixed turbulent flow of methane and air (equivalence ratio 0.9) at 600 K and 60 m/s, owing to a parallel jet of hot gases, at about 2000 K and 120 m/s; the flame develops in a two dimensional duct, allowing easy visualization, as schematized Fig. 2a.

Fig. 2b shows a flame of a high velocity film (4000 exposures per second) obtained with a Shadowgraph technique during the experiments; the turbulent character of the flame is clearly apparent. To the eyes, the flame appears blue, quite transparent, as thick as indicated in Fig. 2a.

Detailed measurements have been done within this flame, and the turbulence in presence of combustion has been measured (Moreau and Boutier [9]); a typical value of $k^{1/2}$, the turbulence characteristic velocity, of 20 m/s has been measured roughly in the middle of the photographs of Fig. 2b; an integral length scale l_t of about 5 cm can be estimated in the same location from the wrinkles: that leads to $Re_T 2.10^4$, giving $\eta \simeq 0.14$ mm, $\tau_t \simeq 2.10-3$ sec, $\tau_K \simeq 10-5$ sec. This flame is very turbulent, compared to others, like [10], with the flame stabilized in grid turbulence with $k^{1/2}$ of the order of 20 cm/sec.

Just from the photographs, we can point out two very important properties of this turbulent times.

First, its equivalent propagation velocity: roughly, it is stabilized in a 60 m/sec flow, and its mean angle α (see the Fig. 2a) is such that $\sin\alpha \simeq 8/60$, for an equivalence ratio of 0.9. Then the velocity of the incoming flow normal to this mean angle is about $60\times 8/60 = 8$ m/sec. Of course this calculation is rather rough but shows that the turbulent flame speed is very high compared to the laminar propagation velocity of methane and air under the same conditions, which is less than 1 m/sec.

Second, the apparent thickness of the flame is quite large compared to the 0.2 mm thickness of the laminar flame in a Bunsen burner. Of course this thickness is a mean thickness, which is well explained, at least partially, by the fluctuating nature of the flame, as seen in Fig. 2b; this mean thickness is of interest for designing the burners, and not the instantaneous thickness.

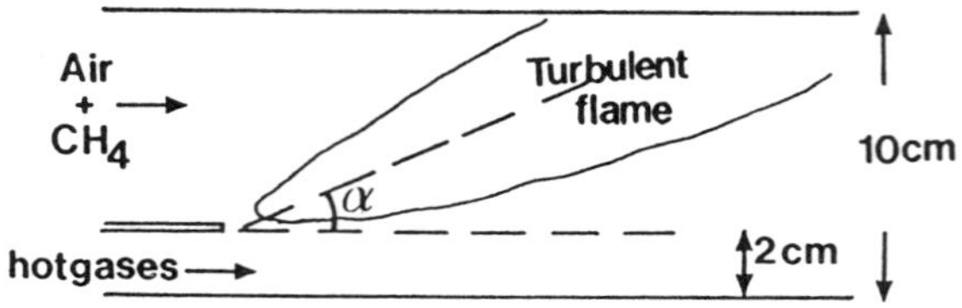

Figure 2. a) A basic experiment for studying premixed turbulent flames.

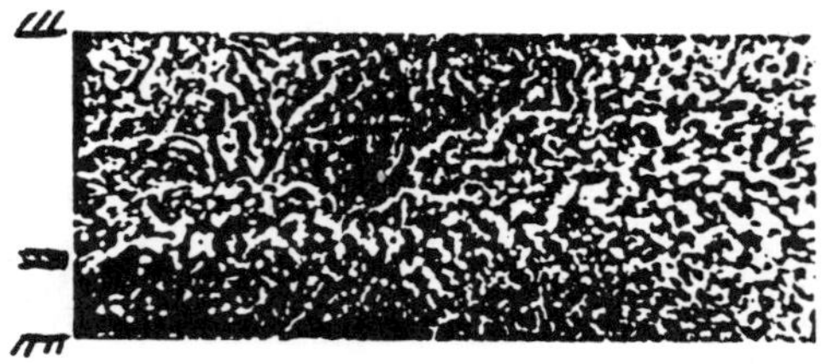

Figure 2. b) Cinematographic shadow-graph record of the flame of [8]. Flow from left to right.

How are the afore mentioned equivalent propagation velocity and thickness varying with respect to the entrance velocities, the equivalence ratio, the entrance temperature? This first question if of great interest to engineers, and is very difficult to answer. In fact, it has been shown by experiments (in the one described here as well as in previous ones [11], [12]) that these variations ar very different compared to the laws followed by laminar premixed flames. One of the purposes of theories and models that will be described in succeeding chapters is to propose a better answer than previously given, while being able to describe with many details the flow and temperature fields.

2.3 The different types of turbulent diffusion flames

As in the case of premixed flames, let us consider now what happens to a diffusion flame subjected to a turbulent flow, whatever may be the source of this turbulence. Is it possible to distinguish several types of turbulent flames, looking again at the plane $(k^{1/2}/u_L, l_t/e_L)$?

u_L and e_L have no physical meaning for diffusion flames. However τ_e and are again the physical parameters relevant for diffusion flames, as well as for premixed ones (assuming that the heat and species diffusitivities are close to viscosity ν). Then u_L and e_L can be taken here, respectively, as a characteristic velocity and a characteristic length of the laminar flame, defined mathematically by (2) and (3), without very precise physical meaning:

$$e_L = (\nu\tau_c)^{1/2} \qquad (2)$$

$$u_L = (\nu/\tau_c)^{1/2} . \qquad (3)$$

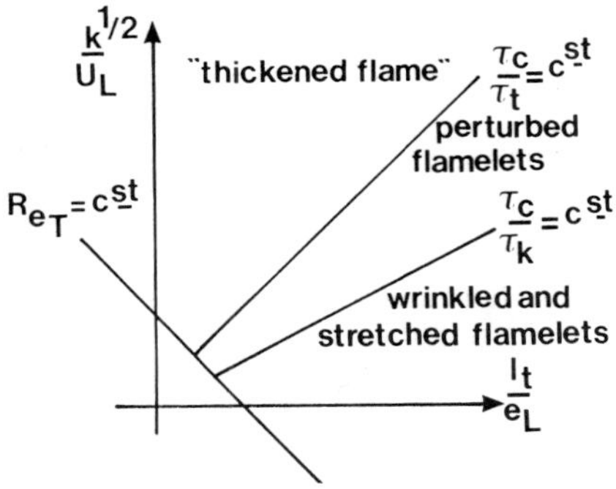

Figure 3. Different types of different flames.

We have just to specify at what equivalence ratio τ_e is taken, because depends strongly on it; as the reaction zone is usually close to the equivalence ratio of one, we can choose, for convenience, this value.

a) Considering now the plane $(k^{1/2}/u_L, l_t/e_L)$, the lines of physical meaning for diffusion flames are $Re_T = c^{st}$ and $\tau_c \alpha \tau_K$. As previously, a line $Re_T = c^{st}$ separates turbulent flames and non-turbulent ones, with of course a transition region, more or less thick. A measure of the mean stretching rate in turbulence can be obtained with τ_K^{-1}, and the limit of local extinction of laminar diffusion flamelets can be written as a line of constant τ_c/τ_K.

We have named, Fig. 3 the regime of "wrinkled-stretched flamelets" when τ_c/τ_K is below a critical value. Within this domain the laminar diffusion flame of has been lengthened, wrinkled and stretched by the turbulent velocity field, more and more when $k^{1/2}$ has been increased.

As a result of this stretching, F.A. Williams has imagined that the turbulent flame, is then composed in many places of the one dimensional stretched diffusion reaction layers [13], identified later by N. Peters [14] as steady counter-flow diffusion flamelets. Just by dimensional analysis, one find, in the case of fast chemistry ($\tau_c/\tau_K \ll 1$) that the thickness of such flamelets is then $e_D \alpha (\nu \tau_K)^{1/2}$, if τ_K^{-1} is a measure of the stretching rate in a turbulent flow; then e_D scales with the Kolmogorov scale itself, and the internal structure of the flamelet is not modified by the details of th turbulence.

However, the fact that a turbulent diffusion flame, in this regime, is mainly composed of these stretched flamelets is disclaimed by R.W. Bilger [15], who says that, due to reverse reactions, the thickness of the reaction zones within flamelets are greater than Kolmogorov microscale, and consequently will be modified by turbulent movements.

On the other hand, it can displays compressed flamelets, or flamelets too close from each other in such a way that they interact. Evidence of compressed and interacting flamelets has been shown in moderate Reynolds number jet diffusion flames in [16], by seeding them with $TiCl_4$. Indeed, more experimental studies are needed in order to know the structure of "wrinkled-stretched" diffusion flames.

b) When τ_c/τ_K is above a critical value, we know that local extinctions of flamelets can occur. But, as in the case of premixed flames, other phenomena have to be considered: interactions of flamelets are frequently present, and the steadiness of flamelets becomes questionable. The structure of the flame then becomes very complicated; we call this type of flame in Fig. 3 "perturbed flamelets", a very imprecise name. We emphasize, however, that the eventual extinctions will be followed by mixing of products, fuel and oxidizer, and later, when the stretching (which is fluctuating) has disappeared, by reignition in a partially premixed medium. The recent study of Stepowski et al. [17] has show the existence, in a turbulent diffusion flames, of diffusion reaction layers with too low maximum temperature for being steady stretched flamelets; they are extinguishing or reigniting flamelets...

c) For larger τ_c, this premixing becomes more efficient because reignition is delayed; there will probably arise a limit, when the large scale turbulent mixing time τ_l is small enough compared to τ_c, above which the diffusion flame resembles a premixed one. On Fig. 3 we have indicated this limit, and we have called the flame behind as the premixed one: "thickened flame".

The final aspect of Fig. 3, at first sight surprising, is that the different regimes for diffusion flames are bounded similarly to those for premixed flames, except for the line $k^{1/2}/u_L = 1$, which is clearly without meaning for diffusion flames. But this is only due to the fact that the same phenomena are relevant for both types of flames. In addition, it is to be noticed that the physical meaning of the limits are not the same in the two cases and consequently, the critical values of the parameters τ_c/τ_K and τ_c/τ_t are not the same.

2.4 On the global properties of turbulent diffusion flames

In order to parallel the way we described premixed turbulent flames, we now present the main properties of turbulent diffusion flames, emphasizing the difference with laminar diffusion flames.

A basic experiment concerning turbulent diffusion flames is the jet flame of Bilger et al. ([18] [19]). A large amount of fine experimental results exists concerning species measurements as well as velocity and turbulence measurements. Fig. 4, for instance, shows measured transverse and axial profiles for species and temperature from [18].

The main property of a turbulent diffusion flame can be seen very simply: the length of the flame, defined on the axis as the length behind which no fuel can be found in a measurable proportion (for instance) is almost independent of the velocity of the jet, while it is proportional to it for laminar diffusion flames. This property has been noticed very early [20].

Another interesting property concerning the maximum temperature is much difficult to point out: in turbulent diffusion flame the maximum temperature is significantly lower

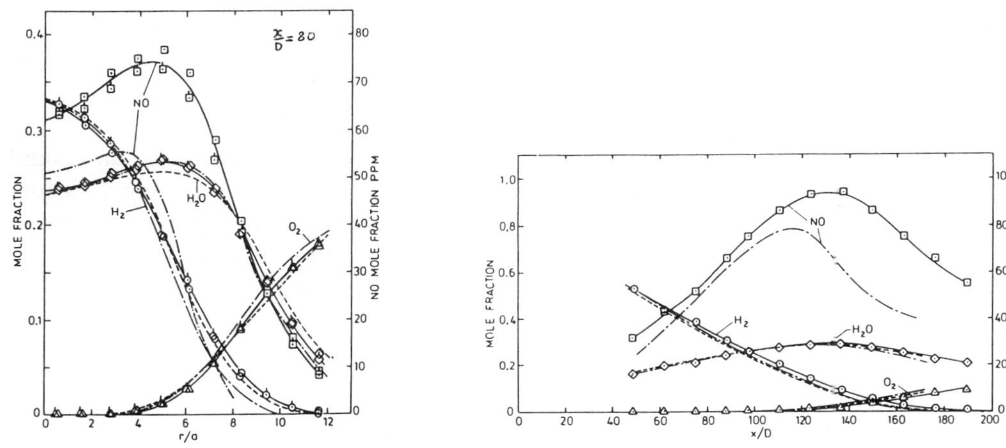

Figure 4. Measured species concentration and temperature in an H_2-Air jet diffusion flame. [18]

than the adiabatic flame temperature, and the thickness of the temperature profiles is larger. This apparently surprising result, that turbulence has the same effect as finite rate chemistry, can now be clearly explained, and can be shown even by the simplest models.

Finally a very surprising result has been found by Bilger et al. [19] concerning the NO_x production: the global NO_x yield by the flame has been found to vary with the square root of a Reynolds number based on the nozzle diameter, when the velocity of the jet is varied. This extraordinary dependence on viscosity, for a fully turbulent jet flame (whose length does not depend on it) has not been explained clearly up to now.

2.5 Some comments concerning the regimes

A few comments are in order after the previous discussions.

a) The first one concern the real nature of turbulent flames. We have discussed here what happens to a laminar flame subjected to a turbulent incoming flow, but, in some cases, the flame is propagating in a non-turbulent flow (in the limit fluid at rest) and becomes "visually turbulent". In fact, a large proportion of flames occurring in practical devices are stabilized in flows already turbulent upstream of the flame. The case of "self turbulizing flame", which arises due to an intrinsical instability of a laminar flame, is quite rare in industry.

Anyway, a flame is always accompanied by some movements of gases, due to the change of density, and when a flame is turbulent, it induces a turbulent flow field. The result then is, if a turbulent flow exists upstream, that turbulence is modified by the flame, and if no turbulence exists in the flow when a turbulent flame is propagating, turbulence is created in the vicinity, even (just) upstream, of the flame.

The previous discussion have then to be understood with the flow turbulence characteristics taken into account along with the presence of the flame, and can also include self turbulizing flames. Of course, for a self turbulizing flame, $k^{1/2}$ and l_t are not independent of u_L and e_L; in particular, $k^{1/2}$ will probably remain of the order of u_L.

b) In practical burners, turbulence is not the same at different locations, and consequently a turbulent flame in a burner can display different regimes at different locations. Furthermore, these burners are not always operated at the same conditions, and a change of conditions will result eventually in a change of flame regime. The implication is that the need of engineers is not for a separate model for each regime, but for a model able to cover two or several regimes. In particular, the domain of interest for turbomachines will change with the altitude (playing mainly on τ_c) and the operating point (playing also on the flow velocity and consequently on the turbulence)

What is the regime expected for turbulent flames in turbomachines combustion chambers? W have first to decide if the flames are diffusion of premixed flames, and this is not always clear. Roughly speaking, the turbulent flames in afterburner devices are close to premixed flames (perhaps with some variations in equivalence ratio from points to points in the flow), while combustion in the main chamber is more similar to diffusion flames.

Strictly speaking, we could consider a single category of non perfectly premixed for both cases, but is is more interesting to take advantage of the numerous studies existing for perfectly premixed flames.

Considering first the after burners devices, the turbulent flame described in 2.2 (Moreau [8]) is a good example; we have seen that it is "wrinkled-thickened", with $u_L/k^{1/2} \simeq 0.05$, $\tau_c/\tau_K \simeq 50$ (with $u_L \simeq 1$ m/sec, $\tau_c \simeq 0.5$ m/sec). At lower equivalence ratio or lower that temperature, the flame will be more thickened, because τ_c will increase; at low velocity, the flame will be more wrinkled, because $k^{1/2}$ will decrease and τ_k increase.

For main combustion chambers, the turbulence is expected to be of smaller scale, depending of the size of the primary zone and of the primary holes. A value of l_t between 1 and 0.1 cm is probably in order, for not too small engines. The turbulence is probably very high, due to the air jets and swirlers; values of 10 to 5 m/s can be achieved, giving τ_K between 10^{-5} and 0.7 10^{-4}, and τ_t between 2.10^{-3} and 2.10^{-4}. For nominal power (and high pressure) τ_c is probably of the order of 0.3 10^{-4} sec, then lower than τ_t and perhaps slightly lower than τ_K for large combustors. For reduced power (and pressure), τ_c is of the order or 10^{-3} sec and may become slightly larger than τ_t.

For reignition conditions, at very low pressure, even if the velocities and turbulence are low, τ_c is probably clearly larger than τ_t.

Indeed, we see that, due to the very different operating conditions, the regimes of turbulent flames cover almost the whole domain, with in some cases wrinkled-stretched flamelets, in most of the cases perturbed flamelets, and for reigniting conditions thickened flames (or distributed combustion).

c) Being able to estimate (and also to compute) the turbulence kinetic energy and length scales is very important, as we have seen.

It may be not expected, in any case, that the turbulence existing in a combustion chamber is the same without combustion and with combustion. The problem of the influence of combustion on turbulence has been discussed from long time, and is not completely solved at this time.

In fact, there are two or three possible actions of the combustion on the turbulence:

i) first, in a given combustor, the heat release due to the combustion decreases the mean density of the gases and increases the mean velocities. Increase of the mean velocities is very often accompanied with the increase of velocity gradients and consequently of turbulence.

ii) Second, the increase temperature of the gases increases the viscosity. If the turbulence Reynolds number is not very large, that can lead to more dissipation of the turbulence. It has been very early seen that a turbulent premixed flame stabilized behind a rod was looking more "stable" than the corresponding cold flow behind the same rod [11]. This has been shown also for diffusion flames by Takagi et al. [21].

iii) Third, it has been also from long time discussed about "flame generated turbulence", independently of the two previous mechanisms. A phenomena of this type, linked with the fluctuations of density and the mean pressure gradient has been clearly put into evidence (see [22], Bray-Libby, [23], Borghi-Escudié). A direct influence also of the non constant mean density can be seen for the turbulent kinetic energy as a "dilation term", which is a sink of k (see [24], Bray et al.). The pressure gradient effect itself, equally, can be a sink of turbulence in certain case, and particularly in jet diffusion flames ([25], Bilger).

In principle, the turbulence models for combusting flows are now able to include these effects, with, perhaps, some improvements needed. At least the two effects first described can be taken into account without problems. The weakest point of these models seems to be the prediction of length scales, l_t, or time scale τ_t.

3. MODELS FOR TURBULENT NON PREMIXED FLAMES

3.6 Turbulent diffusion flames with fast chemistry

The problem with very fast reaction (with a very small reaction time with respect to any convection or diffusion times), has been addressed first. This case has been discussed and almost solved by Toor [26] as early as 1963, followed by Lin and O'Brien [27].

a) The solution is quite clear when we look at the joint p.d.f. of Y_A and Y_B, the two reactants of a single step reaction: $A + r_s B \rightarrow C$. When the reaction is very fast, the coefficients K_i, of $\dot{w}_i = -k_i \cdot Y_A \cdot Y_B$ are infinite (whatever can be their dependence in T) in such a way that the mean reaction rate $\dot{w}_i = -K_i \cdot Y_A \cdot Y_B$ is finite only if $Y_A \cdot Y_B = 0$, i.e. either $Y_A = 0$ or $Y_B = 0$ (and not $\tilde{Y}_A = 0$ $\tilde{Y}_B = 0$!). It follows then that $\tilde{P}(Y_A, Y_B)$ within the phase plane (Y_A, Y_B) is supported only by the axis $Y_A = 0$ and $Y_B = 0$; the p.d.f. is pushed against one boundary of its permitted domain.

The result is then that the p.d.f. is one-dimensionals instead of two-dimensional; in addition if we define $\phi = r_s Y_A - Y_B$, ϕ satisfy a balance equation without reaction term, and, because Y_A and Y_B cannot be non zero at the same time, ϕ represents either $r_s Y_A$, if positive, or $-Y_B$, if negative; one write:

$$\begin{cases} \tilde{Y}_A = \int_0^\infty \frac{\phi}{r_s} \tilde{P}(\phi) \mathrm{d}\,\phi \\ \tilde{Y}_B = -\int_{-\infty}^{0} \phi \tilde{P}(\phi) \mathrm{d}\,\phi \end{cases} \tag{4}$$

and similarly for any moment $\tilde{Y}_A^n$ or $\tilde{Y}_B^m$. The knowledge of $\tilde{P}(\phi)$ allows us to solve the problem completely.

If we assume that the molecular diffusion of A and B are identical, the balance equation for ϕ writes:

$$\frac{\partial}{\partial t}\rho\phi + \frac{\partial}{\partial x_\alpha}\rho u_\alpha \phi = \frac{\partial}{\partial u_\alpha}\left(\rho d \frac{\partial \phi}{\partial x_\alpha}\right) . \tag{5}$$

The problem of reactive turbulent flow is then reduced to the computation of the p.d.f. of the fluctuations of a single inert scalar diffusing into the turbulent flow. That problem, however, was not addressed by specialists of turbulence, at the time of Toor's studies; it has been studied since that, and now we have at our disposal some modeled balance equations for $\tilde{P}(\phi)$ itself.

b) The generalization of (4) to the case of a complex mechanism is quite obvious when the diffusion coefficients of each species involved can be assumed almost equal; in this case one finds that each species Y_i is deterministically related with a single inert ϕ, conveniently defined (see Bilger [25]), and we can write:

$$\tilde{Y}_i = \int_{-\infty}^{+\infty} Y_i^{eq}(\phi)\tilde{P}(\phi)\mathrm{d}\,\phi . \tag{6}$$

Where Y_i^{eq} (ϕ) stands for the equilibrium value for a mixture whose the atomic composition is the one determined by the mixing variable ϕ and at the temperature such that the enthalpy h is also determined by ϕ. Indeed, if all the Lewis numbers are assumed to be unity, we can define a variable ϕ, often called conserved scalar, which his unity in the fuel jet and zero in the surrounding stream, and it can be shown that every non reacting scalar, such as the mass content in any atoms Z_i in a fluid particle, or such as the enthalpy, is linearly related to ϕ:

$$\frac{h - h_{\text{ext}}}{h_j - h_{\text{ext}}} = \phi \,, \qquad \frac{Z_i - Z_{i\text{ext}}}{Z_{ij} - Z_{i\text{ext}}} = \phi \,. \tag{7}$$

This method can also be applied if the streams entering the combustion chamber are not simply of pure fuel or pure oxidizer: fuel and air can be partially premixed in the jet flow, as well as air and fuel can be premixed, with another proportion, in the external flow. The only requirement is that the inlet flow of the chamber would be feeded by two different streams only; if this is not the case more than one inert scalar is needed.

The knowledge of the mean reaction rate is without interest in this case, but the knowledge of $\tilde{P}(\phi)$ is crucial. It is possible to gain it either with a presumed p.d.f. method, or with an equation for $\tilde{P}(\phi)$ itself.

The balance equation for $\tilde{P}(\phi)$ can be found from the primitive equations, i.e. The Navier Stokes equations and the diffusion equation. The derivation is not as easy as the one of the moments equations for $\tilde{\phi}$ or $\widetilde{\phi^{12}}$; there are two ways to obtain it, as explained in [2]; the equation obtained is unclosed, due to two types of things: first, the correlations with the fluctuations of velocity (again the problem of turbulent diffusion), and second the correlations with the fluctuations of ϕ but at another close point: the corresponding physical effect is called the micromixing.

The model that has been first proposed by Dopazo and O'Brien, is:

$$\frac{\partial}{\partial t}\overline{\rho}\tilde{P}(\phi) + \frac{\partial}{\partial x_\alpha}\overline{\rho}\tilde{u}_\alpha\tilde{P}(\phi) = \frac{\partial}{\partial x_\alpha}\left(\overline{\rho}D_p\frac{\partial\tilde{P}(\phi)}{\partial x_\alpha}\right) - \frac{\partial}{\partial c}\left(\overline{\rho}\frac{6d}{\lambda_\phi^2}(\tilde{\phi} - \phi)\tilde{P}(\phi)\right) \,. \tag{8}$$

The turbulent diffusion is modelled through a diffusion coefficient Dp, which has to be related to the turbulence. The micromixing term, the last one, involves λ_ϕ, the Taylor microscale of the fluctuations ϕ, which is also to be prescribed; in turbulence of large Re_T, $6d/\lambda_\phi^2$ is a time scale that can be related to the time scale τ_t of the turbulence itself, eventually with a dependence on the Schmidt number (see [2]).

Another model for the micromixing term can be derived from a mechanistic model, the coalescence-dispersion model or Curl [28]; it gives:

$$\begin{aligned}\frac{\partial}{\partial t}(\overline{\rho}\tilde{P}(\phi) \quad &+ \quad \frac{\partial}{\partial x_\alpha}(\overline{\rho}\tilde{u}_\alpha\tilde{P}(\phi)) = \frac{\partial}{\partial x_\alpha}\left(\overline{\rho}D_p\frac{\partial\tilde{P}(\phi)}{\partial x_\alpha}\right) \\ &- \quad \overline{\rho}\frac{C_D}{\tau_t}\int_0^c 2\tilde{P}(\phi')\left[\int_c^1 \tilde{P}(\phi'')\delta\left(\frac{\phi'+\phi''}{2} - \phi\right)\mathrm{d}\,\phi'' - \tilde{P}(\phi)\right]\mathrm{d}\,\phi'\end{aligned} \tag{9}$$

C_D, here, is a constant relating the micromixing time and the turbulence time. This new model has been improved also by W. Kollman et al. [29], and S. Pope [30].

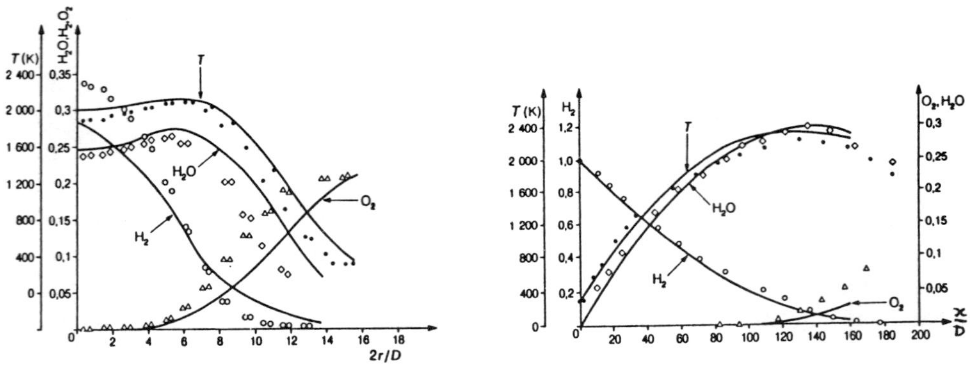

Figure 5. Predictions of a H_2-Air Flame with equilibrium chemistry - from [32].

The closure problem for the turbulent diffusion can be avoided if we look at the p.d.f. of C and u_α; but there are other closure problems, related to the dependence in the velocity. S. Pope [31] has been working along this line from several years and has been able to propose a model for this new p.d.f.; in fact, this new p.d.f. contains also a part of a turbulence model (the k equation, in particular).

c) The paper of Janicka et at. [32] has been the first one in which the prediction of a jet diffusion flame has been performed with an equation for $\tilde{P}(\phi)$. Fig. 5 shows the shapes of $\tilde{P}(\phi)$ found and the agreement obtained for the transverse profiles of species and concentrations.

Of course, jointly with this equation, the Reynolds equations and a turbulence model (for instance the $k-\varepsilon$ model) have to be integrated; they are necessary in order to express $\tilde{u}_\alpha$, D_p and $6d/\lambda^2 = 1/\tau_t$.

3.7 Turbulent non premixed flames with moderately fast chemistry

When the chemical reactions occurring cannot be considered as very fast, three methods have been proposed, up to now, in order to generalize the "conserved scalar approach".

a) The method using the joint p.d.f. equation for the reactive species Y_i is the most direct extension. In fact, it suffices to generalize the equation for $\tilde{P}(\phi)$ given in section 3.1 for a multidimensional p.d.f. $\tilde{P}(Y_1, Y_2 \dots Y_n)$.

If a single, adiabatic, chemical reaction between A and B occurs, two random variables only: Y_A and Y_B, or Y_A and ϕ, are enough. Indeed, the inert scalar ϕ (when $Le_A = Le_B = 1$), defined as $\phi = r_s Y_A - Y_B$, is linearly related to h even when Y_A, Y_B, ϕ and h are fluctuating (see [2]), and this allows us to relate deterministically the temperature to Y_A

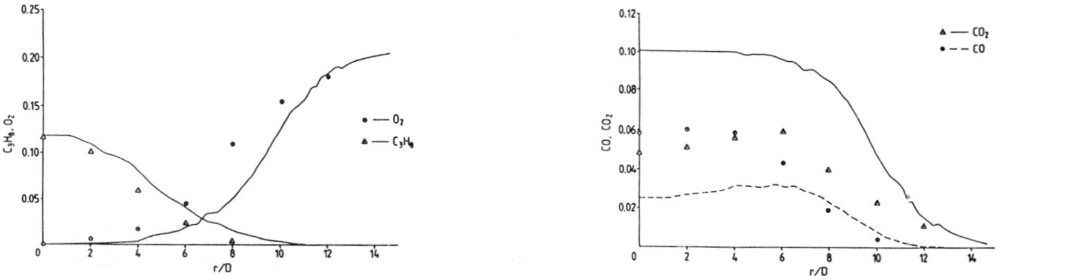

Radial profiles of C_3H_8 and O_2 mole fractions at $x/D = 80$; o, ▵ Measurements

Radial profiles of CO and CO_2 mole fractions a. $x/D = 80$; o, ▵ Measurements

Figure 6. Predictions of a C_3H_8-Air Flame with finite rate chemistry [35]

and Y_B (or Y_A and ϕ). If the medium is not adiabatic, and/or if many chemical species have to be considered involved in a complex kinetical mechanism, a multidimensional $\tilde{P}(\underline{Y})$, $\underline{Y}$ being a vector, is to be used.

Following the same method as or $\tilde{P}(\phi)$, the equation for $\tilde{P}(\underline{Y})$ is just obtained by summing the contribution for each species; for instance:

$$\begin{aligned}\frac{\partial}{\partial t}(\overline{\rho}\tilde{P}(\underline{Y})) \; + \; \frac{\partial}{\partial x_\alpha}(\overline{\rho}\tilde{u}_\alpha\tilde{P}(\underline{Y})) &= \frac{\partial}{\partial x_\alpha}\left(\rho D_p\frac{\partial \tilde{P}}{\partial x_\alpha}(\underline{Y})\right) \\ &- \sum_{j=1}^{n}\frac{\partial}{\partial Y_j}\left[\frac{6d_j}{\lambda_j^2}(Y_j-\tilde{Y}_j)\tilde{P}(\underline{Y})+\dot{w}_j\tilde{P}(\underline{Y})\right] \qquad (10)\end{aligned}$$

with the closure assumptions of Dopazo and O'Brien.

This equation is conceptually as simple as the one of $\tilde{P}(C)$, but is far more difficult to solve numerically, because the number of dimensions of $\tilde{P}$.

Conventional finite differences of finite volume methods cannot be used above 3 variables; such a method has been adopted in [33]. But a larger number of variables obliges to use completely different Monte Carlo statistical simulations. Indeed this method is now widely adopted, from the work of Flagan and Appleton [34], to the ones of S. Pope, W. Jones and W. Kollmann [35]. Fig. 6 shows results of [35] concerning the calculation of a propane jet diffusion flame in air. A moderately complex chemical mechanism, namely:

$$C_3H_8 + \frac{3}{2}O_2 \rightarrow 3Co + 4H_2 \qquad \text{(global reaction)}$$

$$\begin{aligned} H + O_2 + M &\leftrightarrow HO_2 + M \\ H + H + M &\leftrightarrow H_2 + M \\ H + OH + M &\leftrightarrow H_2O + M \\ H + O + M &\leftrightarrow OH + M \end{aligned}$$

The reactions

$$\begin{aligned} H_2+O &\leftrightarrow OH+H \\ O_2+H &\leftrightarrow OH+O \end{aligned}$$

being in partial equilibrium.

That leads to a joint p.d.f. of three variables, chosen as ϕ, the conserved Scalar (called here mixture fraction), n, the total mole number, and $n_{C_3H_8}$, the fuel mole number.

A $k-\varepsilon$ model has been used for computing the turbulent quantities. As we see on Fig. 6, the agreement is only moderate, but the critical factor for this is mostly the chemical scheme, and not the p.d.f. equation.

S. Pope, on his side, has not tried to take into account complex chemistry. But he has studied a multidimensional p.d.f. that include the fluctuations of velocity. The solution of the joint p.d.f. $\tilde{P}(\underline{u}, \underline{Y})$ is possible again with a Monte Carlo simulation. The method is then able to attack the problem of diffusion fluxes and turbulence modifications at the same time as mean reaction rate modelling.

A simpler way to compute multispacies turbulent combustion has been proposed also by Correa et al. [36]; they use a single variable p.d.f. $\tilde{P}(\phi)$ to build a multispecies p.d.f., owen to assumed statistical independence; but the error introduced by this assumption cannot be well appreciated.

b) The "Reacting laminae" models, or "Flamelet models": the idea has been proposed a long time ago by Mao and Toor, in the field of chemical engineering in 1970 [37] and Gibson and Libby in 1972 [38] in the field of combustion. It has become very popular, more recently, and is called "Flamelet approach" by specialists in turbulent combustion.

The method is based on the particular picture of the turbulent reacting flow: assuming that they consist of a collection of reaction-diffusion layers, in quasi-steady state, which are continuously displaced and stretched within the turbulent medium. For the sake of simplicity, the layers are assumed to be one dimensional, that is, almost planar, with a thickness that is very small with respect to their main radii of curvature. It is consequently possible to study separately these layers, or "reacting laminae" or "flamelets", and to compute their internal structure with all chemical details, like in a laminar flow.

After that, it is necessary to build the turbulent field by a convenient collection of such flamelets.

At this stage of the approach, two types of strategies have been worked out. The first one has been proposed by Marble and Broadwell [39], and is based on a key quantity, the mean flame (or reaction) surface per unit volume, Σ. Indeed, the mean reaction rate $\tilde{\dot{w}}_i$ is directly related to Σ, simply by:

$$\tilde{\dot{w}} = V_{DI} \cdot \Sigma \ . \tag{11}$$

Where V_{DI} is a velocity: it is the volumetric flow rate per unit of surface, of the reactant disappearing through the "reacting laminae". V_{DI} is a quantity that can be computed independently of the turbulence, just calculating the diffusion-reaction balance within the "laminae". On the contrary, Σ is a turbulent quantity; Marble and Broadwell have postulated a balance equation for Σ, following the works of Batchelor about surfaces within turbulent flows.

The second type of approach, followed by Liew et al. [40] and Peters [14], consists in extracting from a laminar calculation of the laminae the relations $Y_i(\phi)$ between any species mass fraction and ϕ, an inert scalar diffusing through the laminar. Then the turbulence can be taken into account simply through $\tilde{P}(\phi)$ and one can write for any species

$$\tilde{Y}_i = \int_\phi Y_i(\phi)\tilde{P}(\phi)\mathrm{d}\,\phi \,. \tag{12}$$

Equation (12) has been used directly in [40], but Peters proposed to take into account the actual fact that the flamelets are randomly stretched by the turbulence.

In the Peter's case a library of lamina stretched reaction layers is to be built, with different stretching rates, giving $Y_i(\phi,\gamma)$, and any mean value $\tilde{Y}_i$ is computed through:

$$\tilde{Y}_i = \int_\phi \int_\gamma Y_i(\phi,\gamma)\tilde{P}(\phi,\gamma)\mathrm{d}\,\phi \mathrm{d}\,\gamma \,. \tag{13}$$

For formula (12), only $\tilde{P}(\phi)$ is needed, and we already know the mean to compute it; but now, with (13), we need $\tilde{P}(\phi,\gamma)$, and our present turbulence knowledge is not able to give us this joint p.d.f.

Peters proposed then to assume statistical independence between ϕ and γ, and that $\tilde{P}(\gamma)$ is lognormal, and then to compute $\tilde{P}(\phi)$ as previously.

The extinction effects due to the stretching can be taken into account through the dependence of Y_i on γ.

These "flamelet models", whatever may be their exact form, allow one to take into account the complex chemical mechanism that is actually occurring; they are based on physically clear assumptions. However, in their present form, they neglect important physical phenomena. If we refer to section 2, the picture of the turbulent reacting flow, on which they are based, is not complete: the curvature of laminae is clearly neglected, the re-ignition of extinguished flamelets, as well as their transient evolutions between the several states of strain that are imposed by the turbulence, are not considered; in addition, interactions of adjacent laminae, which are likely to occur, in particular within coherent vertices present in many mixing layers, are also neglected. These approaches are then restricted to the cases where such phenomena are negligible, that is essentially for fast enough reactions, wrinkled flames and not too large turbulence Reynolds number.

Indeed, during their attempt to apply these ideas to a chemically reacting shear layer (with negligible heat release), J.P. Broadwell and P.E. Dimotakis [41] have found that considering only "flamelets" in order to represent the mixing with reactions was not sufficient to explain the experimental results. They have then proposed that two parallel process have to be taken into account in the model, one being the mixing and reactions occurring in flame sheets, the other being mixing and reactions occurring in moleculary mixed homogeneous fluid pockets. They have found that the percentage of each processed depends on the Peclet number, and even if the chemistry is very fast. This theory has not been put into a form such that the mean reaction rate can be found, but has bee written into an integral form. The reader may find details in two different developments of this same idea by P.E. Dimotakis [42] and J.E. Broadwell and M.G. Mungal [43].

c) The Lagrangian models constitute a third type of models. The common feature of such models is to try to represent the evolution of fluid particles, or fluid elements, over their life and trajectory within the reacting medium. Such an approach can, in principle avoid the use of every balance equation, as in Eulerian models, and, in some simple cases, it does so; but in more complex cases, it can be used jointly with Eulerian equations.

The simplest example of Lagrangian models is the JEM model, developed by Villermaux et al. [44], and others in the field of chemical engineering [45]. It consists in replacing the primitive equations of Y_i by:

$$\frac{\mathrm{d}}{\mathrm{d}t}Y_i\left(=\frac{\partial Y_i}{\partial t}+u_\alpha\frac{\partial Y_i}{\partial x_\alpha}\right)=\frac{\tilde{Y}_i-Y_i}{\tau_{\mathrm{IEM}}}+\dot{w}_i\ . \tag{14}$$

The first term on the RHS is a model for the molecular diffusion fluxes and represents them as Exchanges by Interaction with the Mean Value (thus justifying the name). τ_{IEM} is an exchange time, assumed to be simply proportional to the turbulence time τ_t.

This model allows us, by numerical time integration of (14), to take into account quite complex chemistry. The first use of this model has been accomplished for simple homogeneous reactors, in which the residence time distribution $f(t)$ is known. It suffices then to compute any mean value by:

$$\overline{\rho}\tilde{Y}_i=\int_0^\infty \rho Y_i(t,\underline{Y}_0)f(t)\mathrm{d}t\ .$$

An extension of the use of this model to complex flows has been proposed in [46]. The equation (14), integrated for fluid particles entering the combustion chamber from the fuel jet, of from the air jet, allows to build an approximate "skeleton" of the multidimensional p.d.f. $\tilde{P}(Y_1,\dots Y_n)$, where the first variable is an inert scalar (see [2] for more details).

An other extension, called the MIL model (Modèle Intermittent Lagrangien) has been proposed also where the exchange time τ_{IEM} itself is a random variable [47]. In order to simplify the calculations, it has been assumed in addition that the global activation energy

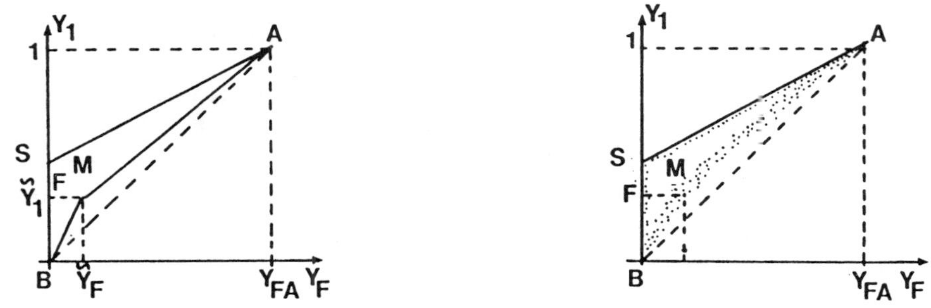

Figure 7. **a)** $\tilde{P}(Y_1, Y_F)$ for the MIL model   **b)** Experimental $\tilde{P}(Y_1, Y_F)$.

of the combustion process is very high, in such a way that the combustion is very "sudden", once ignited.

Let us consider a non-premixed flame; the flow enters the bruner, through two entrances: the fuel through A and the oxidizer through B. If we consider Y_1 as an inert tracer and Y_F as the fuel, and if we look at the plane (Y_1, Y_F) Fig. 7, the evolution of the composition of fluid particles can be represented by trajectories within this plane; some of them are beginning at A, others at B.

From equation (14), but considering a very large value of activation energy and an homogeneous turbulent medium (where $\tilde{Y}_1$ and $\tilde{Y}_F$ are equal everywhere), these trajectories are formed of paths along pure mixing lines (AM or BM) or paths along equilibrium lines (AS or BS); for instance a particle entering the flow through B travels along BM, but jumps at a given time on BS, and follows then BS and SF; for a particle issued from A a similar evolution occurs on AM and AF. As indicated in Fig. 7a, the result is that every fluid particle can be found on the lines AS, SB, BM and MA, but not elsewhere, since the jumps are very fast due to the sudden combustion; so the joint p.d.f. $\tilde{P}(Y_1, Y_F)$ is supported only by BS, SA, AM and MB.

Due to the fact that the I.E.M. model is not exact, and that we are interested in non homogeneous flows, the joint p.d.f. is not actually that simple; instead, it looks like that in Fig. 7b; indeed such a picture has been found experimentally by Masri et al. [48], but in the plane (Nitrogen concentration-temperature).

The trajectories of Fig. 7a allow us to compute directly the mean reaction rate, with the help of the I.E.M. model. This reaction rate requires, as entries, three types of quantities:

i) The knowledge of $\tilde{P}(Y_1)$, where Y_i is an inert scalar.

ii) The knowledge of $\tilde{P}_\tau(\tau_{\text{IEM}})$; this is a new entry. With our assumption of sudden combustion, τ_{IEM} is just an exchange time due to the turbulence before ignition; no influence of chemistry is to be taken into account. A distribution, as a function of τ_t and τ_K, is to be found from theoretical or experimental. basis; the first candidate is a lognormal distribution, as that used for the fluctuations of γ itself.

iii) The knowledge of τ_{ig} (Y_1). It can include complex chemistry effects, but only as an ignition delay, for conditions taken roughly along the AM and MB lines. It could be calculated knowing in full the chemical mechanism, or deduced from experiments. For instance, if we know the $\tau_{ig}(Y_{ox}, Y_F, T)$ for a given oxidizer-fuel mixture, we could calculate τ_{ig} at each given point along the lines AH and BM just using the linear relations:

$$\begin{cases} \frac{Y_F - Y_{FB}}{\tilde{Y}_F - Y_{FB}} = \frac{Y_1 - 1}{\tilde{Y}_1 - 1} = \frac{Y_{ox}}{\tilde{Y}_{ox}} = \frac{h - h_B}{\tilde{h} - h_B} \\ \frac{Y_{ox} - Y_{oxA}}{\tilde{Y}_{ox} - Y_{oxA}} = \frac{Y_1}{\tilde{Y}_1} = \frac{Y_F}{\tilde{Y}_F} = \frac{h - h_A}{\tilde{h} - h_A} \end{cases} \tag{15}$$

and deducing τ_{ig} $(Y_1, \tilde{Y}_1, \tilde{Y}_F, \tilde{Y}_{ox}, \tilde{h})$.

With an arbitrary distribution $\tilde{P}_\tau(\tau_{\text{IEM}}$, it is shown in [2] that this model would lead to a dependence of $\tilde{\dot{w}}_F$ not only on τ_{ig}/τ_t, but also on τ_{ig}/τ_K, as the flamelets models.

This model has been recently compared to experimental results of a methane-air flame close to extinction [47]. See Fig. 8. The chemical model gives again some problems.

Other Lagrangian models have been proposed, in which the Lagrangian equation is intended to represent fluid elements bigger than a fluid particle, with an internal structure; for example the "ESCIMO" model of D.B. Spalding [49], or the "Four environments model" [50] are of this type. The ESCIMO model uses also "reacting laminae", but with upsteady stretching, where the mean stretching rate is related to τ_t. The "Four environments model" is more complicated, involving three types of fluid elements with special exchange processes between them and the surrounding medium.

An exception, there is also another model, neither of flamelet type, nor Lagrangian, for turbulent combustion, namely the "perturbation model" of R.W. Bilger. It is a major extension of the moments approach, in which the calculated variables are the departures (not necessarily small) from the equilibrium values of certain mass fractions. It is described in [51]; it has been applied up to now only to particular chemical systems: H_2 combustion or NO oxidations [52].

d) Some comments.

The three categories of models, the p.d.f. statistical models, the flamelets models, and the lagrangian models, may appear at first sight quite different. However, they can be seen as three different ways to gain a good approximation of the multidimensional p.d.f. $\tilde{P}(Y_1, \ldots Y_n)$.

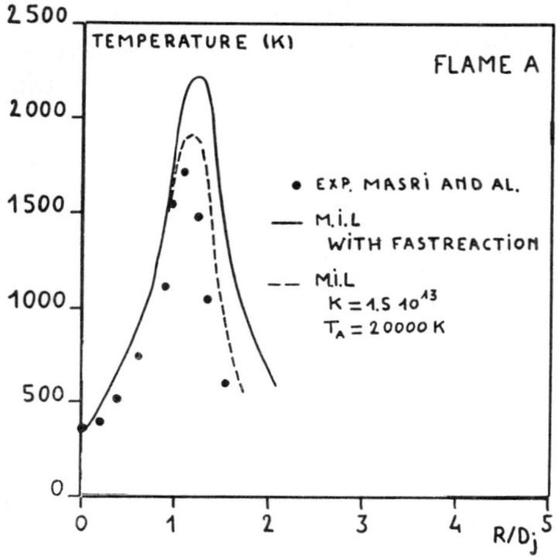

FIGURE 8a

RADIAL PROFILE OF MEAN
TEMPERATURE AT AXIAL LOCATION
x/Dj = 10 ; FLAME A.

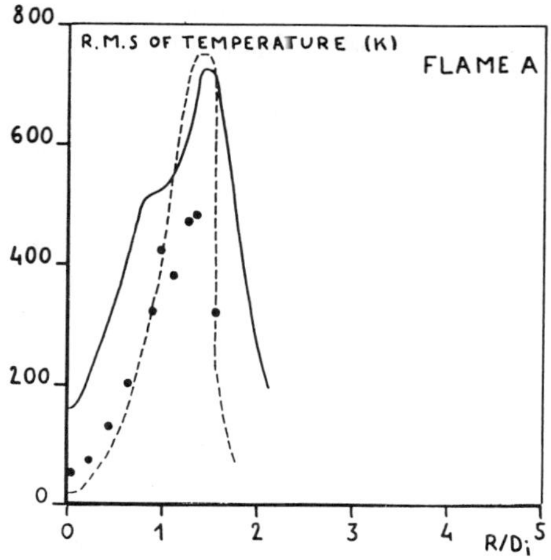

FIGURE 8b

RADIAL PROFILE OF TEMPERATURE
R.M.S. AT AXIAL LOCATION
x/Dj = 10 ; FLAME A.

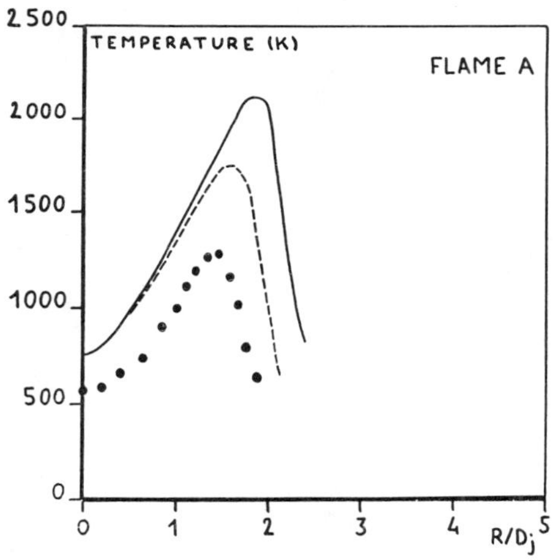

FIGURE 8c

RADIAL PROFILE OF MEAN
TEMPERATURE AT AXIAL LOCATION
x/Dj = 20 ; FLAME A

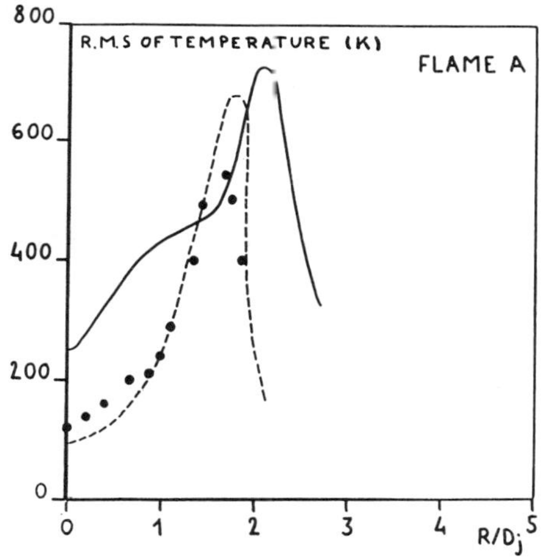

FIGURE 8d

RADIAL PROFILE OF TEMPERATURE
R.M.S. AT AXIAL LOCATION
x/Dj = 20 ; FLAME A

Figure 8.

The p.d.f. statistical model computes this p.d.f. with a modelled equation. In the flamelets models this p.d.f. can be calculated successively in two dimensional spaces (ϕ, Y_i) where ϕ is an inert scalar and Y_i any species; we have (when the stretching is neglected)

$$\tilde{P}(\phi, Y_i) = \tilde{P}(\phi)\delta(Y_i - Y_i^F(\phi)) . \tag{16}$$

Where $Y_i^F(\phi)$ is the relation between Y_i and ϕ inside the flamelet. With the IEM Lagrangian model, the situation is similar; it comes:

$$\tilde{P}(Y_1, Y_i) = \tilde{P}(\ )\delta(Y_i - Y_i^{\text{IEM}}(Y_1)) . \tag{17}$$

Where $Y_i^{\text{IEM}}(Y_1)$ is the relation of Y_i and Y_1 along the trajectories computed with the Lagrangian differential equations. The difference with the flamelets is that $Y_i^{\text{IEM}}(Y_1)$ is influenced by turbulence and by the surrounding medium, because τ_t and $\tilde{Y}_i$ plays in the Lagrangian equations, while for the flamelets, the internal structure leading to $Y_i^F(\phi)$ is independent of the surroundings.

In the Mil model, there is an additional random variable, as for the flamelets models when it includes the stretching; in these cases, we have, respectively

$$\tilde{P}(Y_1, Y_i) = \tilde{P}(Y_1) \int_0^\infty \delta(Y_1 - Y_i^{\text{MIL}}(Y_1, \tau))P_\tau(\tau)\mathrm{d}\,\tau \tag{18}$$

and

$$\tilde{P}(\phi, Y_i) = \tilde{P}(\phi) \int_0^\infty \delta(Y_1 - Y_i^F(\phi, \gamma))P_\gamma(\gamma)\mathrm{d}\,\gamma . \tag{19}$$

4.CONCLUDING REMARKS

The use of turbulent combustion models needs first to identify the combustion regime that is expected to occur. It may depend on the size of the combustion chamber, and of the operating conditions that are chosen. The section 2 has explained the present knowledge that allows this identification.

Section 3 has described the different theories of models that have been proposed so far; they have different validity domains, and are more or less easy to implement. It is to be emphasized that such models have always to be coupled with a turbulence model, that computes the turbulent diffusion and also, at least, one turbulent time scale. It is very important to realistically predict these quantities, because they have a direct influence on the combustion; the turbulence model may need some modifications to be adapted for combusting flows.

The described models are always approximate models, with often some constants that are not well known at present time. This is true even in the simplest case of very fast chemistry. They constitute the present of knowledge, which will be improved further in the near future.

Among the problems that have been attacked with these models, the easiest is the computation of the flame length, or the global heat release, in conditions close to the normal operating point of the engine; in that case, fast chemistry models are often sufficient, and the predictions will give probably a 10%-20% error range, provide that we can be sure enough of the turbulence.

More difficult is the prediction of the limits of operation conditions, such as the stability limits, for instance (in the plance equivalence ratio-mass flow rate); this will need models with finite rate chemistry and the models are more difficult to implement and less validated, but the chemistry itself is not well known. Of the same order of difficulty is the prediction of pollutants, C, N_x or unburned hydrocarbons. This objective needs clearly to use a model with finite rate chemistry, and with a complicated enough chemical scheme. The prediction of pollutants does not constitute a problem that has to be sėt up, and solved, with different and specific models; it can be attacked with the same methodology as explained here for any reactive species; it is just a more complicated case. Concerning these two problems, where chemistry plays a more important role, some progress have been done these last years; but very few comparisons between results of models and experiments have been successful. So we cannot assess now the range of validity of these models; it can be hoped that we will do that in the two or three next years.

REFERENCES

[1] R. Borghi, *On the structure and morphology of turbulent premixed flames*, in *Recent advances in th aerospace sciences*, Ed. Corrado Casci, pp. 117-138, Plenum Publishing corporation (1985).

[2] R. Borghi, *Turbulent combustion modelling*, Prog. Energy Combust. Sci. vol. 14, pp. 245-292 (1988).

[3] K.N.C. Bray, *Turbulent Flows with premixed reactants*, in *Turbulent Reacting Flows*, Ed. P.A. Libby and F.A. Williams. Topics in Applied Physics, vol 44, pp. 115-183, Springer-Verlag, Heidelberg (1980).

[4] G. Damkolher, *Zeitschrift für Electrochemie*, vol. 45, n. 11 (1940).

[5] K.I. Schelkin, *Combustion Hydrodynamics*, Fizika Goreniya i Vzryva, vol. 4, n. 4, pp. 455-468 (1968).

[6] N. Peters, *Laminar flamelets concepts in turbulent combustion*, 21th Symp. (Int.) on Combustion. Munich – Aug. (1986).

[7] V.L. Zimont, *Theory of turbulent combustion of a homogeneous fuel mixture at high Reynolds numbers*, Fizijka Goreniya i Vzryva, vol. 15, n. 3, pp. 23-32 (1979).

[8] P. Moreau, *Turbulent flame development in a high velocity premixed flow*, 16th AIAA Aerospace Science meeting, n. 77/49, Los Angeles (1977).

[9] P. Moreau and A. Boutier, *Laser velocimeter measurements in a turbulent flame*, 16th Symp. (Int.) On Combustion. The Combustion Institute, pp. 17-47 (1977).

[10] R.G. Bill, I. Namer, L. Talbot, R.K. Cheng and F. Robben, *Flame propagation in grid-induced turbulence*, Comb. and Flame, vol 43, pp. 229-242 (1981).

[11] E.E. Zukowski and F.H. Wright, *Flame spreading from bluff-body flameholders*, 8th Symp. (Int.) on Combustion, pp. 933-945, Williams and Wilkins, Baltimore (1962).

[12] N.M. Howe, JR., C.W. Shipman and A. Vranos, *Turbulent Mass Transfert and Rates of Combustion in confined turbulent flames*, pp. 36-47, 9th Symp. (Int.) on Combustion (1963).

[13] F.A. Williams, *Recent advances in theoretical descriptions of turbulent diffusion flames*, in *Turbulent mixing in non reactive and reactive flows*, Ed. S.N.B. Murthy, pp. 189-208. Plenum press, N.Y. (1975).

[14] N. Peters, *Laminar diffusion flamelet models in non premixed turbulent combustion*, Prog. In Energ. Combust. Sci., vol. 10 pp. 319-339 (1984).

[15] R.W. Bilger, *The structure of non premixed flames* – Invited to the 22nd Symp. (Int.) on Combustion. Seattle. Aug. 1988.

[16] W.M. Roquemore, L.D. Chen, L.P. Goss, W.F. Lynn, *The structure of jet diffusion flames*, in *Turbulent reactive Flows*, Ed. R. Borghi, S.N., Murthy, pp. 49-63, Springer-Verlag (1989).

[17] D. Stepowski, K. Labacci, R. Borghi, *Instantaneous radial profiles of OH fluorescence and Rayleigh scattering through a turbulent H2-Air diffusion flame*, pp. 1561-1567 21st Symp. (Int.) on Combustion – The Combustion Institute 1986.

[18] R.W. Bilger, *Turbulent Flows with non-premixed reactants*, in *Turbulent Reacting Flows*, Ed. P.A. Libby and F.A. Williams Topics in Applied Physics, vol. 44, pp. 65-92, Springer-Verlag, Heidelberg (1980).

[19] J.H. Kent and R.W. Bilger, *The Prediction of Turbulent Diffusiôn Flame Fields and Nitric Oxide Formation*, Sixteenth Symposium (Int.) on Combustion, The Combustion Institute, pp. 1643-1656 (1977).

[20] B. Lewis and G. Von Elbe, *Combustion, Flames and Explosions of Gases*, New York Academic Press (1951).

[21] T. Takagi, H.D. Shin, A. Ishio, *Properties of turbulence in turbulent diffusion flames.* Combustion and Flame, 1981, vol. 40, pp. 121-140.

[22] K.N.C. Bray and P.A. Libby, *Interaction effects in turbulent premixed flames*, Phys. Fluids, vol. 19, pp. 1681-1701 (1976).

[23] R. Borghi and D. Escudie, *Assessment of theoretical model of turbulent combustion with a simple experiment.* Combustion and Flame, 56, pp. 149-164 (1984).

[24] K.N.C. Bray, P.A. Libby, G. Masuya and J.B. Moss, *Turbulence production in premixed turbulent flames.* Comb. Sci. Tech., vol. 25, pp. 127-140 (1981).

[25] R.W. Bilger, *Turbulent jet diffusion Flames. Progress in Energy and Combustion Sciences*, vol. 1, pp. 87-109 (1976).

[26] H.L. Toor, *Mass Transfer in dilute turbulent and non turbulent systems with rapid irreversible reactions and equal diffusivities*, AICHE Journal, vol. 8, n. 1, pp. 70-78 (1962).

[27] Ch. Lin and E.E. O'Brien, *Two Species isothermal reactions in Homogeneous turbulence*, Astronautica Acta, vol. 17, pp. 771-781 (1971).

[28] R.I. Curl, *Dispersed phase mixing theory and effects in simple reactors*, AICHE Journal, 9 (2), pp. 175-181 (1963).

[29] J. Janicka and W. Kollmann, *A two variable formalism for the treatment of chemical reactions in turbulent He-Air diffusion flame*, 17th Symp. (Int.) on Combustion, The Comb. Inst. Pittsburg, pp. 421-430 (1979).

[30] S.B. POPE, *P.D.F. methods for turbulent reactive flows. Progress in Energy and Combustion science*, vol. 11, pp. 119-192 (1985).

[31] S.B. POPE, *The Statistical theory of turbulent flames*, Phil. Trans. R. Soc. London Series A, vol. 291, pp. 529-568, (1979).

[32] J. JANIKA, W. HOLBE and W. KOLLMANN, *The solution of a p.d.f. transport equation for turbulent diffusion flames*, Proc. of the 1978 Heat transfer and Fluid Mechanics Institute, pp. 296-312, Ed. C.T. Crowe and W.L. Grosshandler, Stanford University Press (1978).

[33] C. BONNIOT, R. BORGHI, *Joint probability density function in turbulent combustion*, Acta Astronautica, vol. 6, pp. 309-327 (1979).

[34] R.C. FLAGAN, J.P. APPLETON, *A stochastic model of turbulent mixing with chemical reaction*, Fluid Mech. Lab. Pub. 73-10. MIT (1973).

[35] W.P. JONES and W. KOLLMANN, *Multiscalar p.d.f. transport equations for turbulent diffusion flames*, 5th Turbulent shear flows conference. Cornell Univ., Aug. (1985).

[36] S.M. CORREA, M.C. DRAKE, R.W. PITZ and W. SHYY, *Prediction and measurement of a non equilibrium diffusion flame*, Report n. 84CRD171. Corporate Research and Development, G.E. Schenectady, New-York (1984).

[37] K.W. MAO and H.L. TOOR, *A diffusion model for reactions with turbulent mixing*, A.I. Ch. E.J., vol. 16, pp. 49-52 (1970).

[38] C.H. GIBSON and P.A. Libby, *On Turbulent Flows with Fast Chemical Reaction, Part II - The Distribution of Reactants and Products Near a Reacting Surface*, Combustion Science Technol., vol. 8, pp. 29-35 (1972).

[39] F.E. MARBLE and J.E. BROADWELL, *The coherent flame model for turbulent chemical reactions*, Project Squid Tech. Rep. TRW-9-PU, Purdue University, West Lafayette, Indiana (1977).

[40] S.K. LIEW, K.N.C. BRAY and J.B. MOSS, *A flamelet model of turbulent non premixed combustion*, Combustion Sciences Tech.n vol. 27, pp. 69-73 (1981).

[41] J.P. BROADWELL, P.E. DIMOTAKIS, *Implications of recent Experimental results for Modelling reactions in turbulent flows*, AIAA Journal vol. 24 n. 6, pp. 885-889 (1986).

[42] P.E. DIMOTAKIS, *Turbulent shear layer mixing with fast chemical reactions*, in *Turbulent reactive flows*, Ed. R. Borghi, S.N.B. Murthy, pp. 417-485. Lecture notes in Engineering, vol. 40, Springer Verlag (1989).

[43] J.E. BROADWELL, M.G. MUNGAL, *Molecular mixing and chemical reactions in turbulent shear layers*, 22nd Symp. (Int.) on Combustion, pp. 579-587. The Combustion Institute (1988).

[44] J. VILLERMAUX, *Micromixing phenomena in stirred reactors*, Encyclopedia of Fluid Mechanics, Gulf Pub. Co., West Orange, N.J., pp. 707-768 (1986).

[45] H. YAMAZAKI, A. ICHIGAWA, *Int. Chem. Eng.*, vol. 10, pp. 471-478 (1970).

[46] R. BORGHI, E. POURBAIX, *Lagrangian models for turbulent combustion*, 4th Symp. on Turbulent Shear Flows, Karlsruhe, Sept. (1983).

[47] M. GONZALEZ, R. BORGHI, *A Lagrangian intermittent model for turbulent combustion! Theoretical basis and comparisons with experiments*, 7th Symp. on turbulent shear flows, Stanford Un. Aug. 1989.

[48] A.R. MASRI, R.W. DIBBLE and R.W. BILGER, *Turbulent non premixed flames of methane near extinction: Mean structure from Raman measurements*, 21th International Symposium on Combustion, Munich, August 3-8 (1986).

[49] D.B. SPALDING, *A general theory of turbulent combustion*, J. Energy, vol. 2, n. 1, pp. 16-32 (1977).

[50] R.V. MEHTA, J.M. TARBELL, *A four environment model of mixing and chemical reaction. Part II: comparison with experiments*, AICHE Journal, vol 65, pp. 125-133 (1983).

[51] R.W. BILGER, *Perturbation analysis of turbulent non premixed combustion*, Comb. Sci. Tech., vol. 22, pp. 251-261 (1980).

[52] S.S. IBRAHIM, R.W. BILGER and N.R. MUDFORD, *Calculation of mixing and chemical reaction in a turbulent smog chamber*, Physico Chemical Hydrodynamics, vol. 8, n. 3, pp. 265-278 (1987).

R. Borghi

Coira – Ura n° 230 CNRS

Faculté des Sciences de Rouen

Rouen (France)

K N C BRAY AND P A LIBBY

Premixed turbulent combustion

INTRODUCTION

The Bray-Moss-Libby (BML) model [1] - [3] for the aerothermochemistry of premixed turbulent combustion has been found to provide a useful description of the complex processes involved in such combustion under many circumstances of applied interest. It is an extension of the earlier Bray-Moss (BM) model [4] for the thermochemistry of such combustion, an extension which incorporates the velocity components as well as the single scalar, the progress variable defining completely the state of the gas, i.e., its temperature and composition. The purpose of this contribution is to review the development and application of these models and the related literature.

The physical basis for the BM-model resides in the following observation which applies to many flows involving premixed turbulent combustion: If a fast response, highly accurate thermometer is fixed in a turbulent premixed flame, e.g., a flame stabilized on a rod as shown schematically in Fig. 1, it indicates two temperature levels, the low temperature of reactants and the high temperature of products. The transition from one level to the other is abrupt and occurs when a thin structure involving chemical reaction, i.e., transforming reactants to products, passes over the thermometer. There results from this observation a two fluid description of premixed turbulent combustion involving a progress variable $c(x, t)$ with the following values: $c = 0$ in reactants ; $c = 1$ in products and $0 < c < 1$ in fluid within the structure involving chemical reaction. As will be discussed later the thin reacting structures in appropriate circumstances are strained laminar flames, i.e., flamelets.

THE BRAY-MOSS MODEL OF THE THERMOCHEMISTRY

Exploitation of this physical picture calls for consideration of a probability density function (pdf) $P(c; x)$ shown schematically in Fig. 2 and given analytically by

$$P(c; x) = \alpha(x)\, \delta(c) + \beta(x)\, \delta(1 - c) + \gamma(x)\, f(c; x) \tag{1}$$

If the pdf of the interior values of the progress variable is normalized so that

$$\int_{0^+}^{1^-} dc\, f(c; x) = 1 \tag{2}$$

we then have the result that

$$\alpha(x) + \beta(x) + \gamma(x) = 1 \tag{3}$$

However, the physical picture of chemical reaction confined to relatively thin surfaces so that a thermometer indicates essentially only two temperatures, those corresponding to $c = 0$ and $c = 1$, implies that

$$\gamma(x) \ll 1 \tag{4}$$

i.e., that interior values of c are present only a small fraction of the time. Thus mathematically the BM-model can be considered to be based on a series expansion in a representative value of γ, a small quantity. We shall see later that the mean rate of creation of product denoted $\bar{w}(x)$, a quantity which must be proportional to $\gamma(x)$, is retained in the theory because γ is multiplied by a *large* factor, one proportional to a Damköhler number

Before carrying the discussion of the BM-model further we must digress to address the issue of the means for defining the time average of fluid mechanical quantities in turbulent flows involving variable density. In reference [5] Favre introduces the idea of mass weighted averaging in connection with turbulent boundary layers in high speed flows but it is applicable to all turbulent flows involving density inhomogenieties, e.g., low speed binary mixing and turbulent combustion with significant heat release. When first adopted for the latter situation, there were objections on several grounds. Measuring systems, e.g., thermocouples and total head tubes, are generally assumed to yield signals which must be interpreted in terms of unweighted quantities although there may be some ambiguity in this regard. Thus predictions based on Favre averaging cannot be directly compared with experimental results in such systems. In more recent times advanced optical systems, those providing the highest quality and most extensive experimental data on flows involving turbulent combustion, permit the variables determining the state of the gas and the velocity components to be determined at the same point in space and time so that this concern has largely been overcome. A second objection to Favre averaging relates to the modeling of the pressure fluctuation effects in second moment closure theories. Such modeling has been elaborately developed and assessed for constant density flows over many years and a reluctance to reconsider these effects lead some to the use conventional averaging in theories of turbulent combustion. However, it appears at the present time that Favre averaging is generally accepted as providing a convenient approach to the description of turbulent flows involving density inhomogenieties including flows with reaction.

In Favre averaging a velocity component $u_i(x, t)$ is decomposed into

$$u_i(x, t) = \tilde{u}_i(x) + u_i''(x, t)$$

$$= \frac{\overline{\rho u_i}(x)}{\bar{\rho}(x)} + u_i''(x, t) \tag{5}$$

With this averaging $\overline{u_i''} \neq 0$ as is readily found from the first of Eq. (5) but $\overline{\rho u_i''} = 0$.

Central to the notion of a progress variable is the determination of the entire state of the gas if $c(x, t)$ is known. Thus for the low speed, adiabatic flows in which pressure fluctuations are thermochemically insignificant we consider $\rho = \rho(c)$ and $T = T(c)$ according to

$$\frac{\rho}{\rho_r} = \frac{1}{1+\tau c} = \frac{T_r}{T} \tag{6}$$

where the subscript r denotes conditions in reactants and τ is a heat release parameter readily determined if the temperature in products is known; clearly

$$\frac{\rho_p}{\rho_r} = \frac{1}{1+\tau} = \frac{T_r}{T_p} \tag{7}$$

where the subscript p denotes conditions in products. From Eqs. (6) we see that a temperature signal can be immediately interpreted in terms of the progress variable and that the mean density and the mean temperature can be readily calculated in terms of the Favre averaged progress variable, namely

$$\frac{\bar{\rho}(x)}{\rho_r} = \frac{1}{1+\tau\,\tilde{c}(x)} = \frac{T_r}{\tilde{T}(x)} \tag{8}$$

Equation (8) is based on Eq. (6) and is independent of the magnitude of $\gamma(x)$ but if we now calculate the mean progress variable for $\gamma \ll 1$ as in the BM-model, we have

$$\tilde{c}(x) = \frac{\int_0^1 dc\; c\; \rho(c)\, P(c\,;x)}{\int_0^1 dc\; \rho(c)\, P(c\,;x)} \approx \beta(x)\, \frac{\rho_p}{\rho}$$

$$\approx \beta(x)\, \frac{1+\tau\,\tilde{c}(x)}{1+\tau}$$

or

$$\beta(x) \approx \frac{(1+\tau)\,\tilde{c}(x)}{1+\tau\,\tilde{c}(x)} \qquad\qquad \alpha(x) \approx 1-\beta(x) \approx \frac{1-\tilde{c}(x)}{1+\tau\,\tilde{c}(x)} \tag{9}$$

Thus we see that if $\tilde{c}(x)$ is known, so too are the strengths of the delta functions determining the approximate pdf of the progress variable. Furthermore, knowledge of $\tilde{c}(x)$ determines in an approximate fashion the statistical behavior of all state variables. For example, consider the intensity of the temperature fluctuations; according to Favre averaging we have

$$\overline{\rho T''^2} = \overline{\rho T^2} - \bar{\rho}\,\tilde{T}^2$$

$$\approx \alpha(x)\,\rho_r\,T_r^2 + \beta(x)\,\rho_p\,T_p^2 - \bar{\rho}\,\tilde{T}^2$$

$$\approx \bar{\rho}\,T_r^2\,\tau^2\,\tilde{c}\,(1-\tilde{c})$$

Similar calculations can be carried out for all state variables.

We consider next the mean rate of creation of product denoted $\bar{w}(x)$. We have

$$\bar{w}(x) = \int_{0^+}^{1^-} dc\; w(c)\,P(c\,;x)$$

$$= \gamma\,w_{ref} \int_{0^+}^{1^-} dc\; \frac{w(c)}{w_{ref}}\, f(c\,;x) \tag{10}$$

where w_{ref} denotes a reference value of $w(c)$. Equation (10) implies that although γ is a small quantity, w_{ref} is suitably large so that the product $\gamma\,w_{ref}$ is independent of the limit process $\gamma \to 0$. Uncertainity in the value of this product suggests, as indeed is the case, that the BM-model by itself does not lead to an evaluation of $\bar{w}(x)$ and thus that separate considerations are needed to model this important quantity. Indeed we discuss later such considerations as they evolve from the BM-model.

If w_{ref} is defined by the characteristics of an unstrained laminar flame as

$$w_{ref} = \frac{\rho_r\,u_L^0}{l_L^0}$$

where u_L^0 and l_L^0 are the burning velocity and thickness respectively of such a flame, then a Damköhler number

$$D_A = \frac{w_{ref}\,L}{\rho_r\,k_0^{1/2}} = \frac{u_L^0\,L}{l_l^0\,k_0^{1/2}} \tag{11}$$

can be defined in terms of the ratio of a turbulence time $L/k_0^{1/2}$ to a chemical time l_L^0/u_L^0. Here k_0 and L are the turbulent kinetic energy and the integral scale of the turbulence in the stream approaching the flame respectively. We require the latter time to be short compared with the former for the BM-model to apply.

The first application of the BM-model appears in Bray and Libby [6] and involves a classical gradient transport calculation of planar premixed turbulent flames, both normal and oblique to the oncoming reactant stream. As discussed in Bray *et al* [2] the shortcomings of the fluid mechanical treatment in reference [6] within which the BM-model is embedded result in seriously flawed predictions of the influence of heat release on turbulence production, predictions not in accord with experiment. To avoid the gradient transport treatment the BML-model is needed.

THE BML-MODEL FOR THE AEROTHERMOCHEMISTRY AND ITS IMPLICATIONS

The bimodal character of the pdf of $c(x, t)$ leads naturally to conditional sampling of the velocity components and to a useful treatment of the aerothermochemistry of premixed turbulent combustion. Consider a velocity component $u_i(x, t)$ and the bivariate pdf shown schematically in Fig. 3 and given analytically by

$$P(u_i, c; x) = \alpha(x)\, \delta(c)\, f_r(u_i; x) + \beta(x)\, \delta(1-c)\, f_p(u_i; x) + \gamma(x)\, f(u_i, c; x) \tag{12}$$

In this case if $f(u_i, c; x)$ is normalized so that

$$\int_{0^+}^{1^-} dc \int_{-\infty}^{\infty} du_i\, f(u_i, c; x) = 1$$

then again Eq. (9) applies. The quantities $f_r(u_i; x)$ and $f_p(u_i; x)$ are conditional pdf's yielding the statistical properties of the u_i-velocity component within reactants and products respectively. With modern optical diagnostic systems permiting time and space resolved measurements of a state variable, e.g., the temperature, and one or more velocity components it is possible to determine experimentally the conditional pdf's (cf. reference [7]). Earlier experiments involving light scattering from particle loaded premixed flames are due to Moss [8].

The bivariate pdf of Eq. (12) combined with Eqs. (9) leads naturally to the conditional velocity components $\overline{u_{i\,r}}(x)$ and $\overline{u_{i\,p}}(x)$, the mean velocities within reactants and products respectively where the overbar implying conventional averaging is now employed because of the constant density in either reactants or products as appropriate. The connection between the unconditional velocity components which appear in the Favre averaged conservation equations and these conditional velocity components is seen from application of Eq. (12). Consider the following:

$$\begin{aligned}\bar{\rho}\, \tilde{u}_i(x) &= \int_0^1 dc\, \rho(c) \int_{-\infty}^{\infty} du_i\, u_i\, P(u_i, c; x) \\ &\approx \alpha(x)\, \rho_r \int_{-\infty}^{\infty} du_i\, u_i\, f_r(u_i; x) + \beta(x)\, \rho_p \int_{-\infty}^{\infty} du_i\, u_i\, f_p(u_i; x) \\ &\approx \bar{\rho}\left[(1-\tilde{c})\, \overline{u_{i\,r}} + \tilde{c}\, \overline{u_{i\,p}}\right]\end{aligned} \tag{13}$$

and

$$\overline{\rho u_i''^2}(x) = \int_0^1 dc\, \rho(c) \int_{-\infty}^{\infty} du_i\, (u_i - \tilde{u}_i(x))^2\, P(u_i, c; x)$$

$$\approx \alpha(x)\,\rho_r \int_{-\infty}^{\infty} du_i\,(u_i - \bar{u}_i(x))^2 f_r(u_i;x) + \beta(x)\,\rho_p \int_{-\infty}^{\infty} du_i\,(u_i - \bar{u}_i(x))^2 f_p(u_i;x)$$

$$\approx \bar{\rho}\left[\bar{c}\,(1-\bar{c})\,(\overline{u_{ip}} - \overline{u_{ir}}) + (1-\bar{c})\,\overline{u_i'^2}_r + \bar{c}\,\overline{u_i'^2}_p\right] \tag{14}$$

where in the right side of Eqs. (13) and (14) we drop the functional dependencies on x without loss of clarity. Note that Eqs. (13) and (14) reduce properly in the two limits, $\bar{c}(x) = 0, 1$.

Equation (14) introduces the conditional intensities $\overline{u_i'^2}_r$ and $\overline{u_i'^2}_p$ within reactants and products respectively, intensities given by the conditional pdf's $f_r(u_i;x)$ and $f_p(u_i;x)$. We comment on these quantities later. Of interest in Eq. (14) is the indication that the unconditional intensity of the velocity fluctuations is due to two separate mechanisms; one is associated with the first term on the right side and is due to oscillations *between* the conditional velocities within rectants and products while the second is associated with fluctuations *within* the reactants and products. The relative importance of these two mechanisms near the center of premixed turbulent flames appears to depend on flame geometry. Theoretical analysis [2] and experiments [7] suggest that the first mechanism is dominant in flames with negligible Reynolds shear stresses. On the other hand in the experimental studies of reference [9] involving intense turbulence generated by large shear stresses, the conditional intensities dominate.

As a final example of the consequences of Eq. (12) consider the mean turbulent flux

$$\overline{\rho u_i''c''} = \int_0^1 dc\,(c - \bar{c}(x))\,\rho(c) \int_{-\infty}^{\infty} du_i\,(u_i - \bar{u}_i(x))\,P(u_i, c; x)$$

$$\approx -\alpha(x)\,\rho_r\,\bar{c}(x) \int_{-\infty}^{\infty} du_i\,(u_i - \bar{u}_i(x))\,f_r(u_i;x) +$$

$$\beta(x)\,\rho_p\,(1 - \bar{c}(x) \int_{-\infty}^{\infty} du_i\,(u_i - \bar{u}_i(x))\,f_p(u_i;x)$$

$$\approx \bar{\rho}\,\bar{c}\,(1-\bar{c})\,(\overline{u_{ip}} - \overline{u_{ir}}) \tag{15}$$

Equation (15) contains significant implications as may be seen as follows: Consider the planar premixed flame shown in Fig. 4. Conservation of mass requires that the mean velocity within the product stream downstream of the flame be greater than the corresponding velocity within the reactantstream approaching the flame according to the relation

$$\bar{u}_\infty = (1+\tau)\,\bar{u}_0 \tag{16}$$

If the extent of the heat release is small so that these two velocities are essentially the same, the sign of the mean flux $\overline{\rho u_i''c''}$ implied by Eq. (15) is unclear. However, if τ takes on values of applied interest, namely from four to nine, the significantly greater mean velocity within products suggests that $\overline{\rho u''c''} > 0$. However, according to gradient transport

$$\overline{\rho u''c''} \propto -\bar{\rho}\,\frac{\partial \tilde{c}}{\partial x} < 0$$

Accordingly, there is suggested a contradiction between Eq. (16) and the ubiquitous gradient description of turbulent transport when there is significant heat release. Indeed countergradient diffusion in premixed turbulent flames normal to the oncoming reactant stream is found in a series of theoretical papers (references [1] and [2]) and confirmed by experiments (references [8] - [10]).

The physical basis for this phenomena is the interaction of the density inhomogenieties associated with heat release and the weak mean pressure gradient drop across a flame such as that in Fig.4. This interaction causes the light parcels of products to accelerate towards the product side of the flame relative to the heavy parcels of reactants, i.e., differential diffusion results in $\bar{u}_p > \bar{u}_r$ and thus in $\overline{\rho u''c''} > 0$. In more general flow fields involving gradients of mean pressure due to the hydrodynamics of the flow a similar interaction can be expected to be the dominant mechanism of turbulent transport with the sign of the mean flux dependent on the sign of the mean pressure gradient.

Although an interesting phenomena of fundamental and conceptual importance, the practical significance of this interaction in turbulent combustion is as yet unassessed. Nongradient turbulent transport due to the influence of mean force fields on density inhomogenieties can exist only within the combustion zone, i.e., within the flame. Whether in premixed combustion the orientation of the flame and thus the entire flow field is influenced by this interaction is not at present clear. Recent work [11,12] suggests that the *leading edge* of a premixed flame propogates into a uniform turbulent mixture with a turbulent burning velocity which is predictable in terms of gradient transport applied locally. Note that the applicability of such transport at the edges of a turbulent flame is consistent with our earlier discussion that nongradient transport requires the *coexistence* of a mean force field and density inhomogenieties; the latter are absent at such edges. Thus nongradient transport may not be important in determining global features of flows involving premixed turbulent flames. In nonpremixed combustion, in high speed turbulent boundary layers and in high speed turbulent combustion we expect nongradient diffusion to be present but these flows have not been examined from a perspective permiting an assessment of the practical importance of such diffusion.

A related finding of nongradient transport in the tangential direction of oblique turbulent flames by Masuya and Libby [3] leads to different mean streamlines for reactants and products and is due to the force field associated with a gradient of the Reynolds shear stress. There has been no direct experimental verification of this theoretical prediction but supporting suggestions are contained in reference [9].

The bivariate pdf of Eq. (12) with its consequences in Eqs. (13) - (15) results in significant advances in the formulation of a second moment description of premixed turbulent combustion as demonstrated in references [1] -[3]. Since as might be expected there are still uncertainities in such a formulation, we discuss several of these difficulties. The conditional intensities, $\overline{u_i'}_r$ and $\overline{u_i'^2}_p$, in Eq. (14), are examples of more general conditional Reynolds stresses within reactants and products which must be modeled in some fashion to achieve a closed set of equations. In references [1] - [3] various ad hoc models based on physical arguments are used to eliminate these quantities but further research involving careful experimental data are called for to clarify the significance of these stresses and to establish sound models for their treatment.

There is the fundamental problem of combustion induced pressure fluctuations and their influence in theories based on the second moment equations. In references [1] - [3] it is argued that because of the small thickness of the planar turbulent flames studied there, thicknesses on the order of the integral scale of the turbulence in the reactant stream approaching the flame, pressure effects are dominated by the mean pressure gradient but the importance of pressure fluctuations and appropriate descriptions thereof, if they are important, are at the present time unknown. Strahle and coworkers (cf. reference [13]) have made a start on this problem but more research is required.

A final problem arising in second moment closures of the equations for premixed turbulent combustion relates to models for the several dissipation terms. It is argued in reference [14] that chemical reaction within laminar flamelets provides a mechanism for dissipation, one which dominates for planar premixed turbulent flames. However, in applications of the BML-model in turbulent boundary layers with their long flow lengths Champion and Libby [15] and Bray *et al* [16] find it imperative to include a conditional viscous dissipation within reactants $\bar{\varepsilon}_r$ and in products $\bar{\varepsilon}_p$ in a simple additive fashion with the latter given more importance since they estimate that

$$\bar{\varepsilon}_r = \frac{\bar{\varepsilon}_p}{(1+\tau)^4}$$

Since their model is again of an ad hoc nature, additional attention to this matter is indicated although direct experimental support for any model is not likely in the foreseeable future.

FLAMELET CROSSING FREQUENCY

Our earlier description of the response of an idealized thermocouple immersed in a premixed turbulent flame leads to an interesting and useful statistical analysis. The motivation for the analysis resides in the observation in reference [17], namely that the mean rate of creation of product can be described as

$$\bar{w}(x) = \nu(x)\, w_F(x) \tag{17}$$

where the first factor on the right side is the mean crossing frequency of flamelets, the mean number of 0-1 *and* 1-0 changes of the progress variable per unit time with units of reciprocal time, and the second is the mean rate of creation of product per crossing with units of density. According to the laminar flamelet theory of premixed turbulent combustion this second factor depends on the mean rate of strain the flamelet is subjected to by the turbulent velocity field as well as other thermochemical properties of the gas, e.g., the effective Lewis number and activation temperature of the mixture. It is therefore w_F which connects studies of laminar premixed flames and the theory of premixed turbulent combustion. Although later we take up the treatment of $w_F(x)$, here our attention is focused on the first factor on the right side of Eq. (17), i.e., on $\nu(x)$. Consider that an idealized temperature signal is converted to a 0-1 function as shown in Fig. 5, i.e., into a clean, two level time history of the progress variable $c(t; x)$.

The starting point for our discussion [18] is the time autocorrelation of such a function. Generally, we have

$$P_{11}(\hat{\tau}; x) \equiv \overline{c(t; x)c(t + \hat{\tau}; x)} \tag{19}$$

where $\hat{\tau}$ is the time advance taken here to be positive.* When specialized to our idealized progress variable, Eq.(18) yields:

$$P_{11}(\hat{\tau} \rightarrow 0; x) = \overline{c^2(t; x)} - \bar{c}(x) \qquad P_{11}(\hat{\tau} \rightarrow \infty; x) = \bar{c}^2(x) \tag{18}$$

where conventional averaging is found to be more convenient in the present analysis. Moreover, from the theory of two level functions (cf. reference [19]) it is also known that

$$\frac{\partial P_{11}}{\partial \hat{\tau}}(0^+; x) = -\frac{\nu(x)}{2} \tag{20}$$

where $\nu(x)$ is the desired crossing frequency. This last result clearly suggests the utility of considering the P_{11}-function. As a final preliminary we note that an integral time scale which enters the analysis in an important way can be defined as

$$\hat{T}(x) = \frac{1}{\bar{c}(1 - \bar{c})} \int_0^\infty d\hat{\tau}\,(P_{11} - \bar{c}^2) \tag{21}$$

The passage times t_p and t_r shown in Fig. 5 are assumed to be random functions of time and subject to a statistical description in terms of pdf's $P(t_p; x)$ and $P(t_p; x)$. In particular these functions yield mean values which have a simple connections with physically relevant quantities

$$\bar{c}(x) = \frac{\bar{t}_p(x)}{\bar{t}_r(x) + \bar{t}_p(x)} \qquad \nu(x) = \frac{2}{\bar{t}_r(x) + \bar{t}_p(x)} \tag{22}$$

Thus we have two equations among four quantities.

* For two level functions the autocorrelation is nonanalytic at $\hat{\tau} = 0$.

References [18] and [20] provide an analysis permiting calculation of the autocorrelation function from the pdf of the passage times. In the first of these references an exponential distribution is considered for both t_r and t_p, i.e.,

$$P(t_r) = \frac{1}{\bar{t}_r} \exp(-\frac{t_r}{\bar{t}_r}) \qquad P(t_p) = \frac{1}{\bar{t}_p} \exp(-\frac{t_p}{\bar{t}_p}) \tag{23}$$

After a detailed calculation [18] carried out in collaboration with S.O.Rice, the exponential distribution results in a relatively simple form for $P_{11}(\tau; x)$, namely

$$P_{11}[\bar{c}(x), \nu(x)\hat{\tau}] = \bar{c}(1-\bar{c}) \exp\left[-\frac{\nu \hat{\tau}}{2\bar{c}(1-\bar{c})} \right] + \bar{c}^2 \tag{24}$$

where to avoid clutter we drop the x-dependencies of $\bar{c}$ and ν on the right side. From Eq.(21) we have

$$\hat{T}(x) = \frac{g\,\bar{c}(1-\bar{c})}{\nu} \tag{25}$$

where $g = 2$. Thus $\hat{T}$ is the characteristic time scale related to the crossing frequency $\nu(x)$. Initially it was expected that a partial differential equation for $P_{11}(\hat{\tau}; x)$ yielding upon integration $\hat{T}(x)$ and thus completing this analysis could be developed but as demonstrated in reference [21] this expectation has not at present been realized and other approaches to the modeling of $\bar{w}(x)$ are being pursued as discussed later.

The crossing frequency analysis has proven to be useful to experimentalists in presenting and interpreting their results on the dynamics of flamelets in premixed turbulent combustion. Bray *et al* [22] review these data and demonstrate this usefulness.

Bray and Libby [20] revise the analysis sketched here and calculate $P_{11}(\hat{\tau}; x)$ and thus $\hat{T}(x)$ when the pdf of passage times takes on a more realistic form. The exponential distribution has a most probable values of zero, a physically unappealing feature. When a gamma-two distribution is adopted, the result for $P_{11}(\hat{\tau}; x)$ corresponding to Eq. (24) involves much greater complexity but the results corresponding to Eqs. (25) differ only in the replacement of the factor g by unity. In reference [22] it is concluded that the available experimental data, although uncertain because of questions of statistical reliability for long and short passage times, favor the gamma-two distribution. o

MEAN CREATION OF PRODUCT PER FLAMELET CROSSING

We now restrict our attention to aerothermochemical conditions such that a laminar flamelet model is appropriate for the evaluation of the quantity $w_F(x)$ in Eq. (17). If the structures in which chemical reaction occurs are sufficiently thin relative to the Kolmogoroff scale, i.e., to the smallest scale of the turbulence, the internal distribution of their aerothermochemical properties is that same as for strained laminar flames (cf. reference [23]) and laminar flamelet models are then applicable. Peters [24] provides an excellent review of laminar flamelet concepts in both premixed and nonpremixed systems. In other circumstances with lower Damköhler numbers and higher turbulence Reynolds numbers the reactive structures are sufficiently thick as to call for other descriptions.

The quantity $w_F(x)$ we focus on is defined [25] as

$$w_F(x) \equiv <\int w(t;x)\,dt> \tag{26}$$

where $w(t;x)$ is the instantaneous rate of creation of product at the spatial location in question and < > implies conditional averaging within flamelets. The integral is clearly zero except when a flamelet is present so each realization relates to the complete passage of a flamelet. If the speed at which a thin flamelet moves normal to itself across the point x is V_n and if the coordinate normal to the flamelet is z, then Eq. (26) becomes

$$w_F(x) = <\frac{1}{V_n(x)}\int w(z;x)\,dz>$$

Provided we make the plausible assumption that the speed of the flamelet crossing is not strongly correlated with the reaction rate integral, we can write

$$w_F(x) = <\int w(z;x)\,dz\frac{>}{<V_n(x)>} \tag{27}$$

Now the reaction rate integral in this equation is evaluated from solutions for steady strained laminar flames in counterflowing streams [23,26] so that for a given set of aerothermochemical conditions of pressure and reactant composition the instantaneous rate of creation of product depends only on the rate of strain a which is conventionally specified on the reactant side fo the flamelet. Thus we can interpret the integrand in Eq. (27) in terms of $w(z;a)$ and can define an effective laminar burning velocity $u_L(a)$ for each realization from

$$\int w(z;a)\,dz = \rho_r\, u_L(a)$$

Thus the desired mean creation of product per crossing is seen to involve the quotient of two mean velocities, $<u_L(x)>$ and $<V_n(x)>$. The average of the former is evaluated [27] by application of the presumed pdf method applied to the rate of strain a with the parameters in the pdf determined by the distribution of the instantaneous viscous dissipation at the flamelet, i.e., by

$$\varepsilon(t;x) = a^2(t;x)\,\nu(t;x)$$

where ν is the kinematic viscosity which fluctuates from that in reactants to that in products. Roughly

$$\nu_p = (1+\tau)^2 \nu_r$$

so that for degrees of heat release of applied interest there is considerable difference in these two values. We note earlier that the mean values of $\varepsilon(t, x)$ in reactants and products are estimated [15] to obey the relation $\bar{\varepsilon}_p = \bar{\varepsilon}_r/(1+\tau)^4$. Given the uncertainities in this relation and the sensitivity of the instantaneous dissipation to the kinematic viscosity suggested here we must expect large uncertainities in current ad hoc modeling [27] of the pdf of the rate of strain. This matter calls for further attention.

The mean of the integral in Eq. (27) becomes

$$<\int w(z;x)\,dz> = \rho_r <u_L>(x) = \rho_r \int_0^\infty u_L(\varepsilon;x)\,P(\varepsilon;x)\,d\varepsilon \tag{28}$$

which focuses attention on the pdf $P(\varepsilon; x)$. The distribution of ε is assumed following Kolmogoroff [28] to be lognormal with a mean value $\bar{\varepsilon}(x)$ and with a variance on the order of unity. Accordingly, we must add to our equation set an equation for $\bar{\varepsilon}(x)$ with all the uncertainities arising therefrom. Assuming $\bar{\varepsilon}(x)$ is available, we can then make an estimate for $P(\varepsilon; x)$ and can compute $<u_L>(x)$. Thus we can write

$$w_F(x) = \frac{\rho_r\, u_L^0\, I_0(x)}{<V_n>(x)} \tag{29}$$

where u_L^0 is the burning velocity of an unstrained laminar flame and $I_0(x) \equiv <u_L>(x)/u_L^0$. Calculations [12,26,27] and experiment [29] suggest that I_0 can become small compared to unity in intense, small scale turbulence, i.e., near extinction can occur, and thus that a realistic description of premixed turbulent combustion must include the effects of rate of strain on the flamelets.

Equation (28) makes no allowance for effects of flamelet curvature on $<u_L>(x)$. This neglect is justified [29] on the several grounds. When the turbulence Reynolds number is large Klimov [30] finds rate of strain alters flame speed significantly more than curvature. In addition in [29] it is argued that positive and negative curvatures have almost equal probabilities in intense turbulence so that flame curvature effects tend to cancel. However, experiments on extinguishing flames in weak, decaying turbulence show that assymmetry of the distribution of curvature has a profound effect on the rate of heat release.

LAMINAR FLAMELET MODEL FOR THE MEAN REACTION RATE

Combining Eqs. (17), (25) and (28) results in the mean rate of creation of product in laminar flamelet combustion

$$\bar{w}(x) = \frac{\rho_r\, u_L^0\, I_0}{<V_n>} \frac{g\, \bar{c}\,(1-\bar{c})}{\hat{T}} \tag{30}$$

where in the interest of simplicity we supress the dependence of $\bar{c}$, I_0 and $<V_n>$ on x. At this point in the discussion we need a description for either $<V_n>(x)$ or $\hat{T}(x)$ separately or for their product. Uncertainities relative to these quantities lead to alternative formulations for $\bar{w}(x)$.

The mean rate of creation is sometimes expressed in terms of the mean surface-to-volume ratio of laminar flamelets denoted by $\sum(x)$, i.e., e.g., the average number of cm^2 of flamelet surface area per unit volume in cm^3; in this case we have

$$\bar{w}(x) = \rho_r\, u_L^0\, I_o(x) \sum(x)$$

Comparison with Eq. (30) leads to

$$\sum(x) = \frac{g\, \bar{c}\,(1-\bar{c})}{<V_n>\, \hat{T}} \tag{31}$$

where $\bar{c}$, $\hat{T}$ and $<V_n>$ are functions of x. Again we see the two quantities $<V_n>(x)$ and $\hat{T}(x)$.

Alternatively, the integral in Eq. (10) can be evaluated in terms of the laminar flamelet model developed here; for flamelets characterized by a given rate of strain a we write

$$I(a) = \int_{0^+}^{1^-} dc\ w(c\,;a)\, f(c\,;a) \tag{32}$$

Now $f(c\,;a) = (l_L\ dc/dz)^{-1}$ so that

$$I(a) = \frac{1}{l_L(a)} \int_{z_r}^{z_p} w(z\,;a)\, dz$$

$$= \frac{\rho_r\, u_L(a)}{l_L(a)} \tag{33}$$

The coordinates z_r and z_p define the flame thickness in terms of $z_p - z_r$ and identify appropriate locations within the strained flamelet where c is close to zero and unity respectively. Finally, Eq. (33) is weighted with a pdf characterizing the distribution of the rate of strain the flamelets are subject to and integrated as in Eq. (28). The result may be written as

$$\int_{0^+}^{1^-} dc\ w(c\,;a)\, f(c\,;a) = \frac{\rho_r <u_L>(a)}{<l_L>(a)}$$

which defines the mean laminar flamelet thickness $<l_L>(a)$. Equation (10) thus becomes

$$\bar{w}(x) = \frac{\gamma\, \rho_r\, u_L^0\, I_o(x)}{<l_L>(x)}$$

which if compared with Eq. (30) gives

$$\gamma(x) = <l_L>(x) \sum(x)$$

We thus see that $\gamma(x)$ can be interpreted as the *mean volume fraction* occupied by reacting gas. From Eq. (31) we finally obtain

$$\gamma(x) = \frac{<l_L>}{<V_n>} \frac{g \, \bar{c} \, (1-\bar{c})}{\hat{T}} \tag{34}$$

where on the right side $\bar{c}$, $<l_L>$, $<V_n>$ and $\hat{T}$ are functions of x.

Before Eqs. (29),(31) and (34) can be applied, the quantities g, $\hat{T}(x)$ and $<V_n>(x)$ must be specified. As noted earlier experiments confirm that g lies between one and two. Measurements of $\hat{T}$ in a variety of premixed turbulent flames are reviewed in reference [22] while measurements of $<V_n>$ in a bunsen flame are reported by Zhang [32]. Bray [12] suggests that

$$\hat{T}(x) <V_n>(x) \approx C_L \, L \, \bar{\sigma} \left(\frac{u_L^0}{k^{1/2}}\right)^n$$

where C_L and n are empirical constants of order unity [12] and $\bar{\sigma}$ is a mean direction cosine whose value [31] is around one half and where $L(x)$ is an integral length scale of the velocity field and $k(x)$ is the turbulent kinetic energy. As noted earlier the individual turbulence intensities whose sum is $2\,k(x)$ can take different values in reactants and products as can the viscous dissipation $\bar{\varepsilon}(x)$. Consequently the proper evaluation of $k(x)$ and $L(x)$ taking into account conditional values within the two gases is fraught with unresolved difficulties. Earlier studies of boundary layer combustion [15,16] assume that $L(x)$ has the same value in reactant and product parcels but further research is required to assess this assumption. Following the empirical formulation of Bray [12] with $n = 1$ we find from Eqs. (11) and (34) that

$$\gamma D_A = \frac{<l_L>}{l_L^0} \frac{\bar{c} \, (1-\bar{c})}{C_L \, \sigma}$$

which is of order unity as called for in the discussion following Eq. (10). Similarly Eq. (29) become

$$\bar{w}(x) = \frac{g}{C_L \, \sigma} \frac{\rho_r \, I_o \, k^{1/2} \, \bar{c} \, (1-\bar{c})}{L} \tag{35}$$

where I_0, L, k and $\bar{c}$ are functions of x. The characteristic time scale in this expression is the same as that arising in the well-known $k - \varepsilon$ theory, namely $t_T = L/k^{1/2} = k/\bar{\varepsilon}$ but here we see that the mean reaction rate is altered by the quantity $I_0(x)$ which accounts for the influence of the rate of strain on the creation of product within flamelets. Indeed as noted earlier $I_0(x)$ becomes small so that near extinction is predicted in intense small scale turbulence.

The calculation of $I_0(x)$ is therefore of considerable interest. In reference [26] it is shown that for thin flamelets the progress variable c can be unambiguously related to the species concentrations within reactants and products in calculations of strained laminar flames inncorporating detailed chemical kinetic and molecular transport descriptions. Values of I_0 predicted in this manner show satisfactory agreement with values deduced [12] from a correlation of turbulent flame speed data [33].

It is clear from this discussion that despite the progress made toward a rational, well supported model for the mean rate of creation in premixed turbulent flames additional research involving close coordination among theory, computation and experiment is called for. The mean velocity of flamelet passing $<V_n>(x)$ and the integral time $\hat{T}(x)$ need special attention. In this research the flows in which the turbulent flames are embedded should be selected so that such coordination is facilitated. In particular there should not be chosen flows which can be readily established in the laboratory but which involve complex geometries so that theoretical and computational difficulties becloud assessment of the models for $\bar{w}(x)$.

PREMIXED FLAMES IN STAGNATING TURBULENCE

Extinction is one of the interesting characteristics of premixed turbulent flames which can be elucidated in terms of flamelet concepts. The widely held belief which has not been subjected to close scrutiny is that extinction of a turbulent premixed flame occurs when an excessive proportion of laminar flamelets are strained by the turbulent velocity field to such an extent that they are incapable of creating new product, i.e., a large proportion of flamelets involving high temperature products flow counter to cold products but involve little or no chemical reaction. Under these circumstances it is believed a turbulent flame may not be sustained and extinction can occur. From this perspective the onset of extinction may be expected to be dependent on flow geometry such that, for example, a transient spherical turbulent flame may be more susceptible to extinction than a planar flame.

Premixed flames in stagnating turbulence provide a convenient means for examining both experimentally and theoretically the extinction behavior and other important notions relevant to such flames. Such flames are the turbulent counterparts of the laminar flows which have provided important information on the aerothermochemistry of both premixed and nonpremixed systems. Several examples illustrate the flames we have in mind. Figure 6a shows a turbulent flow of reactants from a circular jet impinging on a flat wall and resulting in a flat turbulent reaction zone. This configuration is being studied at the University of California Berkeley (cf. references [10] and [34]) and at Karlsruhe (cf. reference [35]). Additional studies are being planned elsewhere.

A second configuration is shown in Fig. 6b; in this case two identical streams of turbulent reactants flow counter to one another and lead to either one or two reaction zones depending on the aerothermo-chemical conditions of the system. Experiments along these lines are being carried out by Bray and coworkers [36]. The extinction characteristics of these flames are discussed in reference [37].

The advantages of these flows in terms of experiments are that they are relatively easy to set up, that a wide range of experimental conditions can be studied with a single test rig and that access by optical and other instrumentation is readily provided. Of great importance in terms of extinction is the possibility of changing the mean rate of strain the flame experiences by simply changing the mass flow from the jet in the case of a single jet or from both jets in the case of counterflowing streams. Extinction in a single jet configuration is reported in reference [35] and extensively in counterflowing reactant streams in reference [36,37]. Illustrating the flexibity of such test rigs is the observation that with two counterflowing reactant streams the enthalpy of one stream can be altered either by changing its temperature or composition so that the characteristics of nonisenthalpic premixed turbulent flames can be readily investigated. Since from theoretical studies of premixed laminar flames (cf. reference [23]) we know that nonisenthalpic conditions significantly alter the extinction and other characteristics of flamelets, it is of interest to determine the effects of such alterations on turbulent flames. Finally, nonpremixed turbulent flames can be studied in the configuration of Fig. 6b if one turbulent stream carries the oxidizer while the second carries the fuel. Plans to conduct experiments of this sort are underway.

To indicate the possibilities of studying flamelet behaviour in stagnating turbulence we show the results of laser tomography in Fig.7a - 7c. The first two relating to a reactant stream impinging on a wall are kindly supplied by Dr. Robert K. Cheng of the University of California Berkeley. The reactant stream is laden with small oil droplets which evaporate during passage through the flamelet. Thus the photographs show a slice of the laminar flamelet at a particular instant of time. By identifying the location of the flamelet and assembling a series of such photographs there is obtained Fig. 7b which clearly identifies the planar reaction zone discussed earlier. It is within this zone that that the mean rate of creation of product $\overline{w}(x)$ discussed earlier arises. Figure 7c is from reference [36] and corresponds to the two counterflowing reactant streams of Fig. 6b. Again the sharp boundary between light and dark corresponds to the instantaneous position of the flamelets.

The further advantage of studying premixed flames in stagnating turbulence is that complementary theoretical studies are facilitated by the simplicity of the geometry. Bray *et al* [38] show that there can be formed similarity solutions for the mean velocity components, the mean progress variable and the turbulence quantities such as the turbulent kinetic energy and the viscous dissipation. These solutions are analogous to those for laminar flames but with revised characteristic quantities, i.e., the mean rate of strain a and a characteristic turbulent kinetic energy k_∞. The similarity variable is shown to be

$$\eta = \frac{a}{k_\infty^{1/2}} \int_0^y \frac{\overline{\rho}}{\rho_r}\, dy \tag{36}$$

These similarity notions can be extended to the other turbulence quantities. If a wall is involved as in the flow shown in Fig. 6a, there arises the Reynolds number $k_\infty/a\ \nu_0$ where ν_0 is a reference molecular kinematic viscosity. In this case the flow consists of three layers: The viscous sublayer in which molecularity must be taken into account, a shear layer in which turbulent transport is significant and an outer flow whose mean velocity components are given by potential flow theory. This structure is exploited in reference [39] in which asymptotic methods are applied to obtain approximate descriptions of all three layers.

In reference [38] the BM-model is combined either with a simple algebraic transport model or the $k-\varepsilon$ model to treat flows involving counterflowing reactant streams but research is presently underway to apply the BML-model to these flows. Of particular interest in this regard is the finding by Cho *et al* [10] of differential diffusion in their flames.

PREMIXED FLAMES IN CLOSED VESSELS

Another relatively simple and attractive geometry for the study of premixed turbulent combustion involves the transient, one dimensional propogation of a flame in a linear, cylindrical or spherical closed vessel which is initially filled with a combustible mixture in a specified state of turbulent motion. The turbulence may either be allowed to decay after the initiation of combustion or be assumed to be continuously stirred while combustion proceeds. This type of combustion is of practical interest in connection with studies of spark ignition engines and for the assessment of explosion hazards in industrial settings. Also a large amount of experimental data is obtained from combustion in closed vessels and presented in terms of turbulent burning velocities [33]. A successful theory of turbulent combustion should be able to address such data.

The BML-model described earlier requires significant modification [40] before it can be applied to these problems. Although it leads to no fundamental difficulties, transient combustion requires that ensemble rather than the more usual time averaging be used. Moreover, the conditional densities and temperatures of reactants and products are no longer constant but rise as a result of the heat release in a closed vessel while the accompanying variations in the enthalpy require inclusion of a statement of energy conservation in the equation set. Nevertheless the resulting model [27,40] retains key features of the BML-theory including the assumption that chemical reaction is confined to thin surfaces, i.e., that $\gamma(x) \ll 1$. Thus conditional values of aerothermochemical quantities within reactants and products arise naturally and facilitate closure of the system of equations. A second moment turbulence model is employed [40] which contains a mechanism for countergradient turbulent transport via the interaction between density fluctuations and the mean pressure gradient.

The resulting model is used [27,40] to predict the progress of combustion in closed vessels. Countergradient transport is predicted [40] but in the cases considered its magnitude is found to be reduced in comparison with that found in a freely propogating flame. Interestingly, the predicted turbulent burning velocity is unaffected when the mechanism for countergradient transport is suppressed, a finding confirmed by subsequent theoretical analysis [41]. Flame geometry is predicted [41] to have a significant effect on the turbulent burning velocity with spherical flames propogating more slowly than cylindrical flames which in turn propogate more slowly than planar flames. Use of the laminar flamelets model leads to the prediction of important influences of chemical reaction rates on turbulent flames speeds [12,27].

The interesting findings from these studies suggest that further research on these flames is indicated. In due course the phenomenology developed for closed vessels with fixed geometry will be applied to the treatment of the spark ignition engine with the resulting more realistic model for chemical reaction and to the extensive data on premixed turbulent flames in combustion bombs.

Acknowledgements

Preparation of this review was supported by Shell Research Ltd., by the European Community (KNCB) and by the Department of Energy, Office of Basic Energy Sciences, Division of Chemical Sciences under Contract DE-FG03-86ER13527 (PAL)

K.N.C.Bray is Professor of Applied Thermocdynamics in the Department of Engineering at Cambridge University and Paul A. Libby is Professor of Fluid Mechanics in the Department of AMES at the University of California San Diego

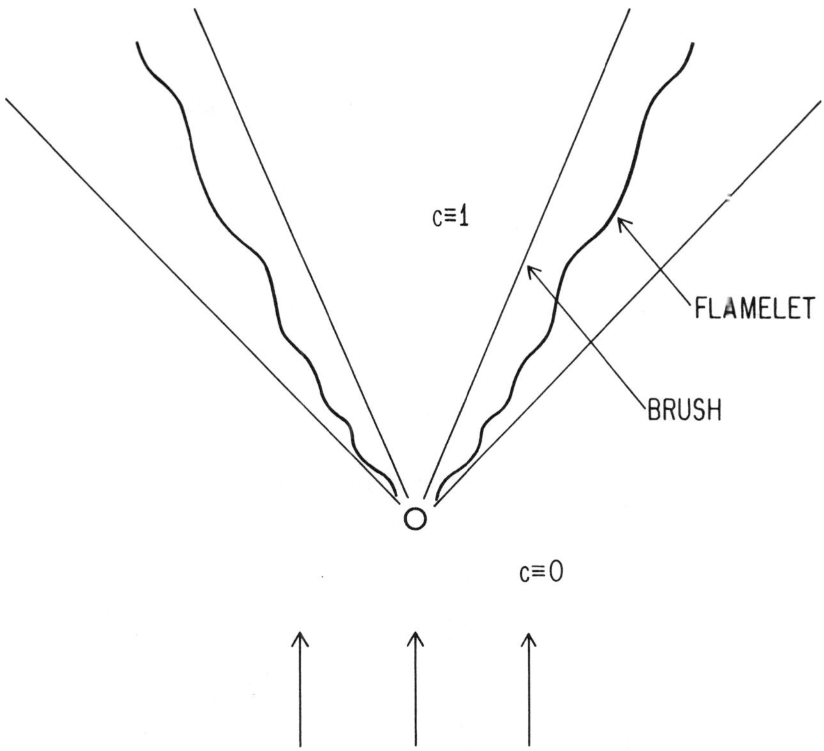

Figure 1. Premixed turbulent flames stabilized on a rod

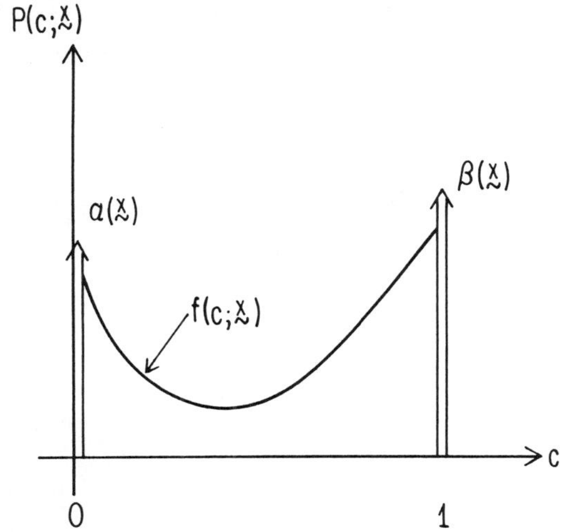

Figure 2. The probability density function for the progress variable

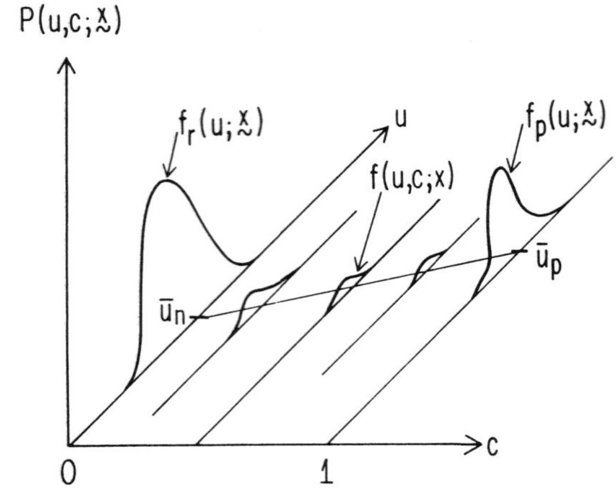

Figure 3. The bivariate probability density function of the progress variable and a velocity component

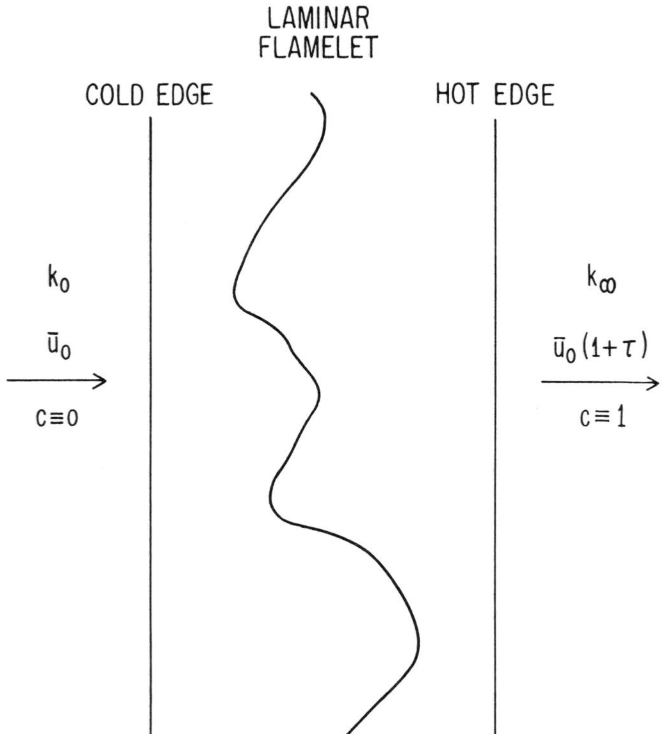

Figure 4. Schematic representation of a premixed turbulent flame orthogonal to an oncoming reactant stream

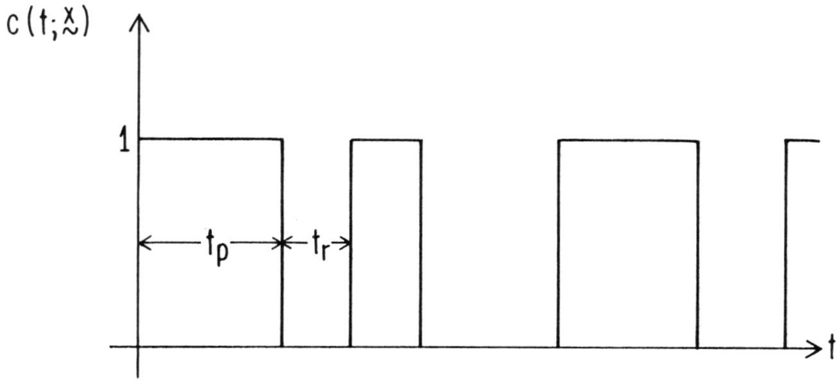

Figure 5. An idealized time history of the progress variable in a premixed turbulent flame

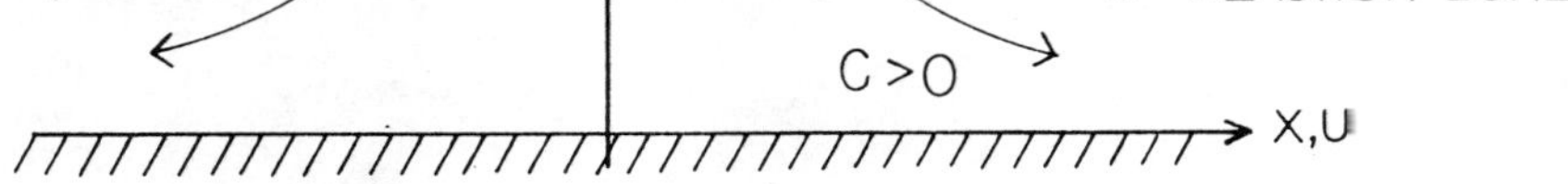

Figure 6. Premixed flames in stagnating turbulence

a. A reactant stream impinging on a wall

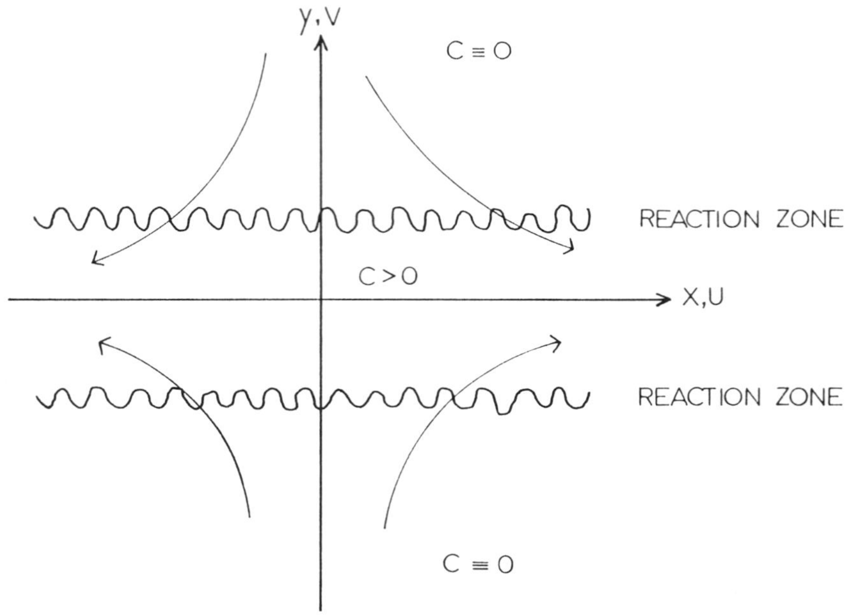

b. Counterflowing reactant streams

Figure 7. Laser tomography of premixed flames in stagnating turbulence
a. A single realization of a flame in reactants impinging
on a wall

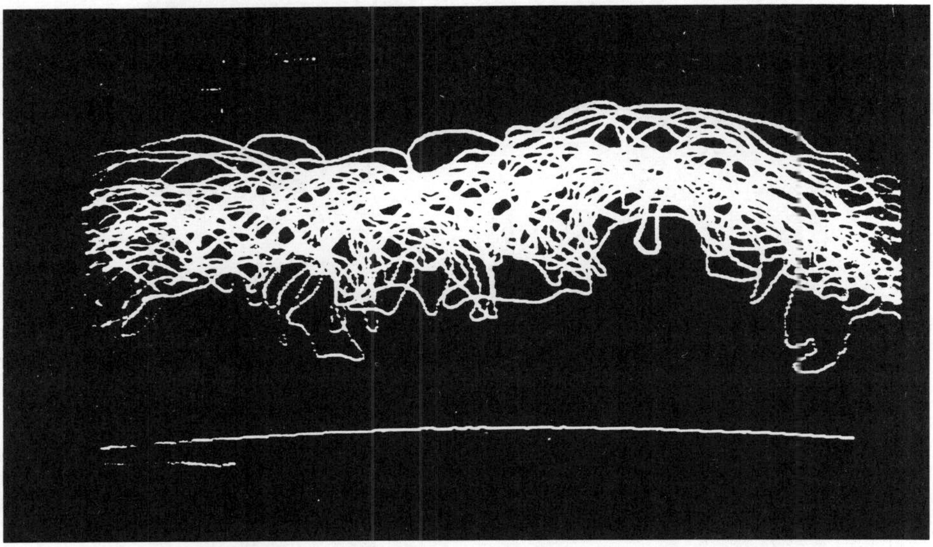

b. Multiple realizations of flamelets in a premixed flame in reactants impinging on a wall

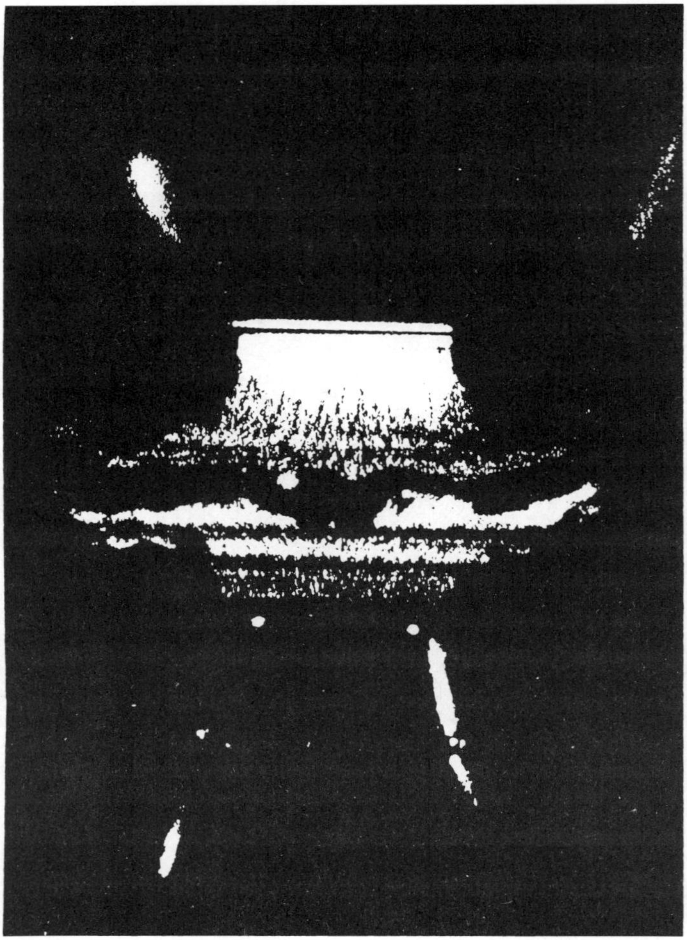

c. Flamelets in counterflowing reactant streams

REFERENCES

1. Libby,P.A. and Bray,K.N.C. (1981). Countergradient diffusion in premixed turbulent flames. *AIAA J.* **19**:205.

2. Bray,K.N.C., Libby,P.A., Masuya,G. and Moss,J.B. (1981). Turbulence production in premixed turbulent flames. *Combust. Sci. Technol.* **254**:127.

3. Masuya,G. and Libby,P.A. (1981). Nongradient theory of oblique turbulent flames with premixed reactants. *AIAA J.* **19**:1590.

4. Bray,K.N.C. and Moss,J.B. (1977). A uniform statistical model of the premixed turbulent flame. *Acta Astronautica* **4**:291.

5. Favre,A. (1969). Statistical equations of turbulent gases. *Problems of Hydrodynamics and Continuum Mechanics*, SIAM, Philadelphia, 231.

6. Bray,K.N.C. and Libby,P.A. (1976). Interaction effects in turbulent premixed flames, *Phys. Fluids* **11**:1687.

7. Cheng,R.K. (1984). Conditional sampling of turbulence intensities and Reynolds stress in premixed turbulent flames, *Combust. Sci. Tech.* **25**:109.

8. Moss,J.B. (1980). Simultaneous measurements of concentrtion and velocity n an open premixed turbulent flame. *Combust. Sci. Tech.* **22**:115.

9. Heitor,M.V., Taylor,A.M.K.P. and Whitelaw,J.M. (1987). The interaction of turbulene and pressure gradients in a baffle-stabilized premixed flame. *J. Fluid Mech.* **181**:387.

10. Cho,P., Law,C.K., Hertzberg,J.R. and Cheng,R.K. (1988). Velocity and scalar fields of turbulent premixed flames in stagnation flow, *Twenty-second Symposium (International, on Combustion*, The Combustion Institute, 739.

11. Gray,H.L. (1988). Theoretical predictions of turbulent burning velocities of steady and accelerating unconfined turbulent flames with premixed reactants. PhD Thesis, University of Southampton.

12. Bray,K.N.C. (1990). Studies of the turbulent burning velocity. Cambridge University Report CUED/A-Thermo/TR 32 (Submitted for publication).

13. Strahle,W.C. and Chandran,S.B.S. (1982). Pressure-velocity correlation in a reactive turbulent flow, *AIAA J.* **20**:129.

14. Libby,P.A. and Bray,K.N.C. (1980). Implications of the laminar flamelet model in premixed turbulent combustion. *Combust. Flame* **39**:33.

15. Champion,M. and Libby,P.A. (1984). Turbulent premixed combustion in a boundary layer, *Combust. Sci. Tech.* **38**:267.

16. Bray,K.N.C., Champion,M. and Libby,P.A. (1987) The turbulent premixed boundary layer with variable enthalpy, *Combust. Sci. Tech.* **55**:139.

17. Libby,P.A. (1975). Studies related to turbulent flows involving fast chemical reactions, in S.N.B.Murthy, Ed., *Turbulent Mixing in Nonreactive and Reactive Flows*, Plenum Press, New York-London, 333.

18. Bray,K.N.C., Libby,P.A. and Moss,J.B. (1984). Flamelet crossing frequencies and mean reaction rates in premixed turbulent combustion, *Combust. Sci. Tech.* **41**:143.

19. Papoulis,A. (1965). *Probability, Random Variables and Stochastic Processes*, McGraw-Hill, New York.

20. Bray,K.N.C. and Libby,P.A. (1986). Passage times and flamelet crossing frequencies in premixed turbulent combustion, *Combust. Sci. Tech.* **47**:253.

21. Bray,K.N.C., Champion,M. and Libby,P.A. (1988). The correlation functions for flamelet crossings in premixed turbulent flames, *Combust. Sci. Tech.* **59**:463.

22. Bray,K.N.C., Champion,M. and Libby,P.A. (1988). Mean reaction rates in premixed turbulent flames. *Twenty Second Symposium (International) on Combustion, The Combustion Institute*, 763.

23. Libby,P.A., Linan,A. and Williams,F.A. (1983). Strained premixed laminar flames with nonunity Lewis numbers, *Combust. Sci. Tech.* **34**:257.

24. Peters,N. (1986). Laminar flamelet concepts in turbulent combustion.*Twenty First Symposium (International) on Combustion, The Combustion Institute*,1231.

25. Bray,K.N.C., Champion,M. and Libby,P.A. (1989). The interaction between turbulence and chemistry in premixed turbulent flames. In *Turbulent Reactive Flows* (R. Borghi and S.N.B.Murthy, Eds.) Lecture Notes in Engineering **40**, Springer-Verlag, New York.

26. Cant,R.S., Rogg,B. and Bray,K.N.C. (1990). On laminar flamelet modelling of the mean reaction rate in a premixed turbulent flame. *Combust. Sci. Tech.* **69**, 53.

27. Cant,R.S. and Bray,K.N.C. (1988). Strained laminar flamelet calculations of premixed turbulent combustion in a closed vessel. *Twenty Second Symposium (International) on Combustion*, The Combustion Institute, 791.

28. Kolmogoroff,A.N. (1962). A refinement of previous hypotheses concerning the local structure of turbulence in a viscous incompressible fluid at high Reynolds number, *J.Fluid*, **13**, 82.

29. Becker,H., Monkhouse,P.B., Wolfrum,J., Cant,R.S., Bray,K.N.C., Maly,R., Pfister,W., Stahl,G. and Warnatz,J. (1990). Investigation of extinction in unsteady falmes in turbulent combustion by 2D-LIF of OH radicals and flamelet analysis. Submitted to *Twenty Third Symposium (International) on Combustion*, The Combustion Institute.

30. Klimov,A.M. (1963). Laminar flames in turbulent flow, *Zhur.Prikl.Mekh.Tekh.Fiz* **3**, 49.

31. Chew,T.C., Bray,K.N.C., Britter,R.E. (1990). Spatially resolved flamelet statistics for reaction rate modelling. *Combust. and Flame* (To appear).

32. Zhang,Y. (1990). Private Communication, University of Cambridge.

33. Abdel-Gayed,R.G., Bradley,D. and Lawes,M. (1987). Turbulent burning velocities: A general correlation in terms of straining rates, *Proc.Roy.Soc. London.* **A414**,389.

34. Cho,P., Law,C.K., Hertzberg,J.R. and Cheng,R.K. (1986). Structure and propogation of turbulent premixed flames stabilized in a stagnation flow. *Twenty First Symposium (International) on Combustion*, The Combustion Institute, 1493.

35. Liu,Y. and Lenze,B. (1988). The influence of turbulence on the burning velocity of premixed $CH_4 - H_2$ flames with different laminar burning velocities. *Twenty Second Symposium (International) on Combustion*, The Combustion Institute, 747.

36. Kostiuk,L.W., Bray,K.N.C. and Chew,T.C. (1989). Premixed turbulent combustion in counterflowing streams, *Combust.Sci.Tech.*, **64**, 233.

37. Bray,K.N.C., Champion,M., Kostiuk,L.W. and Libby,P.A. (1989). Extinction by stretch of turbulent premixed flames in counterflowing streams. *Seventh Symposium on Turbulent Shear Flows* Papere 1-2, Stanford University, August 21-23.

38. Bray,K.N.C., Champion,M. and Libby,P.A. (1990). Premixed flames in stagnating turbulence. Part I - The general formulation for counterflowing streams and gradient models for turbulent transport, *Combust.and Flame* (To appear).

39. Champion,M. and Libby,P.A. (1991). Asymptotic analysis of stagnating turbulent flows. *AIAA J.* (To appear).

40. Cant,R.S. and Bray,K.N.C. (1989). A theoretical model of premixed turbulent combustion in closed vessels. *Combust.and Fl;ame* **76**, 243.

41. Ibrahm,S.S., Cant,R.S. and Bray,K.N.C. (1990). Geometry effects on premixed turbulent propogating flames. *AIAA J.* (To appear).

3 Detonation theory

A K KAPILA

An introduction to detonation theory

1. INTRODUCTION

A combustible or explosive material is capable of sustaining two types of combustion waves: deflagrations or flames, and detonations. Flames are essentially waves of diffusion. As a parcel of material burns, it conducts heat to its neighbor, causing it to burn in turn, thereby advancing the flame. Flame speeds are typically low, of the order of meters per second or fractions thereof, and as a result compressibility of the reactive medium has a minimal influence on flame propagation. A detonation, on the other hand, is a rapid and violent form of combustion, accompanied by a large increase in pressure and travelling at speeds of the order of kilometers per second. At the head of the wave is a strong shock which triggers chemical reaction in the material by compressing it, and the resulting energy release sustains the shock. Diffusion processes, all important in flames, are negligible in detonations.

Flames are discussed extensively elsewhere in this volume and will not be treated here. Instead we shall concentrate on detonations. A thorough understanding of detonation behavior is essential if explosive materials are to be safely stored, transported, and utilized in practical devices. It would not be an exaggeration to say that over the years, most of the knowledge in the field has been acquired as a result of experimental observations. With a few notable exceptions, attempts at theoretical modelling, both explanatory and quantitative, are of relatively recent vintage, and this is especially true of transient phenomena.

This introductory essay, aimed at the nonexpert, has a modest goal: to acquaint the reader with a representative but simple mathematical model on which to base a theoretical study of detonations, and to provide but a glimpse of some of the analytical approaches that are proving increasingly useful in gaining a better understanding of the subject.

We begin with a discussion of the model equations, go on to describe the steady ZND detonation, and end with a brief foray into two topics of current research: an analysis of the early stage of detonation initiation, and an attempt to construct an asymptotic theory of detonation propagation under conditions of varying geometry. A fuller discussion on initiation by Dold [1], and an account of detonation instability by Roytburd [2], appear elsewhere in this volume.

2. EQUATIONS OF REACTIVE FLOW

This section contains an elementary account of flow in a reactive medium. The equations introduced below will form the basis of most of the discussion in the following sections. The treatment is necessarily brief; a thorough and critical discussion of the underlying assumptions and simplifications, as well as generalizations, can be found in the standard literature on the subject ([3],[4]).

In the absence of transport effects the equations of mass, momentum and energy balance for an unsteady flow are

$$\dot{\rho} + \rho \nabla . \mathbf{u} = 0, \tag{2.1}$$

$$\rho \dot{\mathbf{u}} + \nabla p = 0, \tag{2.2}$$

$$\dot{e} + p\dot{v} = 0, \tag{2.3}$$

where ρ is the material density, $v = (1/\rho)$ the specific volume, p the pressure, $\mathbf{u}$ the particle velocity and e the specific internal energy. The dot over a symbol refers to the material derivative

$$\frac{\partial}{\partial t} + \mathbf{u}.\nabla.$$

These are simply the Euler equations of hydrodynamics, to be supplemented by balance equations for the individual species taking part in the chemical reaction. If one idealizes the reaction to the elementary step $\mathcal{A} \rightarrow \mathcal{B}$, then the progress of the reaction can be measured by a single variable λ (for example, the mass fraction of the product species $\mathcal{B}$) governed by the rate law

$$\dot{\lambda} = \mathcal{R}. \tag{2.4}$$

Unreacted material corresponds to $\lambda = 0$ and completed reaction to $\lambda = 1$. Additionally, the constitutive relations

$$e = e(p, v, \lambda) \text{ and } \mathcal{R} = \mathcal{R}(p, v, \lambda)$$

must also be specified. If the two constituents of the reactive mixture, $\mathcal{A}$ and $\mathcal{B}$, are taken to be polytropic gases with identical molecular weights and (constant) specific heats, then, the equation of state of the mixture is

$$e = \frac{pv}{\gamma - 1} - \lambda q, \tag{2.5}$$

where the adiabatic exponent γ is also the ratio of the specific heats, and q is the chemical heat release per unit mass of the reactant. The temperature T is defined by the thermal equation of state

$$pv = RT, \tag{2.6}$$

and the adiabatic sound speed c by

$$c = \sqrt{\gamma p v}. \tag{2.7}$$

The reaction rate will be modelled by the Arrhenius law

$$\mathcal{R} = k(1-\lambda)e^{-T_a/T}, \tag{2.8}$$

where T_a is the activation temperature and k a pre–exponential constant with the dimensions of frequency.

It should be emphasized that the above equations are the result of a very special set of assumptions about the equation of state and the rate law. Even for the simplest of reactive mixtures, chemical kinetics is vastly more complicated than the simple model adopted here. Equation of state can also be complex, especially for condensed materials and in fact, a considerable research effort is currently directed at constitutive characterization of explosives. The present model should therefore be viewed only as an illustrative tool, to be employed for a qualitative understanding of phenomena due to a generic feature of explosives, namely, chemical energy deposition at a strongly state–sensitive rate.

The strong state–sensitivity of the reaction rate can be modelled by a large activation temperature T_a. This choice has substantial implications, mathematical and physical, which are revealed by expressing the rate function in the form

$$\mathcal{R} = (1-\lambda)\exp\left[\frac{T_a}{T_c}\left(\frac{T-T_c}{T}\right)\right]. \tag{2.9}$$

The above expression follows from (2.8) upon replacing the pre–exponential multiplier k by the scale factor $\exp(T_a/T_c)$, where T_c is a characteristic temperature in the flow. Now, if T_a/T_c is large, there are three possibilities. For $T - T_c < 0$ and $O(1)$, the reaction rate is negligibly small; the local temperature in the flow is too low for chemical activity to be significant. For $T > T_c$ and $O(1)$, the reaction-rate is exponentially large, suggesting that either flow gradients are correspondingly large, or that the reaction is essentially complete by virtue of λ being exponentially close to unity. Only when $T - T_c = O(1/T_a)$ is the reaction rate significant in a region of moderate flow gradients. In fact, the prescription $T - T_c = (T_c^2/T_a)\phi$ linearizes the Arrhenius exponent, thereby simplifying (2.9), to leading order, to $\mathcal{R} \sim (1-\lambda)e^{\phi}$. Thus, large activation temperature tends to localize chemical activity, in space and/or time, and the resulting mathematical and conceptual simplification can be substantial.

Much of the following discussion will be devoted to planar, one–dimensional flow. For this simple geometry it is expedient to introduce the mass-weighted, or Lagrangian, spatial coordinate

$$h = \int_{x_0(t)}^{x} \rho(\xi, t)\,d\xi, \tag{2.10}$$

where $x_0(t)$ is the position, at time t, of the particle which was at $x = 0$ at $t = 0$. Then the governing equations, following simple manipulations, reduce to the set

$$v_t - u_h = 0, \tag{2.11}$$

$$u_t + p_h = 0, \tag{2.12}$$

$$p_t + (\gamma p/v)v_t = (\gamma - 1)q\mathcal{R}/v, \tag{2.13}$$

$$\lambda_t = \mathcal{R}, \tag{2.14}$$

$$T = pv/R. \tag{2.15}$$

The material derivative is now simply the derivative with respect to t at fixed h. In this formulation equations (2.11) and (2.12), representing mass conservation and momentum balance, are linear. The energy equation (2.13) reveals the hydrodynamic implications of chemical energy release; it is partitioned into a fraction causing pressure rise, and another leading to volume expansion. Suitable linear combinations with the remaining equations of the above set allow (2.11) and (2.12) to be replaced by the equivalent pair

$$p_t \pm \rho c p_h \pm \rho c(u_t \pm \rho c u_h) = (\gamma - 1)q\rho\mathcal{R}.$$

The three characteristic families of this hyperbolic system can now be identified, as the particle paths

$$\frac{dx}{dt} = u$$

and the sonic lines

$$\frac{dx}{dt} = u \pm c.$$

3. THE STEADY DETONATION

Before launching into unsteady analyses pertaining to initiation, propagation or stability, it is important to have a good understanding of the structure of a steadily propagating detonation wave. The so-called ZND model (due to **Z**eldovich, von **N**eumann and **D**öring) of a planar wave is shown in Figure 1. Ahead of the wave the explosive is quiescent, because the reaction rate at the ambient state is insignificant. Passage through the lead shock puts the material into a highly compressed state at a temperature of several thousand degrees. The ensuing chemical reaction goes to completion very rapidly in a thin reaction zone behind the shock. An inert hydrodynamic flow follows the reaction zone, determined by the rear boundary condition and usually a rarefaction.

For the steadily propagating detonation, the shock velocity is constant and the flow in the reaction zone steady in a frame attached to the shock, while the following rarefaction is necessarily unsteady. For a given shock speed D, and in the planar geometry, the three conservation equations (2.1-2.3) reduce to algebraic (Rankine-Hugoniot) relations,

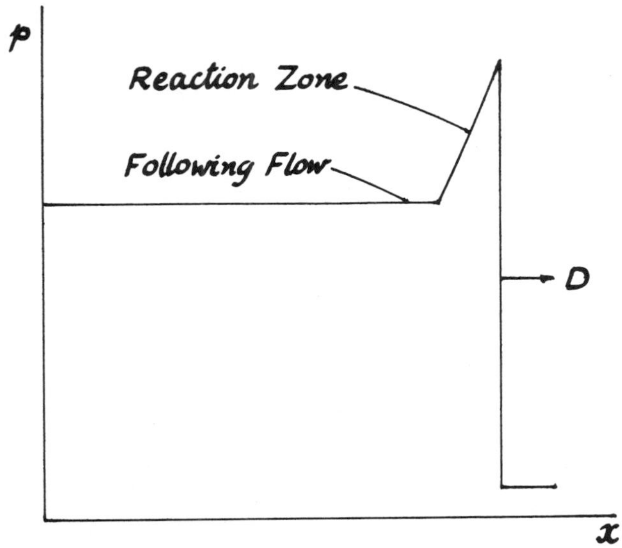

Figure 1: An unsupported detonation

connecting the state at any point within the reaction zone to the state ahead of the shock. These relations are

$$\rho_0 D = \rho(D-u), \tag{3.1}$$

$$p_0 + \rho_0 D^2 = p + \rho(D-u)^2, \tag{3.2}$$

$$e(p_0, v_0, 0) + p_0 v_0 + \frac{1}{2}D^2 = e(p, v, \lambda) + pv + \frac{1}{2}(D-u)^2, \tag{3.3}$$

where the suffix 0 refers to the ambient, unshocked state. Elimination of D from (3.1) and (3.2) yields

$$p - p_0 = -\rho_0^2 D^2 (v - v_0). \tag{3.4}$$

In the pv−plane this the equation of the *Rayleigh line*, a straight line passing through the initial state (p_0, v_0) with a negative slope proportional to D^2. Elimination of u and D from (3.1-3.3) leads to

$$e(p, v, \lambda) - e(p_0, v_0, 0) = \frac{(p + p_0)(v_0 - v)}{2}, \tag{3.5}$$

an equation describing a family of curves in the pv−plane, parametrized by λ and known as the partial-reaction *Hugoniot curves.* As λ increases from zero the curves are displaced upwards. Of special interest are the two extremes: the no-reaction ($\lambda = 0$) curve representing the locus of states immediately behind the lead shock, and the complete reaction ($\lambda = 1$) curve corresponding to the state at the end of the reaction zone. For the polytropic fluid under consideration here, the Hugoniot curves are rectangular hyperbolas

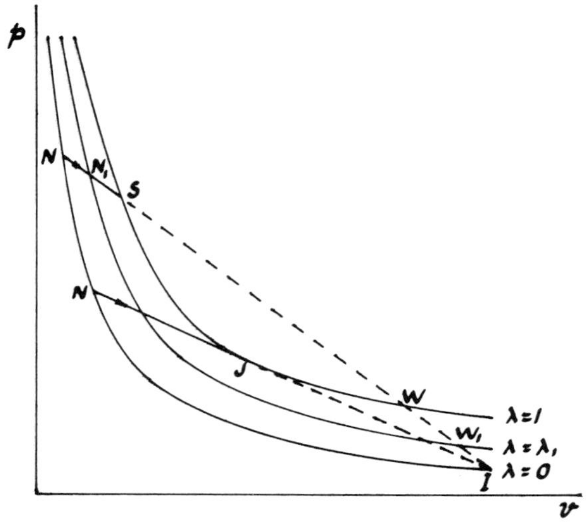

Figure 2: The Hugoniot diagram

$$\left(\frac{p}{p_0}+\Gamma\right)\left(\frac{v}{v_0}-\Gamma\right)=1-\Gamma^2+\frac{2\Gamma\lambda q}{p_0 v_0}, \tag{3.6}$$

where $\Gamma = (\gamma-1)/(\gamma+1)$. The Rayleigh line and the extreme Hugoniot curves are drawn in Figure 2. For a given shock velocity D, all states within the

reaction zone must lie on the corresponding Rayleigh line. Since the Hugoniot relation must also be satisfied, the portion of the Rayleigh line relevant to the ZND structure is that bounded by the extreme Hugoniots and shown solid in the Figure. The history of a particle as it traverses the steady solution can now be traced. Starting at the initial state I, the state of the particle jumps to the point N on the $\lambda = 0$ curve upon passage through the lead shock. As the particle reacts, λ increases, and the state of the particle slides down the Rayleigh line as indicated by the arrow, crossing partial-reaction Hugoniots of increasing λ. At the end of the reaction zone the particle reaches the $\lambda = 1$ curve, at point S or J. (The labels N, J and S stand for the von Neumann point, the Chapman-Jouguet point and a strong point, respectively.)

The lowest possible Rayleigh line, also drawn in Figure 2, is the one tangential to the complete-reaction Hugoniot. The corresponding detonation speed, D_J, is the minimum speed consistent with the conservation laws, and the detonation travelling at that speed is called the Chapman-Jouguet detonation. Physically, this is an unsupported detonation, i.e., one that is not pushed from behind. Higher Rayleigh lines represent supported, or overdriven detonations, forced to run at a higher velocity by being pushed from behind by a piston. The flow at a strong point S is subsonic while the flow at the CJ point J is sonic. Thus an overdriven detonation is subject to degradation by a rarefaction wave

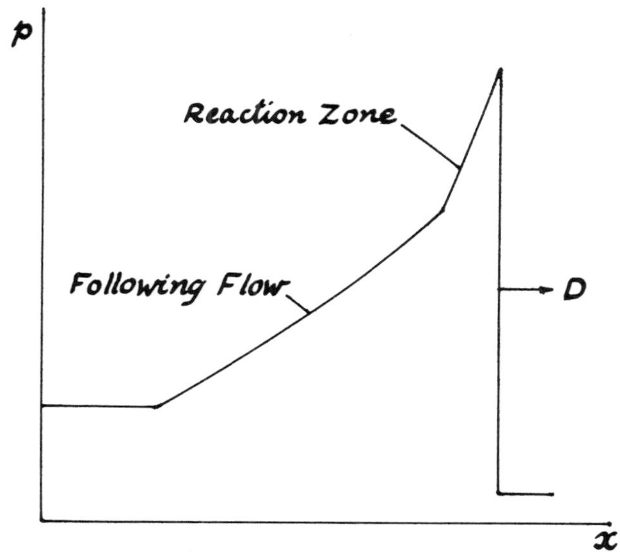

Figure 3: A supported detonation

overtaking it from behind, while the CJ detonation is not.

A particular steady solution, characterized by the speed D, can be set up by appropriate choice of the rear boundary condition. Suppose that the rear boundary is a piston travelling with speed u_p in a laboratory-fixed frame. In order to determine which particular D corresponds to the given u_p, let $U(D)$ denote the particle speed at the end of the reaction zone, with $U_J = U(D_J)$ the corresponding value for the CJ detonation. Let us start with $u_p > U_J$ and examine the change in the nature of the steady solution as u_p is decreased. For $u_p > U_J$ the detonation is overdriven; D is such that $U(D) = u_p$, and the following flow is uniform (Figure 3). The state at the end of the reaction zone is represented by, say, the upper N in Figure 2. As u_p is decreased toward U_J, and at $u_p = U_J$, the same qualitative picture remains, with N dropping lower and S approaching J. A further decrease in u_p will leave the detonation speed and the reaction-zone unchanged since the Rayleigh line can not drop below the limiting tangent. What does change is the following flow, where the previously uniform state is replaced by a rarefaction wave (Figure 1). The smaller u_p, the larger the amplitude of the rarefaction behind the detonation, which is now unsupported and continues to run at the CJ speed, oblivious of the following flow.

There are other types of steady (or at least, quasisteady) solutions consistent with the conservation conditions, quite apart from those of the ZND variety. These include the shockless solutions, with λ increasing along the Rayleigh line from I to W in Figure 2. The label W denotes a point on the weak branch (i.e., the portion below J) of the detonation Hugoniot, and has the property that the flow is supersonic there. These solutions are often rejected as unphysical, because without a lead shock the reaction rate at the initial state would have to be finite to initiate the chemical reaction. However, if an external igniter or trigger were to initiate the reaction, such solutions would be of interest. An extreme case is the constant volume explosion, which does not involve any material motion and can be thought of as an infinite-velocity weak detonation. There are also the hybrid solutions, such as IW_1N_1S in Figure 2. These correspond to a supersonic reaction regime IW_1 from $\lambda = 0$ to an intermediate value λ_1, a shock transition at $\lambda = \lambda_1$ from W_1 to N_1, followed by a subsonic reaction zone N_1S in which λ increases from λ_1 to 1. It has recently been noticed ([5]-[8]) that such hybrid solutions can play a role in the process of detonation initiation.

Once a particular steady solution has been selected, the state of the explosive is completely specified as a function of the progress variable λ. The dependence of λ on the distance behind the shock, or on the time since a fluid element passed through the shock, still remains to be determined. This can be done by integrating the rate law (2.4), which is now a differential equation for $\lambda(t)$. The equation of motion of a particle relative to the shock can then be used to determine $\lambda(h)$. Numerical solutions for the simple model under consideration here are given in Fickett and Davis [3]. Asymptotic solutions in the limit of large activation temperature have also been developed ([9]-[11]). Typically, for a ZND detonation, the asymptotic solution consists of a relatively broad *induction zone* of weak reaction behind the shock, followed by a thin *fire zone* of vigorous reaction. In essence, the structure is that of a subsonic, *fast deflagration* triggered at the lead shock [11].

4. ANALYSIS OF THE EARLY STAGE OF INITIATION

It has already been mentioned that under normal ambient conditions, reaction rates are negligibly small and the explosive exists in a quiescent, metastable state. An external stimulus is therefore needed to provoke ignition, and this can be accomplished in a variety of ways. A common technique, employed in experimental setups as well as practical devices, is to subject the material to rapid compression, say by driving a shock wave through it. Compression leads to drastic reduction in reaction times, thereby switching on weak but significant chemistry. Evolution will be spatially homogeneous, leading to a volume explosion, only in the exceptional case of a high degree of uniformity in the compression process. It is far more likely that small spatial variations will intrude into a nearly uniform state. In the case of shock compression, for example, the finite passage time of the shock across the sample will necessarily produce spatial gradients since each particle begins to react only after experiencing the shock.

For gasdynamics to couple effectively with chemistry (as it must if a detonation is to emerge), the chemical time must be comparable to the characteristic acoustic time in the compressed material. The explosive then finds itself in an unsteady *induction phase*, characterized by a redistribution of chemically liberated energy within the mixture via acoustic transport. The strong state–dependence of the reaction rate leads to a mutual reinforcement of the pressure waves and the intensity of the reaction, and a feedback loop is established. Typically, the induction stage ends with the creation of a site of intense reaction at which a strong reaction wave originates. In fact, under certain conditions the induction process is the sole arbiter of the path along which the reaction wave must travel before it transits into a detonation [1].

In order to analyze the induction phase, we begin with the Lagrangian version of the governing equations. For nondimensionalization we take the uniform state p_i, T_i, and v_i as the reference. The constant-volume induction time at this state, defined by

$$t_i = \left(\frac{q}{kT_a c_v}\right) e^{T_a/T_i},$$

is chosen as the time unit. (In the absence of perturbations, the medium would explode in time t_i, in the asymptotic sense described below.) The mass–weighted acoustic length $\rho_i c_i t_i$ serves as the unit for the Lagrangian spatial coordinate. Then, equations (2.11-2.15) can be written in a shorthand matrix notation as

$$\mathbf{u}_t + \mathbf{A}\mathbf{u}_h = \mathbf{f}, \tag{4.1}$$

where

$$\mathbf{u} = \begin{pmatrix} v \\ u \\ p \\ \lambda \end{pmatrix}, \quad \mathbf{A} = \begin{pmatrix} 0 & -1 & 0 & 0 \\ 0 & 0 & 1/\gamma & 0 \\ 0 & \gamma p/v & 0 & 0 \\ 0 & 0 & 0 & 0 \end{pmatrix}, \tag{4.2}$$

and

$$\mathbf{f} = \epsilon \mathbf{F} W, \tag{4.3}$$

with

$$\mathbf{F} = \begin{pmatrix} 0 \\ 0 \\ 1/v \\ 1/\beta \end{pmatrix}, \quad \text{and } W = (1-\lambda)\exp\{\frac{1}{\epsilon}\left(1-\frac{1}{T}\right)\}. \tag{4.4}$$

The reference state is simply

$$\mathbf{u}_i = \begin{pmatrix} 1 \\ 0 \\ 1 \\ 0 \end{pmatrix}. \tag{4.5}$$

The dimensionless parameters appearing above are the reciprocal activation temperature $\epsilon = T_0/T_a$ and the thermicity parameter $\beta = q/c_v T_0$, c_v being the specific heat at constant volume. Our analysis will proceed under the asymptotic limit $\epsilon \to 0$.

In keeping with the notion of the induction state being a small perturbation about a uniform state, we consider the distinguished limit in which ϵ also defines the perturbation amplitude, and examine the evolution of

$$\mathbf{u} \sim \mathbf{u}_i + \epsilon \mathbf{u}_1(x,t), \tag{4.6}$$

Then the perturbations satisfy

$$v_{1t} - u_{1h} = 0, \tag{4.7}$$

$$u_{1t} + (1/\gamma)p_{1h} = 0, \tag{4.8}$$

$$p_{1t} + \gamma u_{1h} = e^{T_1}, \tag{4.9}$$

$$T_1 = p_1 + v_1, \tag{4.10}$$

$$\lambda_{1t} = (1/\beta)e^{T_1}. \tag{4.11}$$

Suitable initial and boundary data, depending upon the situation at hand, must be appended for a complete specification of the problem. The solution can then be obtained via relatively simple numerics, and this has been done in a number of studies. However, an examination of the equations themselves unveils the underlying mechanisms at play. First, observe that equation (4.11) for reaction progress is uncoupled from the rest of the system, and can be solved after the hydrodynamic variables have been computed. Second, the energy equation (4.9), when combined with the gas law (4.10) and continuity (4.7), can be written in one of two alternate forms:

$$T_{1t} + (\gamma - 1)v_{1t} = e^{T_1}, \tag{4.12}$$

or

$$T_{1t} - \frac{\gamma - 1}{\gamma} p_{1t} = (1/\gamma)e^{T_1}. \tag{4.13}$$

At constant volume, (4.12) yields

$$T_{1t} = e^{T_1}, \tag{4.14}$$

while at constant pressure, (4.13) reduces to

$$T_{1t} = (1/\gamma)e^{T_1}. \tag{4.15}$$

Starting with some prescribed initial value for the perturbation T_1, both of these situations result in logarithmically unbounded growth (thermal runaway) at a finite time. The same processes appear in the full system as well, when equations (4.7-4.11) combine, under appropriate smoothness conditions, to yield

$$\frac{\partial^2}{\partial t^2}\left(T_{1t} - e^{T_1}\right) = \frac{\partial^2}{\partial h^2}\left(T_{1t} - (1/\gamma)e^{T_1}\right), \tag{4.16}$$

or equivalently,

$$\left(\frac{\partial^2}{\partial t^2} - \frac{\partial^2}{\partial h^2}\right) T_{1t} = \left(\frac{\partial^2}{\partial t^2} - \frac{1}{\gamma}\frac{\partial^2}{\partial h^2}\right) e^{T_1}, \tag{4.17}$$

These equations were first derived by Clarke [12]. In version (4.17) we see the appearance of two wave operators, the one operating on T_{1t} describes waves propagating at the frozen adiabatic sound speed of unity, and the other operating on the chemical source e^{T_1} describes waves travelling at the isothermal sound speed of $1/\sqrt{\gamma}$.

Behavior in the induction phase is thus composed of four fundamental processes; energy liberation at constant volume and at constant pressure, coupled with wave propagation at the adiabatic and isothermal sound speeds. As a fluid particle releases chemical energy, its temperature rises and it attempts to increase its volume as a consequence. It is not possible for the element to do this in an unrestricted way, however, since its immediate neighbors are attempting to do the same thing. This leads to interparticle interactions, creating pressure disturbances that propagate through the region via the wave system described above, modifying the local temperature distribution as they do so.

Depending upon the initial and boundary data, the wave equation (4.17) also displays finite-time thermal runaway, but now only locally. The result is the formation of a sharply focussed hot spot, in which T_1 again approaches infinity logarithmically in time. The detailed structures of these hot spots has been elucidated by several authors recently ([13], [14], [15]). All of these works show conclusively that in the core of the hot spot, density ultimately tends to a constant value. The rate of energy release in the core is much too rapid for significant thermal expansion to occur, and the explosive is inertially confined. This behavior has important implications on the post-induction evolution of the system and the ultimate transition to detonation; a detailed discussion of such matters is presented in this volume by Dold [1].

Additional observations are in order about the induction system discussed above. First, the derivation so far has proceeded under the assumption that the perturbations existing in the system are of the long wavelength variety, i.e., their characteristic length is also typified by the overall size of the induction domain. The interaction of short pulses with the induction chemistry can take on quite a different character; in particular, as Clarke has pointed out, weak shocks may appear in the system prior to the formation of hot spots [16]. Recent work by Majda and Rosales [17] and by Almgren [18] has taken the analysis further by introducing perturbations in the form of *high-frequency wave trains*, which lead to a stronger coupling between hydrodynamics of the averaged field and chemistry.

Second, the behavior displayed above is indeed peculiar to the reaction kinetics and the equation of state adopted here. Branched-chain kinetics may have entirely different consequences ([19], [20]). Stewart [21] has shown that the location of the hot spot may depend very strongly on the mole-change character of the kinetics. More recently, Bdzil [22] has demonstrated that the character of runaway can be quite different for an equation of state in which the polytropic exponent and the specific-heats ratio are not the same constant. It is important to note, however, that as long as the model involves high state–sensitivity of the reaction rate, many of these generalizations can be examined in the context of the induction equations because one is looking at the evolution of small perturbations.

5. PROPAGATION OF CURVED DETONATIONS

We have already seen that in a planar geometry, an unsupported detonation propagates at the CJ speed, and that the rear boundary condition only influences the following flow. The lateral boundaries have no role to play in this simple configuration.

Even under circumstances carefully chosen to nurture a plane state, multidimensionality intrudes into the flow in practice. Lateral yielding of the confining walls in condensed explosives, for example, creates rarefaction waves which enter the reaction zone and lower the reaction rate. This causes the planar detonation to slow down, more so near the boundaries than in the center of the charge. If the charge diameter is below a critical value, the detonation fails altogether; otherwise, experiments indicate that a steadily propagating curved detonation is eventually approached.

Still greater complications arise in practical devices where the geometry is inherently nonplanar. The propagation is now necessarily time–dependent, with the detonation front changing speed as well as shape as it advances into the explosive. Computational tools are prohibitively expensive as they must resolve the flow in the reaction zone, even though practical design calculations do not call for such details of the flow.

An analogous problem arises in the unsteady propagation of curved flames. Such a flame may propagate at speeds significantly different from the steady, planar value. If one treats the flame as slender (i.e., thin compared to its radius of curvature) and slowly varying (i.e., the flow perturbations responsible for causing the flame to change its shape have characteristic time scales longer than the time taken by the flame to travel a distance equal to its own thickness), it is possible to develop an asymptotic theory which yields a differential equation for the flame speed as a function of its shape [23]. The flame can then be thought of as a discontinuity in an inert flow and tracked by means of a modified Huygens construction. Whitham's theory of shock dynamics [24], used successfully to explain the diffraction of shock waves around obstacles, has the same key ingredient; the shock velocity is taken to be a function of the local geometry alone, and independent of signals catching up from behind the front.

Guided by these ideas, Bdzil and Stewart ([25]-[27]) have recently laid the foundation of an asymptotic theory for the propagation of unsteady detonations in diverging geometries. They also proceed by recognizing that in many applications of explosives, the reaction zone thickness is small compared to the radius of curvature of the lead shock. As a result, flow in the reaction zone is locally quasiplanar as well as quasisteady, the time dependence being carried entirely by the kinematics of the shock surface. An analysis of the reaction zone provides a relationship between the shock speed and shock curvature parametrized by the details of the kinetics. This relationship, when coupled with shock kinematics, provides the machinery necessary to advance the shock. We now present a brief sketch of the analysis, confining ourselves to a 2-d description only for illustrative purposes; the analysis applies equally well to a fully three dimensional geometry.

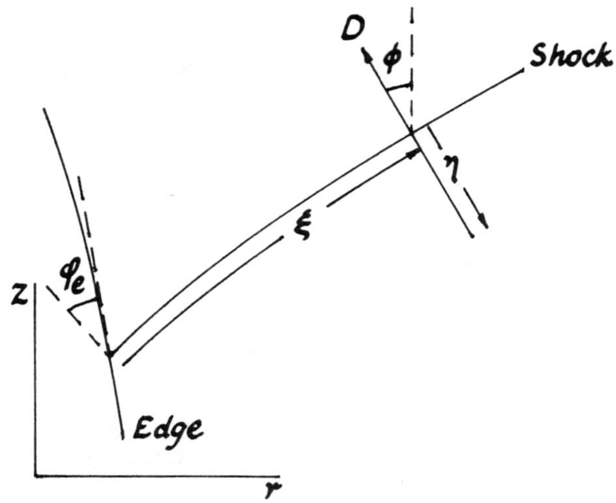

Figure 4: The coordinate system

Imagine a situation in which the explosive medium, semi–infinite in lateral extent, is confined by a single edge, whose effect on the shape of an originally plane detonation is sought. It is useful to employ a shock-attached intrinsic coordinate system (Figure 4) in which the independent variables are time t, the arc length ξ along the shock (measured from the edge), and the distance η measured backwards from the shock along the local normal $\mathbf{n}$. The laboratory coordinates are then given by

$$z_s = z_e - \int_0^\xi \sin\phi d\xi, \quad r_s = r_e - \int_0^\xi \cos\phi d\xi,$$

for a point (z_s, r_s) on the shock, and

$$z = z_s - \eta\cos\phi, \quad r = r_s - \eta\sin\phi,$$

for a point (z, r) off the shock. The subscript e refers to the edge, and ϕ is the angle between the shock normal $\mathbf{n}$ and the $z-$ direction. The shock shape is then described by $\phi = \phi(\xi, t)$, and the curvature by $\kappa = \phi_{,\xi}$, where the comma in the subscript denotes partial differentiation.

The geometric compatibility condition for the shock front can now be expressed as

$$\phi_{,t} + B\phi_{,\xi} = -D_{,\xi}, \tag{5.1}$$

with

$$B = \int_0^\xi \phi_{,\xi} D d\xi + D_e \tan\varphi_e. \tag{5.2}$$

Here, D is the shock speed along its normal and D_e its value at the edge. The angle between the shock normal and the edge tangent is denoted by φ_e.

Equation (5.1) is an equation for evolution of the shock shape, provided one can relate the shock shape to D independently. For this we turn to the dynamics of the reaction zone. In the (ξ, η, t) coordinates, the governing equations (2.1-2.4) are

$$\mathcal{L}(\rho) + \rho(\kappa u_\eta - u_{\eta,\eta} + u_{\xi,\xi}) + \cdots = 0, \tag{5.3}$$

$$\mathcal{L}(u_\eta) - (1/\rho)p_{,\eta} + \cdots = 0, \tag{5.4}$$

$$\mathcal{L}(u_\xi) + (1/\rho)p_{,\xi} - D_{,\xi}u_\eta + \cdots = 0, \tag{5.5}$$

$$\mathcal{L}(e) - (p/\rho^2)\mathcal{L}(\rho) + \cdots = 0, \tag{5.6}$$

$$\mathcal{L}(\lambda) + \cdots = \mathcal{R}. \tag{5.7}$$

Anticipating an analysis in the limit of small κ, all terms not involved in a leading-order analysis have not been explicitly displayed. The operator $\mathcal{L}$ is defined by

$$\mathcal{L} \equiv \frac{\partial}{\partial t} + (D - u_\eta)\frac{\partial}{\partial \eta} + (B + u_\xi)\frac{\partial}{\partial \xi},$$

and (u_ξ, u_η) are the velocity components. These equations, supplemented by the standard jump conditions across the shock, and appropriate initial/boundary data, define the problem for the flow in the reaction zone.

In general the problem is analytically intractable, but considerable progress is possible in the limit of small κ. The Bdzil/Stewart analysis [25] shows that the flow then splits conveniently into three regimes: an outer domain away from the boundary, where edge rarefactions have not yet influenced the planar wave; an inner region where the flow is slowly varying, and an intermediate layer of relatively fast dynamics which links the two. As time evolves, slow variation tends to dominate the region of disturbance. With $\kappa = O(\epsilon^2)$, and $D = D_J + O(\epsilon^2)$, where D_J is the CJ speed of the planar wave, the appropriate scales are

$$\sigma = \epsilon t, \quad \chi = \epsilon^{1/2}\xi, \quad \phi = O(\epsilon^{3/2})$$

for the intermediate layer, and

$$\tau = \epsilon^2 t, \quad \varsigma = \epsilon\xi, \quad \phi = O(\epsilon) \text{ or larger}$$

for the slowly varying regime. The latter has a particularly simple description; in it, equations (5.3-5.7) reduce to

$$[\rho(D - u_\eta)]_{,\eta} + \kappa\rho u_\eta = 0, \tag{5.8}$$

$$(D - u_\eta)u_{\eta,\eta} - (1/\rho)p_{,\eta} = 0, \tag{5.9}$$

$$u_{\xi,\eta} = 0, \tag{5.10}$$

$$e_{,\eta} - (p/\rho^2)\rho_{,\eta} = 0, \tag{5.11}$$

$$(D - u_n)\lambda_{,\eta} = -\mathcal{R}, \tag{5.12}$$

with flow divergence in (5.8) proportional to the curvature. Being quasisteady, these equations are independent of the initial data, and being quasiplanar, the only information needed for their solution is the knowledge of the state immediately behind the shock. The same equations, here derived rationally, were considered by Wood and Kirkwood (see Chapter 5 of [3]) in their ad hoc analysis of slightly divergent flow. It is a simple matter to deduce, as they did, that a solution exists only if D and κ satisfy an *eigenvalue* relation of the form

$$D = D(\kappa). \tag{5.13}$$

The precise form of this relationship depends on the reaction kinetics. For example, for the state-independent rate law

$$\mathcal{R} = k(1 - \lambda)^{\nu}, \tag{5.14}$$

Bdzil and Stewart [26] have shown that

$$D = D_J - \alpha\kappa + o(\kappa) \text{ for } 0 < \nu < 1, \tag{5.15a}$$

and

$$D = D_J + \beta\kappa \ln\kappa - \alpha\kappa + o(\kappa), \text{ for } \nu = 1, \tag{5.15b}$$

where α and β are simply certain kinetics-dependent constants. It is worth pointing out that when detailed kinetic data are unavailable, (the typical situation for condensed-phase explosives), the $D - \kappa$ relation can be determined experimentally from simple steady-state tests.

Substitution of (5.13) into (5.1) leads to an evolution equation for the shock shape. In addition to initial data, an edge condition must also be specified to close the problem. Such a condition for an explosive-vacuum interface is discussed in [29]. For more general confinement conditions at the edge, a rather complex flow needs to be considered at the boundary; such matters are currently being studied.

It should be emphasized that the slowly-varying analysis developed above relies heavily on the assumption that as the detonation advances, it is stable with respect to disturbances on the fast time scale (i.e., the acoustic transit time across the reaction zone). It is well known, however, [3] that strongly state–dependent rates tend to promote fast–time instabilities, which may produce transverse strucures in the reaction zone, thereby destroying its quasisteady and quasiplanar character. Whether the slowly-varying theory is applicable to this situation in an averaged sense is yet unsettled, and is a topic of current research.

6. CONCLUDING REMARKS

Detonations are complex structures, in general three–dimensional, and the manner in which they initiate, propagate and respond to disturbances, depends upon the chemico-thermo-mechanical character of the explosive medium in which they exist. For a given explosive, one can identify at least three fundamental areas of study: initiation, propagation and stability, all of which are currently being pursued vigorously by theoreticians. It is hoped that this elementary treatment will provide the reader with a flavor of some of the analytical techniques that are being employed in these investigations. We have not even mentioned the considerable numerical effort being brought to bear on the subject, nor have we attempted to include multidimensional effects. However, there is still much to do before a complete quantitative description of the various phenomena is achieved, even for the one–dimensional geometry, and for the very simple model introduced here.

ACKNOWLEDGEMENTS

Stimulating and helpful discussions with John Bdzil and Bill Dold are acknowledged with pleasure and gratitude. This article is based on lectures given at the Istituto Mauro Picone, Rome in Summer, 1989, where Professors Marcello Onofri and Alberto Tesei were hosts most gracious. The author's research is supported by the Army Research Office, and by the Los Alamos National Laboratory under Contract No. DOE-LANL-9XG9-4906Y-1.

REFERENCES

1. J. W. Dold: *Emergence of a Detonation within a Reacting Medium*, This Volume.
2. V. Roytburd: *On Detonation Instability*, This Volume.
3. W. Fickett and W. C. Davis: Detonation, University of California Press, 1979.
4. F. A. Williams: Combustion Theory, Benjamin Cummings, 1985.
5. A. K. Kapila and J. W. Dold: *A Theoretical Picture of Shock-to-Detonation Transition in a Homogeneous Explosive*. Paper presented at the Ninth Symposium (International) on Detonation, Portland, Oregon, August 1989.
6. J. B. Bdzil and A. K. Kapila: *Shock-to-Detonation Transition: A Model Problem*. Submitted for publication, 1989.
7. J. W. Dold and A. K. Kapila: *Asymptotic Analysis of Detonation Initiation for One-Step Chemistry:* I–*Emergence of a Weak Detonation*. Submitted for Publication, 1989.
8. J. W. Dold and A. K. Kapila: *Asymptotic Analysis of Detonation Initiation for One-Step Chemistry:* **II**–*From a Weak Structure to ZND*. Submitted for Publication, 1989.

9. W. B. Bush and F. E. Fendell: *Asymptotic Analysis of the Structure of a Steady Planar Detonation*. Combustion Science and Technology, 2, 271 (1971).

10. A. K. Kapila, B. J. Matkowsky and A. Van Harten: *An Asymptotic Theory of Deflagrations and Detonations. I. The Steady Solutions*. SIAM J. Appl. Math., 43, 491 (1983).

11. D. R. Kassoy and J. F. Clarke: *The Structure of a High-Speed Deflagration With A Finite Origin*. J. Fluid Mech., 150, 253 (1985)

12. J. F. Clarke: *On the Theoretical Modelling of the Interaction Between a Shock Wave and an Explosive Gas Mixture*. College of Aeronautics report 7801, Cranfield Institute of Technology, Cranfield, Bedford MK43 0AL, U. K. (1978)

13. T. L. Jackson, A. K. Kapila and D. S. Stewart: *Evolution of a Reaction Center in an Explosive Material*. SIAM J. Appl. Math., 49, 432 (1989).

14. J. W. Dold: *Dynamics of a Self-Igniting Region*. In Mathematical Modelling of Combustion and Related Topics, Eds. C. M. Brauner and C. Schmidt-Lainé, Martinus Nijhoff, Dordrecht, 461 (1988).

15. P. A. Blythe and D. G. Crighton: *Shock-Generated Ignition*. Submitted for Publication, 1989.

16. J. F. Clarke: *Finite amplitude waves in Combustible Gases*. In The Mathematics of Combustion, Ed. J. D. Buckmaster, SIAM, Philadelphia, 183 (1985).

17. A. J. Majda and R. R. Rosales: *Nonlinear Mean Field– High Frequency Wave Interactions in the Induction Zone*. SIAM J. Appl. Math., 71, 149 (1984).

18. R. F. Almgren: Resonant Wave Interaction in a Reacting Gas, Ph.D. Thesis, Princeton University, 1989.

19. R. A. Strehlow: Fundamentals of Combustion, p. 331, International Textbook Company, Scranton, Pennsylvania, 1968.

20. J. W. Dold and A. K. Kapila: *A Comparison Between Shock Initiations of Detonation Using One–Step and Three–Step Chemical Models*. Submitted for Publication, 1990.

21. D. S. Stewart: *Shock-Initiation of Homogeneous and Heterogeneous Condensed-Phase Explosives With A Sensitive Rate* . Combustion Science and Technology, 48, 309 (1986).

22. J. B. Bdzil: Private Communication, 1990.

23. J. D. Buckmaster and G. S. S. Ludford: Theory of Laminar Flames, Chapter 8, Cambridge University Press, 1982.

24. G. B. Whitham: Linear and Nonlinear Waves, J. Wiley, New York, 1974.

25. J. B. Bdzil and D. S. Stewart: *Time-Dependent Two–Dimensional Detonation: The Interaction of Edge Rarefactions With Finite-Length Reaction Zones*. J. Fluid Mech., 171, 1 (1986)

26. D. S. Stewart and J. b. Bdzil: *The Shock Dynamics of Stable Multidimensional Detonation*. Combustion and Flame, 72, 311 (1988).

27. J. B. Bdzil and D. S. Stewart: *Modeling Two-Dimensional Detonations With Detonation Shock Dynamics*. Phys. Fluids A, 1, 1261 (1989)

A. K. Kapila
Department of Mathematical Sciences
Rensselaer Polytechnic Institute
Troy, New York 12180-3590, USA.

J W DOLD

Emergence of a detonation within a reacting medium

1 Introduction

At the heart of this discussion lies the self-evident idea that detonations arise as a result of chemical and compressibility interactions in a reacting medium. For the purpose of mathematical modelling, it is easiest to think of this medium as consisting of a mixture of ideal gases. Of course other media sustain detonations [1], ranging from non-ideal gases to condensed (liquid or solid) phases of explosive mixtures and even multiphase or amorphous mixtures, but for a brief, qualitatively correct exposition, ideal gases provide a useful basis for understanding many of the phenomena and their mathematical representations.

A number of phenomena are discussed in the following pages with the basic purpose of illustrating the way in which a chemically reacting medium can ignite to produce a detonation wave. While the discussion starts with well known results [2], traditional approaches are avoided in order to emphasise more recent work and to provide some fresh interpretation. The exposition is not comprehensive, serving only to outline connected ideas rather than to deal with all possible realisations of them. In this spirit, while the three dimensional behaviour of established detonations is very important, the analysis of the initiation of detonation is presented only for ideal gases in one dimension. The more general aspects of the particulars discussed briefly in this paper should always be borne in mind. They suggest, and often are, subjects for further research.

After introducing an ideal gas model for reacting mixtures, some steadily travelling solution properties are outlined for a chosen (simplest) case where the total number of gas molecules is unchanged by the chemistry. The analysis for changes in the number of molecules is a little more complicated, but would still lead to essentially analogous results. Depending on the ability of the chemistry to proceed at a significant rate, it is shown that reaction waves (or 'fast flames' [3, 4]) of many different speeds are possible.

These may or may not involve shock waves (representing either strong or weak detonations respectively), and may indeed travel at unbounded Mach numbers for suitable behaviour of the chemistry. The word 'propagate' needs to be interpreted carefully in such a context. Over a wide range of wave-speeds, it is shown that solutions describing steady reaction waves are not necessarily unique, having a continuum of possible shock locations separating weak (supersonic) from strong (subsonic) parts of the structure. The analysis of the reaction waves presented here is tied in with more classical steady detonation analyses based on Rankine–Hugoniot relations, as discussed elsewhere in this volume [A.K. KAPILA].

By then focussing attention on a simple one-step model for the chemistry, with a large activation temperature, unsteady behaviour is examined in more detail. It is found that ignition is possible for appropriate ranges of initial and boundary situations, and that the interaction between chemistry and compressibility of the medium during an induction process leads naturally to the emergence of quasi-steadily travelling weak-detonation structures [5]. The movement of these reaction waves arises through spatial variation in the induction process rather than any mechanism for propagation. The paths of such 'induction flames' can be calculated numerically using a fairly simple procedure applied to an asymptotic model [6].

A natural limitation on the possible mass flux through any weak detonation shows that the reaction wave that emerges in this way cannot normally remain shockless—as the induction flame slows down below the speed at which a part of the flow through it becomes subsonic (the Chapman–Jouget speed), a shock wave forms to transform the basic nature of the reaction wave from a weak to a strong, genuinely self-propagating, detonation structure [7]. In this way, a Zeldovich–von Neumann–Döring detonation [8] is born.

2 Basic Modelling

2.1 Diffusive effects

A useful starting point for selecting appropriate model equations to describe detonation phenomena is to note that (on average) only relatively rare, suitably energetic, thermal collisions between the right sorts of molecules can trigger a chemical change. It follows that the Von Kárman number [9], which measures the ratio of a chemical time-scale t_c to the mean molecular collision time t_m, must invariably be large,

$$K\acute{a} = \frac{t_c}{t_m} \gg 1. \tag{1}$$

Typical values for $K\acute{a}$ would be of the order of 10^5 or greater. Over the chemical time-scale, any diffusive processes (such as conduction of heat and diffusion of mass) therefore take place over very many molecular collisions. Elementary arguments, based on the kinetic theory of gases, then show that the Mach number of such diffusive processes must be very small, of the order of $K\acute{a}^{1/2}$.

On the other hand, velocities and changes of velocity in detonations are typically of the order of the sound speed or greater. This makes it clear that where such changes in velocity take place due to chemical effects, diffusive effects can only play a very small role, and where large velocity changes are strongly influenced by diffusive effects (such as viscosity and thermal conductivity, within the internal structure of a shock wave) too few molecular collisions would be involved for significant chemical changes to occur.

It is therefore perfectly realistic to neglect all diffusive effects in examining detonation phenomena, except possibly in the internal structure of shock waves—relative errors being only of the order of $1/K\acute{a}$. Shock waves are adequately treated as discontinuities across which Rankine–Hugoniot jump conditions can be applied at constant chemical composition. These are discussed later.

2.2 Partial inertial confinement

A second point to note is that interactions between chemistry and compressibility are a basic feature of detonations. Let us imagine a chemical reaction releasing energy in a region that happens to be of the spatial size ℓ_c. Resulting disturbances propagate away at a typical (chemically frozen) speed of sound $a_c = (\gamma P_c/\rho_c)^{1/2}$, where P_c and ρ_c simply represent typical values of pressure P and density ρ. In the chemical time t_c these disturbances would propagate a distance of $\ell_a = t_c a_c$.

If this acoustic length scale ℓ_a is much greater than the chemical length scale ℓ_c, then disturbances are rapidly propagated away from the reacting region; the medium is able to expand easily with little overall change in pressure. On the other hand, if ℓ_a is much smaller than ℓ_c, then the velocity and density of the medium remain practically unchanged while the pressure increases. This represents a limit of inertial confinement where the medium is unable to move significantly during the time available.

In a typical detonation, significant changes take place in both pressure and density, as well as velocity and temperature, so that one should naturally expect to find a situation between these two extremes. Thus there must, at least, be a partial degree of inertial confinement at some stage with $\ell_a \approx \ell_c$. Only then is it possible to develop a substantial level of interaction between the effects of exothermic chemistry and the compressibility of the medium.

2.3 Ideal gas equations

If $\alpha \in \mathcal{A}$ is an ideal gas ($\mathcal{A}$ is the set of all chemical species present in a reacting mixture of ideal gases), the difference between the specific heat at constant pressure of α, $C_{P\alpha}$, and the specific heat at constant volume, $C_{V\alpha}$, is

$$C_{P\alpha} - C_{V\alpha} = \frac{\gamma_\alpha - 1}{\gamma_\alpha} C_{P\alpha} = \frac{R}{W_\alpha} \tag{2}$$

where γ_α is the ratio of specific heats, $C_{P\alpha}/C_{V\alpha}$; R is the universal gas constant; and W_α is the molecular weight of α in mass units per mole. We can thus write the useful expression for predicting the value of the specific heat

$$C_{P\alpha} = \frac{\gamma_\alpha}{\gamma_\alpha - 1} \frac{R}{W_\alpha} \tag{3}$$

which shows that $C_{P\alpha}$ generally increases for decreased values of both γ_α and W_α. The classical principle of equipartition of energy may be used to estimate γ_α as being equal to the constant $(f_\alpha + 2)/f_\alpha$, where f_α is the number of degrees of freedom of motion of a molecule of α. For a monatomic, linear or nonlinear gas molecule, f_α is 3, 5 or 6, giving $\gamma_\alpha = \frac{5}{3}$, $\frac{7}{5}$ or $\frac{4}{3}$ respectively. Better estimates, based on quantum or statistical mechanics, are available [10] to show how energy distribution in rotational, vibrational and electronic modes inject some temperature dependence into $C_{P\alpha}$ and γ_α in many real substances.

For a given set of values of the mass fractions y_α ($\alpha \in \mathcal{A}$), one can identify the effective specific heat at constant pressure of the mixture C_P, the number of moles per unit mass

n, the effective molecular weight W and the effective ratio of specific heats γ as follows:

$$C_P = \sum_{\alpha\in\mathcal{A}} y_\alpha C_{P\alpha} = R\sum_{\alpha\in\mathcal{A}} \frac{\gamma_\alpha}{\gamma_\alpha - 1}\frac{y_\alpha}{W_\alpha}, \qquad n = \frac{1}{W} = \sum_{\alpha\in\mathcal{A}} \frac{y_\alpha}{W_\alpha} = \frac{\gamma-1}{\gamma}\frac{C_P}{R}$$
$$\gamma = \frac{C_P}{C_P - nR} = \left(\sum_{\alpha\in\mathcal{A}} \gamma_\alpha \frac{y_\alpha/W_\alpha}{\gamma_\alpha - 1}\right) \Big/ \left(\sum_{\alpha\in\mathcal{A}} \frac{y_\alpha/W_\alpha}{\gamma_\alpha - 1}\right). \tag{4}$$

The values of C_P and γ are now functions of the mixture proportions, and so are not necessarily constant when these change, essentially through chemical reaction. For reactions that preserve the total number of gas molecules, n and W remain unchanged.

The set of conservation equations governing the evolution of pressure P, density ρ, temperature T, velocity $\vec{\boldsymbol{u}}$, specific enthalpy h and the species mass-fractions y_α is:

CONTINUITY OF MASS:	$\dot{\rho} + \rho\nabla\cdot\vec{\boldsymbol{u}} = 0$	
MOMENTUM:	$\rho\dot{\vec{\boldsymbol{u}}} + \nabla P = 0$	
ENERGY:	$\rho\dot{h} = \dot{P}$	(5)
IDEAL GASES:	$P = nR\rho T$	
SPECIES PRODUCTION:	$\rho\dot{y}_\alpha = K_\alpha$	

where K_α is the rate of chemical production of α measured in terms of mass per unit volume per unit time. The superscript 'dot' notation represents a particle-following or Lagrangian time derivative, e.g. $\dot{\rho} = (\partial/\partial t + \vec{\boldsymbol{u}}\cdot\nabla)\rho$.

The specific enthalpy h is composed of the specific enthalpies of the component species of the mixture in the manner

$$h = \sum_{\alpha\in\mathcal{A}} y_\alpha\left(h_{0\alpha} + \textstyle\int_{T_0}^{T} C_{P\alpha}\,dT\right) \tag{6}$$

where $h_{0\alpha}$ is the energy of formation at the reference temperature T_0 and $\int_{T_0}^{T} C_{P\alpha}\,dT$ is the thermal energy of the species α per unit mass. Differentiation leads to the form of the energy equation,

ENERGY:	$C_P\dot{T} = \dot{P}/\rho - \dot{q}$	(7)
where	$\dot{q} = -\sum_{\alpha\in\mathcal{A}} \dot{y}_\alpha\left(h_{0\alpha} + \int_{T_0}^{T} C_{P\alpha}\,dT\right).$	

The 'chemical energy' q that may be released by changes in chemical composition can thus be seen to depend on temperature if $C_{P\alpha}$ is not the same for all species. This is because one species may contain a different amount of thermal energy from another species at the same temperature. However if the overall specific heat C_P remains constant, then the energy per unit mass released by the chemistry, $-[q]$, depends only on changes in the overall energy of formation, $\sum y_\alpha h_{0\alpha}$, of the mixture per unit mass.

It is useful to think of the chemical energy q as being normalised to have a 'base value' of zero at a suitable burnt reference state in chemical equilibrium. Thus the value of q at any stage broadly represents the amount of chemical energy remaining to be released, and $q \to 0$ as the reaction approaches its final completion ($\dot{y}_\alpha \to 0,\ \forall \alpha \in \mathcal{A}$). In general, this picture has to be qualified by recognising that q need not change monotonically, and that the state of chemical equilibrium may change for different values of the thermodynamic variables T, P and ρ.

Differentiation of the ideal gas law gives $\dot{P} = nR\rho T(\dot{\rho}/\rho + \dot{T}/T + \dot{n}/n)$. Thus, using the continuity equation and the energy equation, it follows that

$$\frac{C_P}{nR}\dot{P}/\rho + C_P T \nabla \cdot \vec{u} = C_P \dot{T} + C_P T \frac{\dot{n}}{n} = \dot{P}/\rho + C_P T \frac{\dot{n}}{n} - \dot{q} \tag{8}$$

from which one can obtain an equation relating pressure and velocity changes,

COMPRESSIBILITY:
$$\frac{1}{\gamma - 1}\dot{P}/\rho + C_P T \nabla \cdot \vec{u} = C_P T \frac{\dot{n}}{n} - \dot{q}. \tag{9}$$

The terms on the right hand side of this equation represent forcing terms for either volumetric expansion, $\nabla \cdot \vec{u}$, or an increase in pressure, $\dot{P}$. The balance of chemical activity that goes into either expansion or pressure must depend upon the particular gasdynamical situation.

It may be of interest to bear in mind that non-chemical effects (such as radiative heating or the constriction of a flow in a wind tunnel) could equally well provide a similar forcing in equation (9). Some of the phenomena discussed below might thus have relevance to quite different situations.

2.4 Reaction Models

In order to close the system of equations, a set of chemical rate-laws providing expressions for the species production rates K_α is needed. Starting with a set of reactants $\mathcal{F}$, the progress of the chemistry towards a set of product species $\mathcal{P}$ usually takes place via a set of intermediate chemical 'radicals' $\mathcal{R}$. Although the overall process should release a significant amount of internal energy in a combustible system, some reaction steps may be endothermic and there is no guarantee that the release of chemical energy would proceed monotonically. Likewise, the number of moles per unit mass n may also change non-monotonically as some reactions act to increase, while others decrease, the total number of molecules.

To model all of this in a simple, yet general, way is not a straightforward task. One approach is to assume that the reaction may be characterised by a single, suitably normalised, chemical-state parameter λ (possibly with $\lambda \in [0,1]$). This reduces the total number of chemical variables y_α to the one variable λ by an assumption that $y_\alpha = \widetilde{y}_\alpha(\lambda)$

under all circumstances. The chemical energy q and specific mole-number n are then (possibly non-monotonic) functions of λ :

$$q = \tilde{q}(\lambda) \quad \text{and} \quad n = \tilde{n}(\lambda). \tag{10}$$

With an approach of this sort, the rate of progress of the chemical reaction can typically be expressed in the the form $\dot{\lambda} = -f(\lambda, T, \rho)$.

If only a single chemical reaction-step $F \rightarrow P$ were to take place, converting reactant directly into product, then λ could (for example) be taken to represent a scaled value of the reactant mass-fraction y_F, decreasing from (say) 1 to 0 as the reaction proceeds. In this case, the rate of chemical reaction would be described by an Arrhenius law, such as

$$\dot{\lambda} = -f(\lambda, T, \rho) = -\frac{\lambda}{t_0}\exp\left(-T_A/T\right) \tag{11}$$

where the constants T_A and t_0 are the activation temperature of the reaction and a time that reflects the rate of chemical change at very high temperatures, respectively. This time is generally longer than the mean molecular collision time t_m. A fuller reaction scheme normally involves rate laws for each reaction step that are generalisations of equation (11).

A fairly simple model for the chemistry can now be posed by using equation (11) as a *representative* rate-law for the chemical-state parameter λ, without necessarily assuming that the chemistry proceeds in only one step (for which the functions $\tilde{q}$ and $\tilde{n}$ would be simple linear functions of λ). For equation (11) to reflect the general properties of the progress of most more complex chemical schemes, the activation temperature T_A must be large enough to make $\dot{\lambda}$ a sensitive function of temperature. This is achieved by insisting that

$$T_A \gg T. \tag{12}$$

This property can provide the basis for an asymptotic analysis of equations (3)–(11).

It remains only to specify the functions $\tilde{q}$ and $\tilde{n}$ of equation (10). Since these terms play a similar role in the compressibility equation (9), all qualitative features can be modelled using only $\tilde{q}$. For this reason, we will assume in this discussion that $\dot{n} \equiv 0$, and for a similar reason we will take C_P and γ to remain constant. Different variations of q with λ can lead to different detonation structures and evolutions, so that it is best at this stage to leave $\tilde{q}(\lambda)$ unspecified. While being a good representative model for a detonable medium, the equations presented above may easily be generalised.

2.5 One dimensional Lagrangean coordinate

A 'stream-function' or mass-coordinate in one dimension, that remains fixed for any fluid element or particle, can be defined as

$$\psi = \int_{x_0(t)}^{x} \rho \, dx \tag{13}$$

where the position $x = x_0(t)$ moves with the local fluid velocity, and time is held fixed in the integration. It is also convenient to introduce the specific volume $V = 1/\rho$. Using subscript notation to represent partial derivatives, the model equations then become

$$V_t - u_\psi = 0, \qquad u_t + P_\psi = 0, \qquad VP_t + \gamma P u_\psi = \gamma PV\frac{n_t}{n} - (\gamma - 1)q_t,$$
$$T = \frac{PV}{nR} \quad \text{and} \quad \lambda_t = -f(\lambda, T, \rho) = -\frac{\lambda}{t_0}\exp\left(-T_A/T\right) \tag{14}$$
$$\text{with} \quad q = \tilde{q}(\lambda) \quad \text{and} \quad n_t = 0$$

the latter equation being modelled as shown for simplicity. Usefully, if q_t, n_t and γ are specified, then the first three of these equations become decoupled from the rest so that it is possible to express thermodynamic changes only in terms of changes in q, n and γ, without explicitly requiring a model for the chemistry.

3 Steady reaction waves or 'fast flames'

This fact is particularly useful in seeking solutions representing steady waves. For this purpose, each of the thermodynamic and chemical variables can be considered to behave as a function of a phase coordinate Ξ in the manner

$$f = f(\Xi) \quad \text{with} \quad \Xi = \psi - mt \tag{15}$$

where m denotes the steady mass flux of material passing through a wave travelling from negative to positive values of ψ. With this form of behaviour (and with n and γ constant), the first four of equations (14) are easily solved to give the relations

$$P - P_0 = -m^2\left(V - V_0\right), \qquad u - u_0 = \frac{P - P_0}{m},$$
$$(\gamma - 1)m^2(q_0 - q) = \left(P_0 + V_0 m^2\right)\left(P - P_0\right) - \frac{\gamma + 1}{2}\left(P^2 - P_0^2\right), \tag{16}$$
$$T = \frac{P}{nRm^2}\left(P_0 + V_0 m^2 - P\right) \quad \text{and} \quad \mu^2 = \frac{m^2 V}{\gamma P} = \frac{1}{\gamma}\left(\frac{P_0 + V_0 m^2}{P} - 1\right)$$

for any suitable reference values P_0, V_0, q_0 and u_0. The parameter μ represents the mach number of the flow through the wave at any point and is included for future reference.

The first of these equations represents a Rayleigh line, while elimination of m^2 between the first and third equations leads to the Hugoniot relation

$$\frac{\gamma + 1}{\gamma - 1}(PV - P_0V_0) + P_0V - V_0P = 2(q_0 - q). \tag{17}$$

It follows that the thermodynamic variables in the wave are related throughout by Rankine-Hugoniot relations, as obtained for the appropriate value of the chemical energy q at any point. These are sketched in figure 1. Exactly the same principle would apply for non-constant values of n and γ.

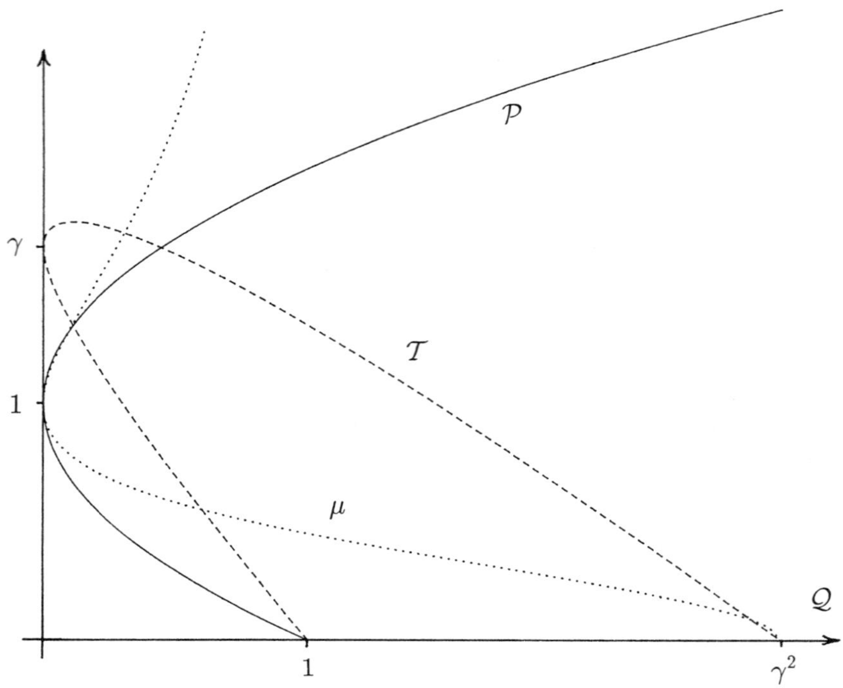

Figure 1: *Rankine–Hugoniot diagram based on equation (18) showing a burnt Hugoniot curve (thick curve) for $q = 0$ and partially reacted Hugoniot curves (thin curves) for different values of q. The dotted curve traces sonic points, $\mu = 1$, and the straight line $O - W$ represents a typical Rayleigh line.*

It may be noted that the chemical energy q and temperature T are given in the relations (16) by simple (but different) quadratics in pressure. A useful way of redrafting these relations is found by defining the dimensionless parameters

$$\mathcal{P} = P \Big/ \frac{P_0\,(1+\gamma\mu_0^2)}{\gamma+1}, \qquad \mathcal{T} = T \Big/ \frac{P_0^2\,(1+\gamma\mu_0^2)^2}{(\gamma+1)^2 nRm^2}$$

and

$$\mathcal{Q} = \mathcal{Q}_b + (q - q_b) \Big/ \frac{P_0^2\,(1+\gamma\mu_0^2)^2}{2(\gamma^2-1)m^2} \tag{18}$$

where

$$\mathcal{Q}_b = 1 + \frac{\gamma+1}{(1+\gamma\mu_0^2)^2}\left(\gamma - 1 - 2\gamma\mu_0^2\right) - \frac{2(\gamma^2-1)}{(1+\gamma\mu_0^2)^2}\,\frac{(q_0-q_b)m^2}{P_0^2}$$

where μ_0 is the flow Mach number through the wave at the reference values, and q_b is the final burnt value of the chemical energy q. The relations (16) now become simply

$$\mathcal{T} = \mathcal{P}\,(\gamma+1-\mathcal{P}), \qquad \mathcal{Q} = (\mathcal{P}-1)^2 \qquad \text{and} \qquad \mu^2 = \frac{\gamma+1-\mathcal{P}}{\gamma\mathcal{P}}. \tag{19}$$

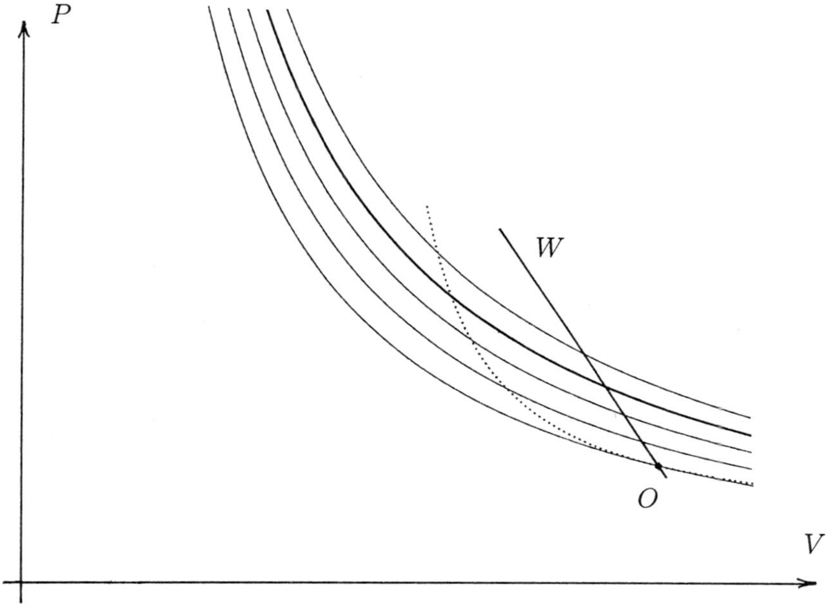

Figure 2: *Temperature (dashed curve), pressure (solid curve) and Mach number (dotted curve) variations with chemical energy through a steady reaction wave.*

Values for $\mathcal{P}$, $\mathcal{T}$ and μ^2 are plotted against values of $\mathcal{Q}$ in figure 2. It can be seen that two branches of solution exist. The subsonic upper (deflagration or strong detonation) branch, with $\mu^2 < 1$, involves higher pressures and temperatures for any given value of $\mathcal{Q}$, while the supersonic lower (or weak detonation) branch, with $\mu^2 > 1$, involves lower pressures and temperatures. Reaction wave solutions would generally follow one or other of these branches as $\mathcal{Q}$ changes through the wave towards a final stable chemical equilibrium value. In equations (18) this is defined as $\mathcal{Q}_b$, a final value that is, in general, a function of the wave speed or mass flux m through the reaction-wave. Thus different speeds generally lead to different final values. Clearly no speed is possible that would make $\mathcal{Q}_b < 0$, since real solutions only exist for $\mathcal{Q} \geq 0$.

A variety of solutions are then possible, provided only that a chemical reaction can actually proceed to drive $\mathcal{Q}$ towards $\mathcal{Q}_b \geq 0$, not necessarily monotonically, in the range $\mathcal{Q} \geq 0$. A chemical change that releases energy would cause an overall decrease in $\mathcal{Q}$ even though local increases may occur in some parts of the wave. It is interesting to note that temperatures on the strong detonation branch do not increase monotonically as $\mathcal{Q}$ decreases, a temperature maximum occurring at the isothermal sound speed, $\mu^2 = 1/\gamma$.

Figure 2, together with the scalings (18), provides a more detailed unravelling of possible

Rayleigh line variations occuring in a more conventional Rankine–Hugoniot diagram (as illustrated by the line $O - W$ in figure 1).

There are two ways in which solutions can swop between the deflagration and weak detonation branches. If $\min\{\mathcal{Q}\} = 0$ with $\mathcal{Q}_b > 0$, then as $\mathcal{Q}$ decreases to zero and increases again, solutions can move smoothly from one branch to the other, passing through a sonic point at $\mathcal{Q} = 0$. This exothermic–endothermic change between branches can generally only happen for a unique 'eigenvalue' mass flux, which we denote by m_S (in order to emphasise the presence of a sonic point in the flow).

The other means of changing from one branch to the other is via a shock-wave (across which $\mathcal{Q}$ remains constant). This, of course, is only able to *increase* pressures so that it can only cause a change from the weak detonation to the deflagration or strong detonation branch. Thus a solution can start by moving along the lower (weak) branch until, at some value of $\mathcal{Q}$, the solution jumps to the upper branch where the solution continues to evolve as $\mathcal{Q}$ continues to change. Such solutions would remain on the deflagration branch for all mass fluxes that give $\min\{\mathcal{Q}\} > 0$. The condition $\min\{\mathcal{Q}\} = 0$ (or $m = m_S$) would identify the lower bound for such mass fluxes. Curiously, if $\mathcal{Q}_b > 0$, an exothermic–endothermic change could then, potentially, return the solution once again to the lower branch from which yet another shock wave increase could not be ruled out.

Unless the reaction-rate actually is zero at some unburnt stable equilibrium point on the weak detonation branch, all such solutions involving a shock-jump from the weak branch to the strong branch are non-unique. There is no criterion that fixes the value of $\mathcal{Q}$ at which the shock-wave would lie, so that a continuum of steady reaction-wave structures involving both weak detonation and deflagration parts becomes possible for all mass fluxes in the range $m \geq m_S$.

According to these arguments, both weak and strong detonations are possible for all mass fluxes bounded below by m_S provided only that chemical changes can realistically take place at 'weak' temperatures and pressures. Moreover, mixed reaction waves are also possible with shock-waves separating weak from strong (or deflagration) parts of the structure.

At this point it is worth commenting on the role of mechanisms for propagation in these waves. In the strict interpretation of the word, 'propagation' does not simply denote movement, but requires some action on the part of the object in question that brings about a reduplication of the object at some other point and time. Since information in this compressible system cannot travel faster than the local speed of sound, the supersonic weak detonation waves do not genuinely self-propagate. One may imagine these 'very fast flames' as representing either convected explosions [4] or spatially non-uniform explosion processes [11]. An example of the latter is considered in the next section.

4 Unsteady behaviour

4.1 Induction

We will now invoke the reaction-rate model (11), requiring only that it should hold true for relatively small temperature changes, of the order of $T_i^2/T_A \ll T_i$, about some reference temperature T_i. This and the other reference values, P_i, V_i, u_i, q_i and $\lambda_i = 1$, may be taken to typify the state of a medium behind an initiating shock-wave that triggers a chemical reaction into life, or it may represent peak initial values in the conditions produced by bulk compression or radiant heating of an explosive medium, either in a container or in an unbounded region. At this state, the medium would explode in a time of order t_i (to be specified below).

A dimensionless asymptotic form of the chemical and compressible model is now usefully expressed in terms of the dimensionless time and mass-coordinates

$$\tau = \frac{t}{t_i} \qquad \text{and} \qquad \chi = \frac{\psi}{t_i\sqrt{\gamma P_i/V_i}}. \tag{20}$$

With these as independent variables, the thermodynamic parameters may be considered to satisfy the following asymptotic relations

$$\begin{aligned} \frac{T}{T_i} &\sim 1 + \frac{T_i}{T_A}\,\phi(\tau,\chi) \qquad & u &\sim u_i + \sqrt{\gamma P_i V_i}\,\frac{T_i}{T_A}\,v(\tau,\chi) \\ \frac{P}{P_i} &\sim 1 + \gamma\frac{T_i}{T_A}\,p(\tau,\chi) \qquad & \lambda &\sim 1 - \frac{C_P T_i}{\tilde{q}'(1)}\frac{T_i}{T_A}\,\ell(\tau,\chi). \end{aligned} \tag{21}$$

After substituting these rescalings into equations (14) and neglecting lower order terms, one obtains the set of equations

$$\phi_\tau = (\gamma-1)p_\tau + \ell_\tau, \qquad p_\tau + v_\chi = \ell_\tau, \qquad v_\tau + p_\chi = 0 \qquad \text{and} \qquad \ell_\tau = e^\phi \tag{22}$$

after selecting the following definition for t_i

$$t_i = t_0\,\frac{C_P T_i}{\tilde{q}'(1)}\frac{T_i}{T_A}\,\exp(T_A/T_i). \tag{23}$$

The perturbation model (22) describes the small-scale behaviour of thermodynamic variables in the presence of (at least) a partial inertial confinement.

In general, these equations would need to be solved subject to an appropriate specification of initial and boundary conditions. There are clearly many possible forms for this, representing different physically imposed conditions as suggested earlier. For the present discussion, it is simplest to suppose only that initial conditions

$$\phi(0,\chi) = \phi_0(\chi), \quad u(0,\chi) = u_0(\chi), \quad p(0,\chi) = p_0(\chi) \quad \text{and} \quad \ell(0,\chi) = \ell_0(\chi) \tag{24}$$

are imposed over an unbounded domain $\chi \in (-\infty, \infty)$.

Assuming appropriate differentiability, equations (22) can be condensed into a single differential equation for ϕ, namely Clarke's equation [12]

$$\left(\phi_\tau - \gamma e^\phi\right)_{\tau\tau} = \left(\phi_\tau - e^\phi\right)_{\chi\chi}. \tag{25}$$

By associating different terms in this equation, one can observe some of the rich structure of possible interactions and processes taking place as the medium reacts. These include: the presence of the adiabatic sound speed of unity; the isothermal sound speed of $\gamma^{-1/2}$; constant pressure blowup if $\partial_\chi \gg \partial_\tau$; constant density blowup if $\partial_\tau \gg \partial_\chi$; and possible transition from the former to the latter [13].

Alternatively, integrating along characteristics τ and $\chi \pm \tau$, and eliminating p and v, leads to a single integral equation for ϕ,

$$\begin{aligned}\phi \;=\; &\int_0^\tau \left[e^{\phi(s,\chi)} + \frac{\gamma-1}{2}\left(e^{\phi(s,\chi-\tau+s)} + e^{\phi(s,\chi+\tau-s)}\right)\right] ds \\ &+ S_0(\chi) + \frac{\gamma-1}{2}\left[R_0(\chi-\tau) + L_0(\chi+\tau)\right]\end{aligned} \tag{26}$$

where the functions $R_0 = p_0 + v_0$, $L_0 = p_0 - v_0$ and $S_0 = \phi_0 - (\gamma-1)p_0$ represent right travelling, left-travelling and stationary (with respect to any fluid element) thermodynamic waves, respectively. Under conditions where p_0, v_0 and ϕ_0 are uniformly bounded from below within a suitable range of characteristics, this equation shows that ϕ would remain bounded from below by the singular function $\hat{\phi} = -\ln(\hat{\tau}_I - \tau)$ for some appropriate constant $\hat{\tau}_I < \infty$. It follows that ϕ would become unbounded, $\phi(\tau_I, \chi_I) \to \infty$, at some location χ_I in the finite time $\tau \to \tau_I \leq \hat{\tau}_I$.

This is highly significant for the emergence of detonation in the context of the present discussion. However, it is equally important to note that ϕ does not only become unbounded at an isolated point; it also grows to infinity at other locations. In fact, so long as the integrations in (26) are suitably well-defined, there must exist a path or continuum of singularity times $\tau = \tilde{\tau}(\chi)$ at which ϕ becomes unbounded:

$$\phi(\tau, \chi) \to \infty \quad \text{as} \quad \tau \to \tilde{\tau}(\chi). \tag{27}$$

The time τ_I simply represents the first time of singularity, $\tau_I = \min\{\tilde{\tau}(\chi)\}$, at $\chi = \chi_I$.

This singularity path has certain important properties. Firstly, since the integrals in (26) can only increase along any characteristic, $\chi = \chi_0 \pm (\tau - \tau_0)$, it follows that the speed of the singularty path $1/\tilde{\tau}'$ must be supersonic, that is greater than unity:

$$\tilde{\tau}'(\chi) < 1. \tag{28}$$

Also, towards the first point of singularity, the gradient $\tilde{\tau}'(\chi)$ tends to zero, so that the path of singularity first appears with *infinite* speed. Using results found for the shape of the function ϕ as $\tau \to \tau_I$ [13, 14], and the local solutions (30) below, the singularity path can typically behave in one of at least the following two ways near its initial point:

$$\tilde{\tau} \sim \tau_I + \kappa_p \chi^{(2\gamma-1)/\gamma} \qquad \text{or} \qquad \tilde{\tau} \sim \tau_I + \kappa_i(\chi - \chi_I)^2 \tag{29}$$

for some constants κ_p and κ_i. The former behaviour arises when the singularity evolves from an initial temperature maximum at a rigid boundary with non-zero gradient (as typically found if the reaction is triggered by a strong piston-driven or reflected shock wave). The latter behaviour arises from an initial temperature distribution with zero

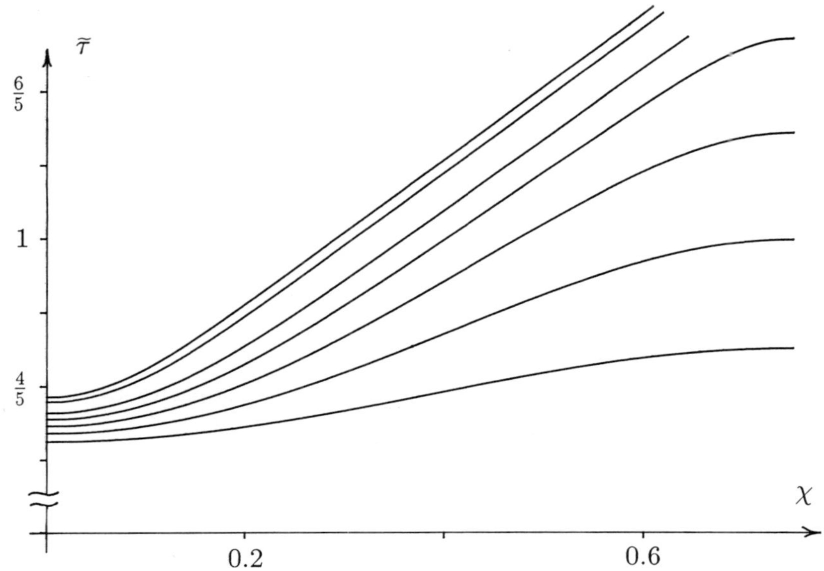

Figure 3: *Calculated curves of* $\tilde{\tau}(\chi)$ *for* $\gamma = 1.4$ *with the initial conditions* $p_0 \equiv v_0 \equiv \ell_0 \equiv 0$ *and* $\phi_0(\chi) = \frac{\kappa}{2}\left[\cos\left(\pi\chi/\frac{3}{4}\right) - 1\right]$ *for* $\kappa = 0.2,\ 0.4,\ 0.6,\ 0.8\ 1.0,\ 1.4$ *and* 1.6.

gradient at its peak value. Some numerical results for the flame-path $\tilde{\tau}$, are reproduced in figure 3 for a sample set of periodic initial conditions.

Because e^{ϕ} becomes very large close to $\tau = \tilde{\tau}$, the behaviour close to the singularity is locally quasi-steady. As a result, the equations are readily solved in this region to yield the asymptotic structures of ϕ, v, p and ℓ close to the singularity

$$\phi \sim -\ln\left(\frac{\gamma - \tilde{\tau}'^2}{1 - \tilde{\tau}'^2}\varpi\right) + F\varpi\ln\varpi, \qquad \phi \sim \left(\gamma - \tilde{\tau}'^2\right)p + a + A\varpi\ln\varpi,$$
$$\ell \sim \left(1 - \tilde{\tau}'^2\right)p + b + B\varpi\ln\varpi \qquad \text{and} \qquad v \sim \tilde{\tau}'p + c + C\varpi\ln\varpi \tag{30}$$

where $\varpi = \tilde{\tau} - \tau$, and

$$A = B = \frac{\gamma + 3\tilde{\tau}'^2}{(\gamma - \tilde{\tau}'^2)^2}\tilde{\tau}'', \qquad C = -\frac{2\tilde{\tau}'\tilde{\tau}''}{(\gamma - \tilde{\tau}'^2)^2} \quad \text{and} \quad F = -\frac{\gamma - 1}{2}\frac{A}{1 - \tilde{\tau}'^2}. \tag{31}$$

The functions $a(\chi)$, $b(\chi)$ and $c(\chi)$, as well as the function $\tilde{\tau}(\chi)$, must depend on the set of boundary and initial conditions that are relevant to the particular situation being modelled.

4.2 Induction flame; weak detonation

Naturally, as the perturbation quantities ϕ, p, v and ℓ become too large, they lose their physical meaning. However, the quasi-steady behaviour (30) near the path of singularity remains significant in that it shows the emergence of a chemical reaction wave involving large-scale variations in thermodynamic and chemical variables. Provided only that the reaction rate $f(\lambda, T, \rho)$ remains much larger than its value during the induction stage of the evolution, the relations (16) describe the quasi-steady variation of temperature, pressure etc. through the wave. Because this strongly active chemical wave and its movement arise as a direct result of a spatially varying induction process, it is natural to describe it as an 'induction flame.'

Selecting $P_0 = P_i$, appropriate values for the reference parameters V_0, q_0 and u_0, and the mass flux m, that determine the thermodynamic and chemical variation through the flame, are obtained by asymptotic matching with the local singularity structure (30) in the limit $T_i/T_A \to 0$. This gives

$$\begin{aligned} P_0 &= P_i, & & m(\psi) \sim \sqrt{\gamma P_i/V_i}\Big/\tilde{\tau}'(\chi), \\ V_0(\psi) &= \frac{nRT_0}{P_i} & \text{where} \quad & T_0(\psi) \sim T_i\left(1 + \frac{T_i}{T_A}a(\chi)\right), \\ q_0(\psi) &\sim q_i - C_P T_i \frac{T_i}{T_A} b(\chi) & \text{and} \quad & u_0(\psi) \sim u_i + \sqrt{\gamma P_i V_i}\,\frac{T_i}{T_A} c(\chi) \end{aligned} \tag{32}$$

in which m would generally vary significantly as a function of ψ. The remaining functions, V_0, T_0, q_0 and u_0, are in fact constant to leading order as $T_i/T_A \to 0$.

Since the mass flux through the induction flame, m, starts off being infinite (where $\tilde{\tau}'$ is zero), and therefore must be very large at least for a short while, the induction flame initially has the structure of a weak detonation. In figure 2, it would be described by thermodynamic variations along the lower curves. On the Rankine–Hugoniot diagram, figure 1, it would be described by changes that take place along a continuum of Rayleigh-lines, such as $O-W$, that gradually decrease in slope after initially being vertical. Although steady solutions that begin on the (supersonic) weak detonation branch can, in principle, be lifted by a shock-wave to the (subsonic) deflagration branch, there is no mechanism in the early progress of the induction flame that would create such a shock.

Thus, although thermodynamic changes generally increase the local sound speed as the combustion proceeds, the induction flame passes through the medium at a rate that is *uniformly* supersonic. In the sense discussed earlier, it does not strictly 'propagate' itself forwards since information cannot travel faster than the flame. The only way in which compressibility and chemistry can interact to determine the speed and structure of the induction flame is during the induction process, before the flame has actually appeared at any location.

One vital feature that is worth highlighting before proceeding further is the thermodynamic state of the medium at the point at which the compressible forcing terms in equation (9) approach zero (possibly by changing sign). Since we are modelling n as remaining constant, the chemical energy q approaches a minimum value $\hat{q}$ at this stage. For a reaction in which q decreases monotonically, $\widehat{q}$ would be the same as the final burnt value q_b of q (which may, possibly, be normalised to zero). Provided the chemical reaction approaches this point much more rapidly than the rate of reaction during the induction process, the thermodynamic state is determined by the intersection between the appropriate Rayleigh line and the weak detonation part of the Hugoniot curve corresponding to $\widehat{q}$ in figure 1. From equations (16), it is determined by the lower root of the equation determining the chemical energy q, as q approaches $\widehat{q}$. This gives

$$P = \widehat{P} = \frac{P_0 + V_0 m^2}{\gamma + 1}\left(1 - \sqrt{1 + \frac{(\gamma^2 - 1)P_0^2 - 2(\gamma + 1)m^2(\widehat{Q} + P_0 V_0)}{(P_0 + V_0 m^2)^2}}\right) \qquad (33)$$

where $$\widehat{Q} = (\gamma - 1)(q_0 - \widehat{q}).$$

Other thermodynamic quantities are then determined by the remainder of equations (16).

As the mass flux m decreases, one finds that $\widehat{P}$ increases until the point is reached where the two roots become equal—there are no real roots for lower values of m. Equation (33) and the last of equations (16) make it clear that this happens precisely at a sonic point ($\widehat{\mu}^2 \to 1$). As in section 3, we use the subscript S to denote the state at this sonic critical point, obtaining values for P_S, V_S, u_S, T_S and m_S.

4.3 Shock formation

At this point, the induction flame is supersonic with respect to the sound speed in the medium ahead of it, where an unsteady induction process still prevails. Thus the critical mass flux m_S can be attained for some perfectly realistic value of $\widetilde{\tau}' = \widetilde{\tau}_S' < 1$. In general, it is very likely that an induction flame would slow down below this critical mass flux at some stage.

When this happens, the steady solution descriptions presented in section 3 are no longer tenable over the entire range of chemical energy q. Once $m < m_S$, the steady wave description is only feasible, in terms of the scaled chemical energy $\mathcal{Q}$ (in equations (18), (19) and figure 2), within the domain where real solutions can be found, namely where $\mathcal{Q} \geq 0$. Something else must happen wherever $\mathcal{Q} < 0$ (at the least).

This breakdown in the quasi-steady weak detonation description of the induction flame can be analysed by focussing attention on the point where a sonic flow first appears. For simplicity, we will consider cases in which the chemical energy q decreases monotonically so that the sonic point only appears towards the very end of the chemical reaction. It is also convenient to describe the reaction-rate at this point using the simple model formula

$$\lambda_t = -\frac{\lambda}{t_c} \qquad (34)$$

where the 'fast' chemical time t_c is very much shorter than the induction time t_i identified in equation (23). In terms of the reaction-rate model (11), one should expect to find

$t_c = t_0 \exp(T_A/T_S) \ll t_i$, although a different chemical mechanism and model from that governing the induction stage might well describe the final stage of the reaction.

By identifying the 'acoustic impedance' $\alpha = \sqrt{\gamma P/V}$, the momentum and compressibility equations from (14) can be combined to describe evolutions along the characteristics $\partial\psi/\partial t = \pm\alpha$:

$$P_t \pm \alpha P_\psi \pm \alpha(u_t \pm \alpha u_\psi) = \gamma P \frac{n_t}{n} - (\gamma - 1)\frac{q_t}{V}. \tag{35}$$

For an induction flame that slows down through a critical mass flux $m_S > 0$, it is worth noting that only the equation that follows forward-travelling characteristics is likely to reveal significant unsteady behaviour. Also, with $t_c \ll t_i$, all of the remaining equations can be shown still to lead to good quasi-steady descriptions in coordinates that shadow the induction flame. It follows that only the third of equations (16) runs into difficulty near $m = m_S$, all of the other relations remaining useable. Moreover, the appearance of the square-root singularity of P in equation (34) indicates that changes in pressure dominate over changes in mass flux in these relations. All of these assertions can be verified using formal asymptotic techniques [5, 7].

In particular one obtains

$$\alpha \sim \sqrt{\frac{\gamma P}{V_S}\left(1 + \frac{P - P_S}{V_S m_S^2}\right)} \sim m_S + \frac{\gamma + 1}{2}\frac{P - P_S}{V_S m_S} \quad \text{and} \quad u_t \pm \alpha u_\psi \sim \frac{P_t \pm \alpha P_\psi}{m_S} \tag{36}$$

after using the fact that $\mu_S = 1$. The resulting form of equation (35) that is relevant for following the induction flame path over small ranges of $P - P_S$, is a chemically-forced Burgers' equation

$$P_t + \frac{\gamma + 1}{2}\frac{P - P_S}{V_S m_S} P_{\psi - \tilde{\psi}} \sim \frac{m_S}{2}\left(\frac{\gamma - 1}{V_S} q_{\psi - \tilde{\psi}} - \frac{\gamma P_S}{n_S} n_{\psi - \tilde{\psi}}\right) \tag{37}$$

in which the time differentiation is taken along paths of constant $\psi - \tilde{\psi}$, where the induction flame path is represented by $\psi = \tilde{\psi}(t)$. It may be convenient to think of $\tilde{\psi}(t)$ as tracing the path of some fixed 'small' value of λ, or perhaps the path of maximum chemical activity. To a certain degree, the exact definition of $\tilde{\psi}(t)$ is a matter of choice.

Extending equations (33) by simply dropping the superscript '$\hat{}$' throughout, the behaviour of pressure close to P_S, takes on the asymptotic form

$$\frac{\gamma + 1}{2}(P - P_S)^2 \sim [(P_S - P_0)V_0 - Q)]\, 2m_S(m - m_S) + (\gamma - 1)m_S^2(q - \hat{q}) \tag{38}$$

for small enough values of $m - m_S$ and $q - \hat{q}$. This expression describes variations in pressure over those parts of the domain where unsteady effects governed by equation (37) are still unimportant. It therefore provides a matching or initial/boundary condition for equation (37) once the variations in mass flux $m - m_S$ and chemical energy $q - \hat{q}$ are known.

Near the critical point, the mass flux through the induction flame, $m = \tilde{\psi}'(t)$, is given by

$$\tilde{\psi}'(t) - m_S \sim - m_S \frac{t - t_S}{t_i} \frac{\tilde{\tau}_S''}{\tilde{\tau}_S'^2} + \cdots \tag{39}$$

which changes significantly over the induction time-scale t_i. The time t_S represents the moment at which $\tilde{\psi}'(t_S) = m_S$. Also, close to the critical point which we take to exist at the end of the chemical reaction, the model rate equation (34) has the quasi-steady solution

$$\lambda = \tilde{\lambda} \exp\left(\frac{\psi - \tilde{\psi}}{m\, t_c}\right) \tag{40}$$

for a suitable constant $\tilde{\lambda}$ that depends on the definition of $\tilde{\psi}(t)$.

The unsteady development of pressure can now be cast into an asymptotic form of analysis by identifying the rescalings

$$\psi - \tilde{\psi} = m_S t_c(\xi - \xi_0) \qquad \text{and} \qquad t - t_S = \hat{t}_S \zeta \qquad \text{with} \qquad P - P_S \sim \hat{P}_S\, \theta(\zeta, \xi) \tag{41}$$

in which

$$\begin{aligned} \hat{P}_S^3 &= V_S m_S^4 \left[(P_S - P_0)V_0 - \hat{Q}\right] \frac{8}{\gamma + 1} \frac{\tilde{\tau}_S''}{\tilde{\tau}_S'^2} \frac{t_c}{t_i} = O\left(\frac{t_c}{t_i} P_S^3\right), \\ \hat{t}_S &= \frac{2}{\gamma + 1} \frac{V_S m_S^2}{\hat{P}_S} t_c \ll t_i \quad \text{and} \quad \xi_0 = \ln\left((\gamma - 1) \frac{\hat{t}_S}{t_c} \frac{\tilde{q}'(0)\tilde{\lambda}}{\hat{P}_S V_S}\right). \end{aligned} \tag{42}$$

The problem can now be expressed in the simplified form

$$\theta_\zeta + \theta\theta_\xi = \tfrac{1}{2}e^\xi \qquad \text{with} \qquad \theta \sim -\sqrt{e^\xi - \zeta} \quad \text{as} \quad \begin{cases} \zeta \to -\infty \\ \text{or} \quad \xi \to \infty. \end{cases} \tag{43}$$

A more general form of this problem can be produced by replacing the first occurrence of e^ξ by $f'(\xi)$ and the second by $f(\xi)$, for any differentiable function f that ultimately depends on the reaction-rate model and the behaviour of n and q. For example, the simple case $f = \xi^2$ would model a situation in which the compressibility forcing terms of equation (9) change sign before the end of the chemical reaction.

By defining a characteristic η such that

$$\zeta - \eta + \int_\xi^\infty \frac{d\xi}{\theta} = 0 \tag{44}$$

it becomes a straightforward task to solve equations (43) to give

$$e^\xi - \theta^2 = \eta = e^\xi \sin^2\left(\frac{(\zeta - \eta)\sqrt{\eta}}{2}\right) \tag{45}$$

for positive values of η. When η is negative, $\sqrt{\eta}$ can be replaced by $\sqrt{-\eta}$, with $\sin^2$ being replaced by $-\sinh^2$. Solutions must be selected such that θ is negative and satisfies

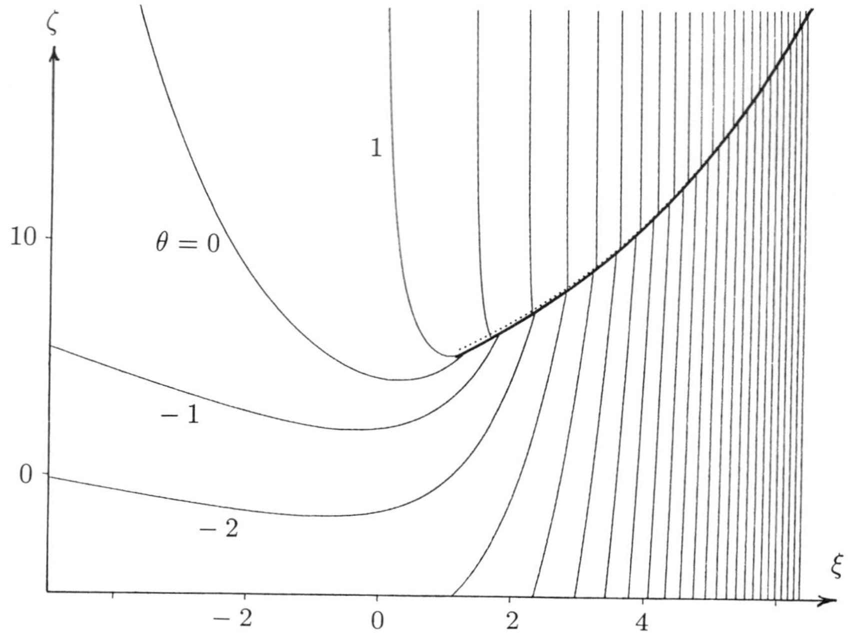

Figure 4: *Pressure perturbation contours θ, showing the formation and movement of a shock wave (thick curve). Contours are plotted at unit increments of θ from -24 to $+24$.*

the asymptotic boundary condition in (43) for all sufficiently large positive values of ξ and/or $-\zeta$. Resulting pressure profiles at different times ζ are plotted in figure 4.

An examination of the solution (45) reveals that an infinite pressure gradient first arises when

$$\zeta = \zeta_* \approx 5.076205, \quad \xi = \xi_* \approx 1.142412 \quad \text{and} \quad \theta = \theta_* = 1/\sqrt[3]{3}. \tag{46}$$

This identifies the moment of shock-formation. Subsequently, if the path of the shock-wave is denoted by $\xi = X(\zeta)$, then the shock-wave propagates ahead at the speed [15]

$$X'(\zeta) = \frac{\theta(X^+, \zeta) + \theta(X^-, \zeta)}{2} \tag{47}$$

the initial value of which is $X'(\zeta_*) = \theta_* = 1/\sqrt[3]{3}$, with $X(\zeta_*) = \xi_*$. The superscripts $^+$ and $^-$ denote evaluation of θ immediately ahead of and behind the shock discontinuity respectively. As the shock moves forwards, it stabilises asymptotically about the path

$$X(\zeta) \sim 4\ln(\zeta/4) \tag{48}$$

shown as a dotted line in figure 4. Thus it can be seen that the shock wave slows down dramatically as it moves forwards, approaching the shock-speed

$$X'(\zeta) \sim 4/\zeta. \tag{49}$$

4.4 Development of a strong-detonation

The existence of this shock wave (or indeed any positive value of θ) marks a fundamental change in the nature of the reaction wave. It shows that a part of the flow through the wave has become subsonic, and that genuine propagation mechanisms can at last play a role in its evolution. As will be seen, these mechanisms ultimately dominate and lead to a truly 'propagating' detonation wave.

The physical significance of the behaviour (49) of the shock wave becomes clearer as soon as one refers back to dimensional coordinates. If $\psi = \Psi(t)$ represents the path of the shock-wave in terms of ψ and t, then the mass flux through the shock is given by

$$\Psi'(t) - \tilde{\psi}'(t) \sim \frac{\gamma+1}{2}\frac{\widehat{P}_S}{V_S m_S^2} m_S X'(\zeta) = O\left((t_c/t_i)^{1/3} m_S X'(\zeta)\right) \tag{50}$$

from equations (41) and (42). This should be compared with the difference between the induction flame mass flux and the critical (sonic) mass flux m_S, which can be seen from equations (39), (41) and (42) to be

$$\tilde{\psi}'(t) - m_S \sim -\frac{2}{\gamma+1}\frac{V_S m_S^2}{\widehat{P}_S}\frac{t_c}{t_i}\frac{\tilde{\tau}_S''}{\tilde{\tau}_S'^2} m_S\zeta = O\left((t_c/t_i)^{2/3} m_S\zeta\right). \tag{51}$$

Since $(t_c/t_i)^{1/3} \gg (t_c/t_i)^{2/3}$ it follows that the initial unsteady shock speed (relative to the critical sonic speed) is much faster than the imbalace between the induction flame-speed and the sonic speed. However, as the time-perturbation ζ increases this imbalance decreases, until for

$$\zeta = O\left((t_i/t_c)^{1/6}\right) \quad \text{or} \quad t - t_S = O\left((t_i/t_c)^{1/2} t_c\right) \tag{52}$$

the two speeds become similar but still relatively small, at the order of $(t_c/t_i)^{1/2}$ compared with the sonic speed. At this point, the subsonic part of the reaction wave that lies behind the shock-wave involves a pressure excursion of the order of $\widehat{P}_S\sqrt{\zeta}$ or $P_S(t_c/t_i)^{1/4}$. This is generally small compared with $P_S(T_S/T_A)$ which confirms, in retrospect, that the model reaction rate (11) does indeed reduce to the simple linear form (34) for the purpose of analysing the early shock behaviour.

During the time-scale identified by (52), it is no longer justifiable to neglect changes in the convective mass-flux in equations (36) and (37). Without entering into the details (see reference [7]), analysis of this stage reveals that two connected quasi-steady evolutions emerge. Ahead of the shock-wave, the induction flame continues to behave as previously described. It takes the form of a quasi-steady weak detonation travelling forwards at the relatively slowly-varying mass flux m.

However, the evolution behind the shock-wave also becomes quasi-steady [7] with thermodynamic changes taking place in the manner of a quasi-steady strong detonation as described in section 3. There is, however, one crucial feature of this wave. Examining the Rankine–Hugoniot diagram in figure 1, it can be seen that as the induction flame slows down along the weak detonation part of Rayleigh lines (such as $O-W$) of decreasing slope, the pressure that remains behind the wave increases. In the subsequent unsteady,

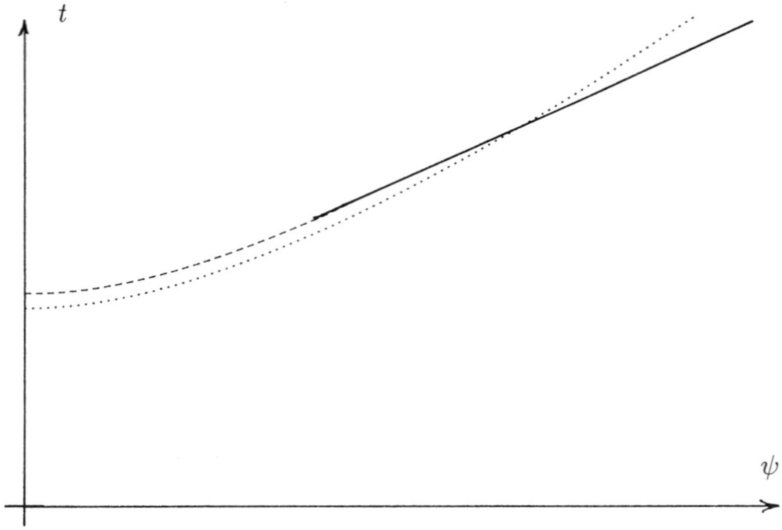

Figure 5: *The movement of an induction flame, between the dotted perturbation singularity path and the dashed path of maximum chemical activity, until it is transformed into a strong detonation following the solid shock path.*

now chemically inert, evolution that takes place it follows that an expansion wave must follow behind the chemical reaction wave. This ensures that a sonic flow is always found very close to the point at which the chemical energy q reaches its minimum value. It thus, effectively, fixes the mass flux traversed by the strong detonation part of the reaction wave to be very close to the critical sonic mass flux m_S.

It may be noted that an analogous pair of quasi-steady structures would be found in the more general case where the number of moles per unit mass n is able to change. The sonic point then arises when the chemical forcing term on the right hand side of the compressibility equation (9) approaches zero. Again this may happen either at the end of the chemical reaction or, if the chemical forcing is generated by non-monotonic or competing changes, before the reaction is complete.

Since $m_S > m$ at this stage, the strong detonation structure actually travels faster than the induction flame and simply overtakes it as sketched in figure 5. In this, the induction flame path $\psi = \tilde{\psi}(t)$ (that may, most usefully, be thought of as representing a path of maximum chemical activity) is asymptotically close to the singularity path $\tau = \tilde{\tau}$. Once the induction flame slows down below the critical sonic mass flux m_S, the quasi-steady shock and deflagration or strong detonation complex (formed after the brief unsteady evolution over the time-scale (52)) continues to move at a mass flux that remains very close to m_S. It therefore *propagates* steadily forward through the slowing induction flame.

As it does so, the shock wave passes into less and less burned parts of the induction flame. Correspondingly, the value of the chemical energy q at the shock location generally

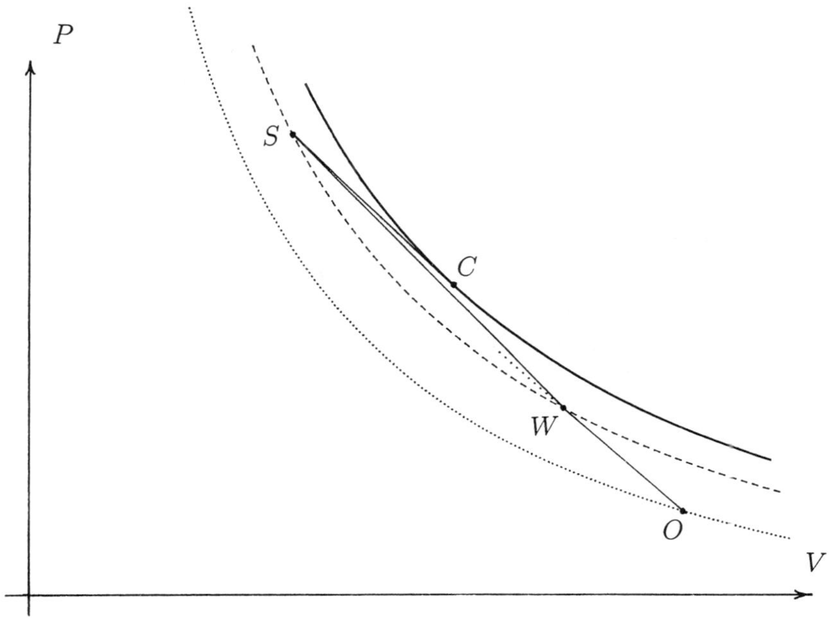

Figure 6: *Rankine–Hugoniot diagram showing weak detonation, shock and strong detonation Rayleigh lines. The solid curve, dashed curve and dotted curve represent fully burnt, partially burnt and fully unburnt Hugoniot curves respectively.*

increases (although not necessarily monotonically). In terms of figure 2, this would correspond to a shock location (joining weak and strong detonation branches of solution) that gradually moves into increasing values of $\mathcal{Q}$. It should be noted that, because the scalings (18) that define $\mathcal{Q}$, $\mathcal{P}$ and $\mathcal{T}$ depend on the mass flux m which is slightly different for the two wave structures, the value of $\mathcal{Q}$ also changes very slightly through the shock wave. This does not, however, significantly alter the overall picture.

On a Rankine–Hugoniot diagram, the variation is not quite so clearly demonstrated. Since the difference between m and m_S is relatively small, the evolution all lies close to a single Rayleigh line. The difference between the Rayleigh lines illustrated in figure 6 is exaggerated for greater clarity. Starting at the point O, the induction flame increases along a Rayleigh line $O - W$ that has too small a slope to be able to reach a fully burnt Hugoniot curve. At some point W where the reaction has only partially proceeded, a shock wave raises the pressure along the Rayleigh line $W - S$ to the point S on a suitable Hugoniot curve for the same fixed chemical state. Finally, the pressure decreases again along the deflagration part of another Rayleigh line $S - C$ that reaches the corresponding fully burnt Hugoniot curve at the tangent point C. This corresponds to the sonic point or Chapman–Jouget point that is found at the end of the chemical reaction. Since the shock wave travels slightly faster than the deflagration that follows it (in order to be able to progress ahead into less reacted regions), the slope of the line $W - S$ is slightly greater than $S - C$. As the shock wave advances, the Hugoniot curve containing the points W and S gradually shifts so as to correspond to less advanced chemical change.

The strong detonation ultimately propagates ahead of the induction flame into the region where unsteady induction processes still prevail. It then has the typical form of a Zeldovich–von Neumann–Döring detonation wave [2], propagating into a weakly nonuniform medium.

5 Conclusions

Using relatively mild assumptions about the nature of a compressible reacting medium and the way in which it undergoes an exothermic reaction, this article has demonstrated a number of features of both steady and unsteady behaviour. Steady combustion waves of many different speeds and configurations are possible. Weak detonations (that do not involve a shock wave) are found to fall within this spectrum, and should only really be excluded if chemical changes are unable to proceed significantly at the appropriate 'weak' values of the thermodynamic parameters.

Indeed they are found to describe the structure of 'induction flames' that emerge in a natural way from a wide class of unsteady ignition processes in compressible media. Since the flow through such a weak detonation is uniformly supersonic, these waves do not genuinely self-propagate but rely on conditions for reaction runaway that are set by events ahead of the waves. Importantly, a transition from such a weak detonation to a strong detonation takes place automatically once the induction flame slows down below a speed at which a sonic point first appears in the flow.

An unsteady evolution then describes the formation of a shock wave and the development of a strong detonation. The flow through this particular reaction wave is subsonic, so that information can be transmitted ahead and thus contribute to a genuine propagation of the wave. As a result, the strong detonation is able to move ahead and completely engulf the induction flame.

No attempt has been made to be comprehensive in describing these events. For example, questions about the stability of steady reaction wave structures has been left completely aside. However, the reader is encouraged to bear in mind more general questions that are hardly touched on in this discussion.

Acknowledgements

Many people contributed substantially towards formulating the arguments that are presented in this article. John Clarke and Ashwani Kapila (through contacts facilitated by a NATO collaborative research grant) were most instrumental in helping to develop the ideas concerning unsteady behaviour, while the document also benefitted from many other useful discussions, especially with Amable Liñán and Paul Clavin.

Above all, this article is based on a series of lectures delivered at the Instituto per le Applicazione del Calcolo in Rome, so that many thanks are due to Marcello Onofri and Alberto Tesei (and all at the I.A.C.) for their warm hospitality.

References

[1] F.A. WILLIAMS, 'Combustion Theory,' 2nd ed., Benjamin/Cummings, California, 1985.

[2] W. FICKETT and W.C. DAVIS, 'Detonation,' University of California Press, Berkeley, 1979.

[3] D.R. KASSOY, *Fast flames,* Ann. Rev. Fluid Mech., **17** (1985) 267–287.

[4] J.F. CLARKE, *Fast flames, waves and detonation*, Manuscript, 1988.

[5] J.W. DOLD and A.K. KAPILA, *Asymptotic analysis of detonation initiation for one-step chemistry: I–emergence of a weak detonation,* submitted.

[6] J.W. DOLD, *Induction period generation of a supersonic flame,* in 'Numerical Combustion,' A. Dervieux and B. Larrouturou, Eds., Lecture Notes in Physics **351**, Springer Verlag, 1989, pp. 245–256.

[7] J.W. DOLD and A.K. KAPILA, *Asymptotic analysis of detonation initiation for one-step chemistry: II–from a weak structure to ZND,* submitted.

[8] YA.B. ZELDOVICH and A.S. KOMPANEETS, 'Theory of Detonation,' Academic Press, 1960.

[9] T. VON KÁRMAN, *Models in thermogasdynamics*, in 'I Modelli nella Technica,' Atti del Convegno di Venezia, **1** (1955) 643–651.

[10] M.W. CHASE, C.A. DAVIES, J.R. DOWNEY, D.J. FRURIP, R.A. MCDONALD and A.N. SYVERUD, *JANAF Thermochemical Tables, Third Edition*, J. Phys. Chem. Ref. Data **14**, Suppl. 1 (1985) 17–18.

[11] YA.B. ZELDOVICH, *Regime classification of an exothermic reaction with nonuniform initial conditions,* Combustion and Flame, **39** (1980) 211–214.

[12] J.F. CLARKE, *Finite amplitude waves in combustible gases*, in 'The Mathematics of Combustion,' J.D. Buckmaster, Ed., SIAM Publications, Philadelphia, 1985, pp. 183–245.

[13] J.W. DOLD, *Dynamic transition of a self-igniting region,* in 'Mathematical Modelling of Combustion and Related Topics,' C-M. Brauner and C. Schmidt-Laine, Eds., Martinus Nijhoff, Dordrecht, 1988, pp. 461–470.

[14] T.L. JACKSON, A.K. KAPILA and D.S. STEWART, *Evolution of a reaction center in an explosive material*, SIAM J. Appl. Math., **49** (1989) 432–458.

[15] G.B. WHITHAM, 'Linear and Nonlinear Waves,' Wiley, New York, 1974.

Author:

J. W. Dold, School of Mathematics, University of Bristol, Bristol, United Kingdom.

V ROYTBURD

On detonation instability

1. INTRODUCTION

The simplest mathematical model of detonation is provided by the reactive Euler equations, a set of hyperbolic conservation laws for mass, momentum and energy augmented by a balance equation for the consumption of reactant within the fluid. In this framework, the classical theory of von Neumann, Zeldovich, and Doering (the ZND theory) describes detonation waves as steadily traveling waves with a quasi one-dimensional structure of an ordinary fluid dynamic shock followed by a reaction zone.

Given the quiescent upstream conditions and the downstream conditions, one can determine the corresponding wave speed and, provided the chemical kinetics is known, the wave structure. The ZND theory predicts that steadily traveling detonations can propagate with speeds above a certain minimal value called a Chapman-Jouguet (CJ) speed. In order to travel faster than the CJ speed, the wave should be overdriven, i.e., some support (for example, a piston moving with appropriate velocity) should be provided behind the wave. Unsupported detonations have the structure of the CJ wave followed by a rarefraction. The ZND theory is presented in many books, see, for example, Fickett & Davis [10] and also the companion paper by A. Kapila.

The ZND theory does not give any clues concerning stability of such traveling waves. As it turns out these waves are, in many cases, dynamically unstable — small perturbations of the wave structure grow in time causing the original planar front to wrinkle, the originally uniform wave speed to change with time etc.

Phenomena of detonation instabilities have attracted a substantial experimental and theoretical effort. In this paper we give a brief review of (or rather a reader's guide to) some parts of this immense material. We begin with a brief description of some phenomena involved and of some relevant mathematical theories. In somewhat more detail we sketch a recent weakly nonlinear theory of one-dimensional instabilities. This theory is developed in a joint work with A. Majda (see [6] for a full exposition). We end with brief remarks how the weakly nonlinear theory can be extended to the multidimensional case. This version of the theory is the most relevant to detonation instability in solid explosives where the scales of transverse and longitudinal instabilities are comparable.

2. PHENOMENA OF DETONATION INSTABILITY

There is conclusive experimental evidence that in many regimes the gaseous detonation front almost immediately develops a cellular structure with secondary shocks and even detonation waves propagating transverse to the leading detonation wave (see, for example, J.H.S. Lee [14]). Experiments show that although on average the detonation front still propagates at a constant velocity close to the theoretical CJ value, the local shock strength fluctuates $\sim \pm 50\%$ from the CJ strength.

The transverse shocks sweeping back and forth, colliding with each other are extremely important. Some recent experiments demonstrate that when the transverse waves are artifically damped (by lining the wall of the shock tube with special attenuating material) the detonation velocity drops dramatically by $\sim$50%. The precise mechanism, how the transverse waves facilitate propagation of the detonation front is not quite clear. Roughly speaking transverse shocks and companion shear layers together with turbulent flows inside the cells enhance turbulent mixing and combustion behind the blast wave.

Cellular instabilities are fully 3-dimensional and extremely complex. Therefore, it is of considerable interest that in some experimental situations it is possible to excite quasi one-dimensional instabilities (see Alpert & Toong [3], Lehr [16]). In the experiments, a steady overdriven detonation loses stability going over into an oscillatory galloping detonation. The galloping detonation is produced by shooting a blunt body through a detonable gas mixture. On schlieren photographs from these experiments, one can clearly observe a regular periodic pattern of density variations left behind the oscillating detonation wave.

3. MATHEMATICAL THEORIES OF DETONATION INSTABILITY

The mathematical investigation of detonation instabilities has started in the 1960's. As with many other complex physical phenomena, there are two mutually complementing approaches: through analytical studies, and through direct numerical simulations.

The most complete formal treatment of the hydrodynamic stability of traveling wave solutions of reactive Euler's equations is due to J.J. Erpenbeck (see [9] for a summary of his work or [10] for a complete set of references). The linearized analysis of Erpenbeck's is based on a study of the Laplace transform on t for the initial value problem. A formally equivalent approach is through the normal mode analysis of the linearized problem. By using the latter approach H.I. Lee and D.S. Stewart [15] have recently extended and greatly clarified Erpenbeck's results.

The linearized analysis preducts periods for the modes which are exponentially growing for the linearized problem. In a surprising development Fickett and Wood [11] have shown that in the case of one space dimension, fully nonlinear numerical simulations produce instabilities with periods very close to the linearized ones. Analogous computations (see, for example, [1]) were performed later on using different numerical techniques and gave rather similar results. It should be mentioned that numerical simulations of galloping detonations are very sensitive and provide a severe test for numerical schemes. In a recent work with A. Bourlioux and A. Majda [5] we compare different numerical methods for unstable detonations and discuss interrelations with our nonlinear asymptotic theory which is described later on.

The one-dimensional linearized analysis invloves numerical solution of an inhomogeneous eigenvalue problem for a system of ordinary differential equations (cf. (9)). In order to avoid numerics and also to elucidate the salient mechanisms responsible for generation of linear instabilities, G.E. Abouseif and T.Y. Toong [1] arrived at an approximate system that can be treated more or less analytically. The essential part of their analysis is the large activation energy asymptotics. Note though that Erpenbeck's results demonstrate the possibility of linearized instabilities for low and even zero activation energy. Approximate results of [1] are in a good accord with Erpenbeck's linearized computations as well as with direct numerical simulations. The latter is not surprising since the simulations are very consistent with the linearized theory anyway.

A first attempt of a nonlinear perturbation analysis of detonation instabilieis was also made by Erpenbeck. But due to some deficiencies of this method (he never described systematically the adjoint problem, he relied on a rather arbitrary separation between "stable" and "unstable" modes), the result is unsatisfactory because, as noted by Erpenbeck, in his derivation the same physical instability could lead to several different equations governing the instability's evolution.

The previous discussion was concerned mostly with one-dimensional instabilities while the most interesting and practially important questions arise for multidimensional instabilities. The transverse instabilities of detonation fronts were demonstrated in several direct numerical simulations (see, for example, [21, 12, 2]). They appear to be extremely robust against numerical perturbations introduced by different numerical methods. In contrast to the numerical simulations which are fully nonlinear, the available mathematical theories investigate just the initial stages of development of transverse instabilities in regimes not far from the linearized one. At the present time, the mathematical understanding of strongly nonlinear regimes seems to be beyond the reach of modern applied mathematics.

The first group of theories is based on ideas of linearized hydrodynamic stability. Through numerical evaluation of eigenvalues and eigenfunctions, Erpenbeck obtained the value of the shortest unstable wave number in the transverse direction for different regimes. This number is postulated to determine the transverse wave length for a cellular detonation. The theory does not explain why or how harmonic oscillations in the y-direction produce the complicated shock structure. being a linearized theory it treats only small perturbations.

In development of their "one-dimensional" paper Abouseif and Toong [3] applied the same linearized method in the limit of large activation energy. In this approximation the governing linearized equations simplify tremendously. They are reduced to a wave equation for the pressure perturbation with the source term determined by chemistry plus a coupled chemical rate equation. By dropping the source term in the induction zone (near the shock), Abouseif and Toong were able to illuminate the role of incident and relected waves in obtaining an eigenmode. Also by a rather crude and not quite justifiable approximation, they decoupled the rate equation from the hydrodynamic fields. It allows to find the source term and to solve the wave equation "almost explicitly". In any event, the results are very similar to those of Erpenbeck: the shortest unstable transverse wave is found and postulated to determine the spacing.

Another high activation energy attack was attempted recently by J. Buckmaster [7] (in colaboration with G.S.S. Ludford in the beginning of this project). In contrast with "irrational" asymptotics of Abouseif and Toong, a rational asymptotic analysis is employed. Buckmaster's principal result is that some distrubances lead to a thermal runaway of the detonation structure at discrete points distributed along the detonation plane. Spacing of the blow-up points is determined by the initial disturbance. In this context the runaway means exponentially large transverse pressure gradients. It should be noted that in some cases, for the validity of this approach the activation energy must be unrealistically large. R. Strehlow, H. Barthel and coworkers (see, for example [4]) apply to the problem of transverse waves ideas of geometrical acoustics. The main point is that the acoustic refraction index changes dramtically through the region of rapid reaction (the "fire") which traps some rays between itself and the shock. The trapped rays, bouncing back and forth between the fire and shock, need to get amplified at the reflections in order to survive. If sufficient amplification is provided, the "bouncing" is sustained and eventually leads to periodically spaced transverse waves. Barthel–Strehlow's method treats only the acoustic wave fronts following them by ray-tracing. Their work provides only a very intuitive evaluation of the amplification factor. A systematic and mathematically compelete attack along these lines, based on the ideas of weakly nonlinear geometrical optics, has been attempted recently by Majda [17].

There is a fundamental limitation restricting the acoustic method to nonsolid explosives only: the characteristic property of the refraction index essentially used in this theory is valid for γ - gas law materials only if γ is less than 2 while solids are approximately described by ideal gases with $\gamma > 2$. Our work with Majda [18] (see also [6]) complements the results obtained with the acoustic method. It describes generation of low frequency transversal instabilities on the small scale which is characteristic for condensed phase explosives. We follow the ideas of classical theory of hydrdynamic stability which were developed for finite bandwidth instabilities (see W. Eckhaus [8], A. C. Newell and J.A. Whitehead [19]). A typical example of such an instability is Benjamin–Feir instabilities of a Stokes' wave train on water. For unstable detonations, we need to develop the analysis in the context of a complex free-boundary problem for a nonlinear hyperbolic system.

In the case of one-dimensional, longitudinal instabilities the analysis is somewhat more tractable. It is simialr to the classical Landau's derivation of the 1940's (see also J.T. Stuart [20]).

4. WEAKLY NONLINEAR THEORY: ONE-DIMENSIONAL CASE

4.1 Formulation

It is convenient to use a general setting of hyperbolic conservation laws with source terms. The specific properties of the reactive Euler equations will be invoked when they are needed. The reactive Euler equations can be written in the form

$$u_t + F(u)_{\tilde{x}} = S(u) \tag{1}$$

with the state vector u representing the density, velocity, specific energy and fuel mass concentration.

We will be concerned with solutions of (1) having the structure of a wave propagating into a quiescent equilibrium medium:

$$u(\tilde{x},t) = \begin{cases} u(\tilde{x},t), & \tilde{x} < \varphi(t) \\ u_q \quad , & \tilde{x} > \varphi(t) \end{cases} \tag{2}$$

with the Rankine-Hugoniot conditions at the leading shock front $\tilde{x} = \psi(t)$ (cf. Kapila's paper in this volume). In the shock attached coordinate system, $x = \tilde{x} - \psi(t)$, one obtains the system

$$u_t + F(u)_x - \psi_t u = S(u), \quad x < 0, \quad t > 0. \tag{3}$$

The boundary conditions take the form:

$$F(u)\mid_{x=0} - F(u_q) - \psi_t(u\mid_{x=0} - u_q) = 0, \tag{4}$$

where the constant vector u_q charcterizes the quiescent medium.

4.2 Eigenvalue Problem

We linearize (3-4) about the steadily traveling ZND solution,

$$u^0(x,t) = u^0(x;f), \quad \psi^0(t;f) = tD. \tag{5}$$

The linearization of (3) has the following form:

$$\frac{\partial v}{\partial t} + L(v,\psi) \equiv v_t + (F^1(u^0)v)_x - Dv_x - \psi' u_x^0 - S^1(u^0)v = 0 \tag{6}$$

$$t > 0, \quad x < 0$$

where F^1 and S^1 are the first order terms of the Taylor's series for the flux and the source; v and ψ are perturbations of the basic flow. Similarly, the boundary conditions at $x = 0$ linearize to become

(7) $$M(v, \psi) \equiv [F^1(u^0)v - Dv - \psi'(u^0 - u_q)]_{x=0} = 0.$$

As usual, we seek a normal mode solution of (6),

(8) $$v(x, t) = g(x)e^{\alpha t}.$$

Then for $g(x)$ we obtain the nonhomogeneous system of ordinary differential equations:

(9) $$((F^1 - D)g)_x - S^1 g + \alpha g - \alpha u_x^0 = 0.$$

It is not difficult to show than any solution of (9) is asymptotically represented by a linear combination of 4 exponentials as $x \to -\infty$. For Re $\alpha > 0$, all these exponentials, except one, are decaying. Following Lee and Stewart [15] we introduce a radiation boundary condition at $-\infty$ which eliminates the growing exponential. The radiation boundary condition has the following interpretation in terms of the linear hyperbolic system (6): there are three characterisitcs approaching $-\infty$ and one characteristic emanating from $-\infty$, the boundary condition prohibits the information transfer from $-\infty$ towards the shock.

The system in (9) together with the radiation boundary condition at $-\infty$ and 4 boundary conditions at 0 (which arise from the time dependent boundary conditions in (7)) comprise an eigenvalue problem for g and α. This eigenvalue problem is solved numerically by shooting. Qualitatively, the solution depends on the parameter D or on the degree of overdrive $f = D^2/D_{CJ}^2$. We summarize the results of the numerical solution as follows:

(10) There exists a degree of overdrive, f_{cr} so that for $f_{cr} > f > f_{cr} - \delta$ the problem has exactly two complex conjugate eigenvalues $\alpha = \tau(f) \pm i\omega(f)$ with real part $\tau(f) > 0, \lim \tau(f) = 0$ as $f \to f_{cr}$, and $\partial\tau/\partial f \mid_{fcr} = -\chi, \chi > 0$. The corresponding eigenfunction, $e(x; f)$, decays exponentially in x.

Thus, our intuition suggests that a Hopf bifurcation occurs for $f < f_{cr}$. However, the potential Hopf bifurcation is not of classical type because for $f > f_{cr}$ the corresponding "eigenmode" grows expontially in space. The situation here reminds the classical acoustic scattering (see Lax & Phillips [13]) where improper eigenmodes which grow exponentially in space correspond to resonant scattering states — such modes typically describe the long time behavior on bounded sets in x.

4.3 Modulation Theory

In order to study the nonlinear evolution of the eigenmode $e(x; f)$ we consider a solution of the nonlinear problem which, to the leading order, is a slow modulation of the eigenmode:

$$
\begin{aligned}
(11) \qquad u(x,t) &= u^0(x; f) + \epsilon[A(\epsilon^2 t)e(x; f)e^{i\omega(f)t} + C.c.] \\
&\quad + \epsilon^2 u^{(2)} + \ldots \\
\psi(t) &= Dt + \epsilon[A(\epsilon^2 t)e^{i\omega(f)t} + C.c.] + \epsilon^2 \varphi^{(2)} + \ldots
\end{aligned}
$$

where C.c. stands for the complex conjugate. The small parameter is given by $\epsilon = (f_{cr} - f)^{1/2}$. We note that the modulated function $e(x; f)e^{i\omega(f)t}$ is bounded, it differes by the factor $e^{\tau(f)t}$ from the eigenmode. For brevity, from now on we do not indicate explicitly the dependence of all parameters on f.

We substitute the ansatz (11) into the nonlinear equation regarding $\epsilon^2 t$ as an independent variable (the two-timing method). We make use of the Taylor series expansions of the flux and source vectors about $u^0(x)$

$$
\begin{aligned}
F(u^0 + \epsilon v) &= F(u^0) + \epsilon F^1(u^0)v + \epsilon^2 F^2(v, v \mid u^0) + \epsilon^3 F^3(v, v, v \mid u^0) + \ldots \\
S(u^0 + \epsilon v) &= S(u) + \epsilon F^1(u^0)v + \epsilon^2 S^2(v, v \mid u^0) + \epsilon^3 S^3(v, v, v \mid u^0) + \ldots
\end{aligned}
$$

where F^1 and S^1 are Jacobians, F^2 and S^2 are bilinear vector-valued forms, F^3 and S^3 are trilinear vector-valued forms, etc. By matching like powers of ϵ, one gets a series of equations of the type

$$
(12) \qquad \partial_t u^{(j)} + L(u^{(j)}, \psi^{(j)}) = R_j, \quad j = 1, \ldots
$$

with the corresponding boundary conditions. The solubility criterion for the equations in (12) is given by the Fredholm's alternative: the inhomogeneous part, R_j, must be orthogonal to the solutions of the adjoint differential equation.

If the inner product is introduced as the following limit

$$
(13) \qquad (u, v) = \lim_{T \to \infty} \frac{1}{T}(u, v)_T \equiv \lim_{T \to \infty} \frac{1}{T} \int_0^T \int_{-\infty}^0 u(x,t) \cdot v(x,t)\,dx\,dt
$$

then upon the integration by parts, one can compute the adjoint operator

$$
(14) \qquad (\partial_t + L)^* u = -(\partial_t u + (F^1 - D)^t u_x + (S^1)^t u)
$$

where the superscript "t" stands for adjoint matrices. In order to define an operator, the expression in (14) should be supplemented by the appropriate boundary conditions. It is possible to find the adjoint eigenmode in the form

(15) $$u(x,t) = h(x)e^{\alpha t}.$$

Now we return back to the consideration of the equations for the terms of perturbation series (12). In the order ε, the right-hand side is 0 since the solution u^1, ψ^1 is proportional to the linearized eigenfunction. In the order ε^2, R_2 have no contribution from differentiation of the amplitude. The temporal harmonics in R_2 are of the form $e^{\pm 2i\omega t}$ or 1, for this reason they are non-secular (i.e. orthogonal to the adjoint solution). Thus, the inhomogeneous equation in (12) is solvable for $j = 2$. One can show that there exists a solution of the form

(16) $$u^{(2)} = A^2 e^{2i\omega t} e_2(x) + A\bar{A} e_0(x) + C.c$$

$$\psi^{(2)} = C_2 A^2 e^{2i\omega t} + C.c.$$

Remark. It is important to note that although any solution of the homogeneous problem can be added to the particular solution in (16), we **select** the solution in (16). Thus, we study the nonlinear response to the self-interaction terms represented by R_2 and ignore the possibility of perturbations coming from other sources.

In the order ε^3, R_3 contains nontrivial secular terms which arise from the three sources: the interaction terms, the results of differentiation of the amplitude, and the remainder terms coming from the fact that α, not $i\omega$, is the eigenvalue (see the remark immediately below (11)). As the solvability condition we get the following equation for $A(T)$:

(17) $$c\frac{dA}{dT} = c\chi A + dA^2\bar{A}, \quad T \equiv \epsilon^2 t,$$

where

$$c = \int_{-\infty}^{0} \left(u_x^0 - e\right) \bar{h}(x)dx$$

$$d = \int_{-\infty}^{0} [S^2(e, e_0) + S^2(\bar{e}, e_2) + S^3(e, e, \bar{e}) - F^2(e, e_0)_x - F^2(\bar{e}, e_2)_x$$

$$-F^3(e, e, \bar{e})_x + 2i\omega\bar{e}_x - i\omega(e_2)_x + i\omega(e_0)_x]\bar{h}(x)dx.$$

Here h is the adjoint eigenfunction, e is the linearized eigenfunction, and e_0 and e_2 are solutions of certain 2nd order inhomogeneous ordinary differential equations. The coefficients in (17) are determined through numerical quadratures, their explicit values are not needed here except for the fact that $Re(d/c) < 0$.

The asymptotic theory which we have sketched in this section leads to several predictions regarding the time-dependent nonlinear development of instabilities for $f < f_{cr}, f_{cr} - f << 1$. First, perturbed amplitudes of the hydrodynamic variables are coupled with the amplitude of the perturbation in the shock front location through the amplitude function A(T). Furthermore, since $Re(d/c) < 0$, the explicit solution of (17) predicts the temporal oscillations with spatial magnitude $O(\varepsilon)$ which develop on time scales $O(\varepsilon^{-2})$. We note that the asymptotic expansions (11) for the solution are uniformly valid for times of order ε^{-2}. As follows from (11), the spatial structure of the waves is given by the ZND profile perturbed by the linearized eigenvector. These wave structures are qualitatively similar to the regular wave pattern observed in experiments [16]. The predictions of the weakly nonlinear theory find their confirmation in direct numerical simulations (see [5,6]).

5. A GINZBURG-LANDAU EQUATION FOR TWO-DIMENSIONAL INSTABILITIES

In this section we just briefly indicate the ways in which the two-dimensional theory differs from the one-dimensional theory of the previous section. First of all, the Euler's equations gain one more independent variable and one more conservation law (for the y-component of momentum). The linearized normal mode for perturbations of the front position takes the form

$$u_k(x,y,t) = e(x)e^{iky+\alpha(k)t}, \psi_k = e^{iky+\alpha(k)t}, \tag{18}$$

where $e(x)$ and $\alpha(k)$ are the eigenvector and eigenvalue which depend on the transverse wave number k. Below we summarize the properties of the spectrum of the linearized problem which give rise to our weakly nonlinear asymptotic theory (cf. (10)):

(19) (i) For any $k \geq 0$ there exists $f = f(k)$ so that in some interval $f(k) > f > f(k) - \delta(k)$ the eigenvalue problem has exactly one eigenvalue $\alpha = \tau(f,k) + i\omega(f,k)$, $\tau(f,k) > 0, \tau \to 0$ as $f \to f(k)$. Note that automatically $\tau - i\omega$ is an eigenvalue corresponding to $-k$

. (ii) As a function of k, for $k > 0$, the stability boundary $f = f(k)$ has a global maximum f_{cr} attained at some $k = k_{cr}$.

The statement in (19) has been verified numerically for some regimes.

It follows from (19) that for $f < f_{cr}$ the whole band of wave numbers around k_{cr} loses (linear) stability. Our goal is to derive an amplitude equation analogous to the Laudau-Stuart equation in (17) which accounts for the mode interaction. The **crucial observation** here is that if

$$e(x \mid f_c)e^{i\omega_c t + ik_c y} \tag{20}$$

is the "most unstable" mode then for a harmonic part of any mode in the band we have

(21) $$e^{iky+\alpha(k)t} = e^{ik_cy+i\omega_ct}exp\{i(k-k_c)y+(\alpha(k)-i\omega_c)t\}$$

$$= e^{ik_cy+i\omega_ct}exp\left\{i(k-k_c)\left[y+t\frac{\partial\omega}{\partial k}\right]\right\}$$

$$\times exp\left\{\frac{1}{2}\left(\frac{\partial^2 t}{\partial k^2}+i\frac{\partial^2\omega}{\partial k^2}\right)(k-k_c)^2t\right\}e^{0(k-k_c)^3t},$$

all the derivatives are evaluated at critical values. Here we used the fact that $\frac{\partial\tau}{\partial k}\mid_{kcr}=0$, where $\alpha=\tau+i\omega$. Since $k-k_c\equiv\sigma\epsilon=O(\epsilon)$, the term $e^{i(k)y+\alpha(k)t}$ is a slow modulation of the critical mode (20) with the modulation factor of the form

(22) $$A(\epsilon t,\epsilon y,\epsilon^2 t).$$

It is also apparent from (21) that ϵt and ϵy should enter in the combination $\epsilon(t\partial\omega/\partial k+y)$. The observation in (21) give rise to the following ansatz for the solution. We look for a solution of the nonlinear problem with the form

(23) $$\psi = Dt+\epsilon A(Y,T)exp(ik_cy+\omega_ct))+\epsilon^2\psi^{(2)}+C.c.+...,$$

$$u=u^0(x\mid f)+\epsilon A(Y,T)e(x\mid f_c)e^{i(k_cy+\omega_ct}+\epsilon^2u^{(2)}+C.c.+...,$$

where $Y=\epsilon\left(y+t\frac{d\omega}{dk}\mid_{k=kc}\right), T=\epsilon^2 t$. Note that the basic flow is taken at the current value of D while the normal mode is taken at critical. The ansatz postulates that the spatial modulation of the state variables is identical to the modulation of the front.

The subsequent strategy is to insert the solution (23) into the equation, match like powers of ϵ and obtain an equation for the amplitude A as a Fredholm's solvability condition. Note that A is differentiated according to formulas

$$\frac{\partial A}{\partial y}=\epsilon\frac{\partial A}{\partial Y},\ \frac{\partial A}{\partial t}=\epsilon\frac{d\omega}{dk}\frac{\partial A}{\partial Y}+\epsilon^2\frac{\partial A}{\partial T}.$$

As a compatibility condition in the 3rd order of perturbation theory we obtain the following Landau-Ginzburg type equations:

$$\frac{\partial A}{\partial T}=\chi A+\frac{d}{c}\bar{A}A^2+\frac{\partial^2(\tau+i\omega)}{\partial k^2}\frac{\partial^2 A}{\partial Y^2}$$

where the constants χ and d/c are the same as for the one-dimensional case. It is well known that the Landau-Ginzburg equation describes an amazing variety of possible dynamical behaviors, from solitons to chaotic dynamics. In a work in progress with A. Majda we investigate different relevant detonation regimes.

ACKNOWLEDGEMENTS

The lectures this paper is based upon were prepared while the author was on sabbatical leave at Princeton University. The research was partially supported by the National Science Foundation Grant DMS 8911888 and through research grants of Professor A. Majda, ARODAAL03-89-K-0013 and ONR N00014-89-J-1044; the support of A. Majda is gratefully acknowledged. The author is thankful to Italian organizers and participants for their warm hospitality and enthusiasm. Special thanks are due to Marcello Onofri for his friendly help in a variety of matters (including lending a portable crib for author's son).

REFERENCES

1. Abouseif, G.E., and Toong, T.-Y., Theory of unstable one-dimensional detonations, Combust. Flame 45: 67-94 (1982).

2. Abouseif, G.E., and Toong, T.-Y., Theory of two-dimensional detonations: genesis of the transverse waves, Combust. Flame 63: 191-207 (1986).

3. Alpert, R.L., and Toong, T.-Y., Periodicity in exothermic hypersonic flows about blunt projectiles, Astronautica Acta 17, 539 (1972).

4. Barthel, H.O., and Strehlow, R.A., Wave propagation in one-dimensional reactive flows, Phys. Fluids 9, 1986-1907 (1966).

5. Bourlioux, A., Majda, A., and Roytburd, V., Theoretical and numerical structure for unstable one-dimensional detonations, SIAM J. Appl. Math. 51, 303-343, (1991).

6. Bourlioux, A., Majda, A., and Roytburd, V., Nonlinear theory of unstable one-dimensional detonations, to appear in I.M.A. volumes in Mathematics and Its Applications, Proceedings of the Workshop on Dynamic Issues in Combustion Theory, preprint (1990).

7. Buckmaster, J., Pressure transients and the genesis of transverse shocks in unstable detonations, Combust. Sci. and Tech. 61, 1-20 (1988).

8. Eckhaus, W., Studies in Nonlinear Stability Theory, Springer, Berlin (1965).

9. Erpenbeck, J.J., Theory of detonation stability, 12th Symposium on Combustion, 711-721, Pittsburgh: The Combustion Institute (1969).

10. Fickett, W., and Davis, W., Detonation, U. of California Press (1979).

11. Fickett, W., and Wood, W., Flow calculations for pulsating one-dimensional detonations, Phys. Fluids 9, 903-916, (1966).

12. Guirguis, R., Oran, E.S., and Kailasanath, K., Numerical simulations of the cellular structure of detonations in liquid nitromethane, Combust. and Flame 65, 339-366 (1986).

13. Lax, P.D., and Phillips, R.S., Scattering Theory, Academic Press, N.Y. (1967).

14. Lee, J.H.S., On the universal role of turbulence in the propagation of deflagrations and detonations, IMA Volumes in Mathematics and Its Applications, Computational Fluid Dynamics and Reacting Gas Flows, B. Engquist et al. Eds, 169-193, Springer, New York (1988).

15. Lee, H.I. and Stewart, D.S., Calculation of linear detonation instability. Part I: one-dimensional instability of plane detonation, J. Fluid Mech., 216, 103-132, (1990).

16. Lehr, H.F., Experiments in shock induced combustion, Astonaut. Acta 17, 589-597, (1972).

17. Majda, A.J., Criteria for regular spacing of reacting Mach stems, Proc. Nat. Acad. Sci. USA, 84, 6011-6014, (1987).

18. Majda, A., and Roytburd, V., Multidimensional instabilities of reacting shock waves, Preprint (1991).

19. Newell, A.C., and Whitehead, J.A., Review of the finite bandwidth concept, Proc. IUTAM Symp. Inst. Cont. Syst., H. Liepholz ed., 284-289, Springer, Berlin (1971).

20. Stuart, J.T., On the nonlinear mechanics of wave disturbances in stable and unstable parallel flows I, J. Fluid Mech. 9, 353-370, (1960).

21. Taki, S., and Fujiwara, T., AIAA J. 16, 73-77 (1978).

V. Roytburd
Mathematical Sciences Department
Rensselaer Polytechnic Institute
Troy, NY 12180-3590

C BRAUNER, J BUCKMASTER, J W DOLD AND C SCHMIDT-LAINE

On an evolution equation arising in detonation theory

1. INTRODUCTION

Detonation waves are, for the most part, unstable [1] and it is important to understand the origins and consequences of the instability. Since activation energy asymptotics is such a valuable tool in small Mach number combustion [2], it is natural to try it on the detonation problem, a high Mach number phenomenon, and there have been several attempts of this nature [3-7]. Some of this work will be reviewed here.

In a certain distinguished limit, described later, an evolution equation can be deduced for the shock displacement, specifically [7]

$$(1)\qquad \begin{aligned} g_t &= -\frac{1}{2}(g_y)^2 + ln[\frac{e^{cgg_y}-1}{cg_{yy}}],\ y\varepsilon[o,l] \\ g_y(0,t) &= g_y(l,t) = 0. \end{aligned}$$

This equation has been solved numerically [7] and solutions have been found that exhibit singular behavior after a finite time. The stability of the stationary solution $g \equiv 1$ is of interest and a linear analysis provides the qualitative framework. However, a higher order approximation of the *invariant manifolds*, performed in Section 4, validates the numerics. This describes the nonlinear effects at the saddle point. In the degenerate case $l = l_c = \pi\sqrt{\frac{c}{2}}$ a center-manifold analysis has to be performed.

The field equations that govern the structure of a detonation wave, and whose solutions we shall discuss, are :

$$(2)\qquad \begin{gathered} \frac{\partial\rho}{\partial t} + \vec{\nabla}.(\rho\vec{q}) = 0\ ,\ \rho\frac{D\vec{q}}{Dt} = -\vec{\nabla}p, \\ \rho\frac{DI}{Dt} - \frac{DP}{Dt} + Q\Omega\ ,\ \rho\frac{DY}{Dt} = -\Omega, \\ \Omega = BYe^{-E/RT}\ ,\ I = \frac{\gamma}{\gamma-1}\frac{p}{\rho},\ T = \frac{p}{\rho R}. \end{gathered}$$

We are concerned with solutions valid in the limit $\theta \equiv \frac{E}{RT_s} \to \infty$, where T_s is the post-shock temperature.

2. STRUCTURE AND STABILITY OF THE DETONATION WAVE

The steady detonation structure consists of a shock wave, followed by an induction zone in which reaction is weak, followed by a zone in which vigorous reaction and heat release occurs. Behind the reaction zone is uniform burnt gas. In the limit $\theta \to \infty$ this structure reduces to the well-known square-wave [3] in which conditions are uniform in the induction zone (of length L) and the reaction zone is a discontinuity. Perturbations in the induction zone are $0(1/\theta)$ and are characterized by a logarithmic singularity as the reaction zone or fire is approached, i.e., the singularity is located at $x = L$ and in this way the end of induction is precisely identified.

We can now pose the general question : Given the large θ structure how will perturbations to this structure evolve ? The answer, of course, depends on the nature of the perturbations and in our asymptotic context requires identification of appropriate distinguished limits. Of particular interest are long wavelength disturbances that evolve slowly.

Reference [3] is concerned with corrugations to the lead shock that have a characteristic length θL and are associated with disturbances that evolve on the time scale $\theta L/u_s$ (u_s is the post-shock velocity), i.e., the lead shock position

$$x = h(y,t)$$

can be written as

$$\frac{x}{L} = \widehat{h}\left(\frac{y}{\theta L}, \frac{t\,u_s}{\theta L}\right) \tag{3}$$

where $\widehat{h}$ is an $0(1)$ quantity. Then the perturbed induction zone is quasi-steady, quasi-one-dimensional, and its asymptotic description is straightforward.

The motion of the lead shock associated with the displacement (3) generates $0(1/\theta)$ perturbations behind the shock, defined by the Rankine-Hugoniot conditions. Of particular interest is T_{1_s}, the perturbation temperature, for this affects the length of the perturbed induction zone. Indeed this length is

$$L\, e^{-\theta T_{1_s}/T_s}. \tag{4}$$

These large variations in the wave thickness imply that there is significant transient storage of mass, momentum, and energy within the induction zone, and this affects the $0(1/\theta)$ perturbations behind the thin reaction zone. These perturbations can be expressed in terms of the shock displacement h by applying Rankine-Hugoniot conditions between the unshocked gas and the burnt gas, modified to allow for the transient storage. This description in inherently nonlinear, although the perturbations are small, because of the nonlinearity of the formula (4). For the purpose of a linear stability analysis, however, it can be linearized (i.e., (4) is replaced by $L(1-\theta T_{1_s}/T_s)$).

There remains a discussion of the burnt gas perturbations. These must be examined on the length scale θL and consist of two acoustic waves, an entropy wave, and a vorticity wave. For a single Fourier mode with dependence $e^{iky+\alpha t}$, the four perturbations p_1, ρ_1, u_1 and v_1 may then be expressed as a linear combination of four exponential functions, each corresponding to one of these waves. We now eliminate the upstream propagating acoustic signal, consistent with a late time description of perturbations with initial compact support. This may be called the *radiation* condition, analogous to the Sommerfield radiation condition [8]. The four perturbation functions are now represented by a linear combination of three functions, and so cannot be independent. They must satisfy a *compatibility* condition obtained by eliminating the three functions amongst the several equations, and this is [3]

$$\alpha u_1 - iku_b v_1 - \frac{p_1}{\rho_b c_b}\,[\alpha^2 + k^2(c_b^2 - u_b^2]^{1/2} = 0\,. \tag{5}$$

When the expressions for u_1, v_1 and p_1 determined from the structure are substituted into this formula it becomes the dispersion relation describing the eigenvalue α as a function of the wave number k.

Two limits are of interest. For one-dimensional disturbances ($k = 0$), equation (5) simplifies to

$$(p_1 - \rho_b c_b u_1) = 0\;; \tag{6}$$

and for small wavelengths ($k \to \infty$) we have (since v_1 is proportional to k whereas u_1 and p_1 are k independent)

$$\frac{v_1}{k} = 0. \tag{7}$$

Equation (6) yields the unstable root

$$\frac{L\theta\beta}{T_s}\alpha = \frac{(1-\mu_b^{-1})}{(\sigma_b^{-1}-1)}\cdot\frac{1+M_b-(\mu_b-1)(\gamma-1)M_b^2}{1+M_b+(1-\sigma_b)(\gamma-1)M_b^2}\,, \tag{8}$$

whereas equation (7) yields the larger root

$$\frac{L\theta\beta}{T_s}\alpha = \frac{(\mu_b-1)}{(1-\sigma_b)}\,, \tag{9}$$

where $\mu = \rho/\rho_f$, $\sigma = \rho/\rho_s$, M_b is the burnt-gas Mach number, γ is the specific heat ratio, and β is defined by the perturbation Rankine-Hugoniot conditions so that

$$T_{1_s} = -\beta\, h_t,\ \beta > 0. \tag{10}$$

α is a monotonically increasing function of k.

Reference [5] also is concerned with long wavelength slowly varying disturbances, but now the lead shock displacement is

$$\frac{x}{L} = \widehat{h}\left(\frac{y}{\sqrt{\theta}L}, \frac{t\,u_s}{\theta L}\right), \tag{11}$$

i.e., the corrugations have length $\backsim \sqrt{\theta}\,L$. This introduces two changes into the analysis.

First, the compatibility condition (5) is modified. Since the time scale is unchanged but the length scale is much smaller than the previous one, the short wavelength approximation is appropriate, namely

$$v_1 = 0. \tag{12}$$

Second, the induction zone analysis is modified by stream-tube divergence. The term $\frac{\partial v}{\partial y}$ in the continuity equation generates an $(0(1/\theta)$ perturbation in the induction zone, and this plays as important a role as the $0(1/\theta)$ temperature perturbations generated by the shock displacement. Now the induction zone length is

$$\frac{L}{\kappa}\; ln[1 + \kappa\; e^{-\theta T_{1_s}/T_s}] \tag{13}$$

where $\kappa\alpha\; \frac{\partial v}{\partial y} \;\alpha\; \frac{\partial^2 h}{\partial y^2}$ is a measure of the shock curvature. This result reduces to (4) in the limit $\kappa \to 0$.

When these modifications are incorporated into a linear analysis we find

$$\frac{L\theta\beta}{T_s}\alpha = \frac{(\mu_b - 1)}{(1-\sigma_b)} - Ck^2 \tag{14}$$

where C is a constant. This matches with the result (9) as $k \to 0$, monotonically decreases with increasing wave-number, and shows that sufficiently short waves are stable.

These linear analyses identify the importance of the two wavelength scales $\sqrt{\theta}\,L$ and $\theta\,L$. A third, $\theta^{2/3}\,L$, lying between these two, can be shown to encompass the wavelength for which α assumes its maximum value [5]. The *ratio* of these three scales $\theta^{1/2}L : \theta^{2/3}L : \theta L$ is $1 : \theta^{1/6} : \theta^{1/2}$ so that the *wave number* ratios are $1 : \theta^{-1/6} : \theta^{-1/2}$. Thus a graph of α vs. wave number, drawn on the scale of the largest wave numbers, is monotone decreasing with the exception of a thin boundary layer near the origin (Fig. 1).

3. THE NONLINEAR EVOLUTION

The analysis of [5] is inherently nonlinear, cf. eqn. (13), and yields an evolution equation for the shock displacement that can be written in the form [7]

$$\begin{gathered} g_t + \frac{1}{2}\,g_y^2 = ln[\frac{e^{cg_{yy}} - 1}{cg_{yy}}],\; y\varepsilon\; [0,l],\; g > 0 \\ g_y(0,t) = g_y(l,t) = 0 \end{gathered} \tag{15}$$

corresponding to propagation in a duct of finite width. c is a positive constant and the undisturbed wave corresponds to $g \equiv 1$.

Solutions of the linearized equation are

$$(g-1) \backsim \cos\left(\frac{n\pi y}{l}\right) e^{\alpha t}, \quad n \text{ an integer} \tag{16}$$

where

$$\alpha = 1 - \frac{c}{2} \cdot \frac{n^2\pi^2}{l^2} \tag{17}$$

corresponding to the result (14). For each mode characterized by n there is instability if

$$l > n\pi\sqrt{\frac{c}{2}} \tag{18}$$

and for channels for which

$$l < l_c = \pi\sqrt{\frac{c}{2}} \tag{19}$$

only the planar mode ($n = 0$) is unstable. However, the growth rate $\alpha = 1$ at $n = 0$ is incorrect (Fig. 1), and (15) needs to be modified for long wavelength disturbances. We have not yet considered such modifications but have, instead, suppressed the planar mode in numerical simulations by suitable choice of the initial data. The hope then is that if one or more nonplanar modes grow more rapidly than would the planar mode of a uniformly valid equation, this nonlinear growth will be adequately modeled by (15).

Solutions obtained in [7] reveal a typical behavior in which g vanishes at some point after a finite time has elapsed, and becomes nonregular. This corresponds to an unbounded perturbation flame speed and a corresponding unbounded perturbation shock pressure (pressure spot formation). It is suggested in [7] that this pressure spot might eventually give rise to weak transverse shock waves, a familiar feature of unstable detonations [1].

The remainder of our discussion is concerned with the stability characteristics of (15).

4. STABILITY ANALYSIS

Equation (15) falls within the framework of fully nonlinear parabolic equations as studied by Da Prato and Grisvard [9]. Here we will consider the stability of the steady stade $g \equiv 1$. The *bifurcated branches* are not tackled in this paper.

Classical results on stability for autonomous semilinear evolution equations (e.g., [10], [11]) have recently been generalized [12]. As in [9] these authors have to consider some specific interpolation spaces including the "little Hölder" spaces :

For $0 < p < 1$,

$$(20) \qquad h^p([0,l]) = \{v \in C^0([0,l]),\ \max_{|x-y|\le t} \frac{|u(x)-u(y)|}{t^p} \to 0 \text{ as } t \to 0\}.$$

However, we shall avoid technical discussion here.

We write (15) in the abstract form

$$(21) \qquad g_t = \mathcal{F}(g),$$

write

$$(22) \qquad g = 1 + v,$$

and, since $\mathcal{F}$ is a sufficiently smooth function, expand it up to 2nd order :

$$(23) \qquad \mathcal{F}(g) = Lv + \frac{1}{2}\, H(v,v) + \mathcal{R}(v).$$

Here L is the linear operator defined in the space $E = C^0([0,l])$ by

$$(24) \qquad \begin{cases} Lv = \dfrac{c}{2}\, v_{yy} + v \\ D(L) = \{v\ \varepsilon\ C^2([0,l]),\ v'(0) = v'(l) = 0\}. \end{cases}$$

As noted earlier the spectrum of L consists of the eigenvalues $\lambda_j = 1 - j\frac{c\pi^2}{2l^2}$ with corresponding eigenvectors $w_j = \cos\, j\frac{\pi y}{l}$, $j = 0, 1, \cdots$ There is at least one unstable mode (the planar mode) corresponding to $\lambda_0 = 1$. In the case $0 < l < l_c = \pi\sqrt{\frac{c}{2}}$, all the other eigenvalues are negative. The special case $l = l_c$, corresponding to $\lambda_1 = 0$ leads to a center manifold, discussed at the end of this paper. However, for the general case, we will assume that $\lambda_0, \lambda_1, \cdots \lambda_n$ are strictly positive while the rest of the eigenvalues $\lambda_{n+1}, \lambda_{n+2}, \cdots$ are strictly negative. Therefore, the spectrum of L is split in two parts $\sigma_s = \{\lambda_j\}_{j=n,n+1\cdots}$ and $\sigma_u = \{\lambda_0, \lambda_1 \cdots \lambda_n\}$, (in an obvious notation).

We denote by P the projection operator on the dimension $n+1$ subspace E^u generated by $\{1,\ w_1, \cdots, w_n\}$ and $Q = I - P$ the projection operator on the codimension $n+1$ eigenspace E^s. One has $E = E^s \oplus E^u$. For example if $n = 0$, $E^s = \{v\ \varepsilon\ E,\ \int_0^l v\ dy = 0\}$.

Clearly, this is a *saddle point* configuration. The existence in a neighborhood of the equilibrium point of a codimension $n+1$ stable manifold W^s and of a dimension $n+1$ unstable manifold W^u follows from [12], th.2.4. They may be (locally) characterized as the graph of some continuous functions (Fig. 2), and W^s (resp. W^u) is tangent to E^s (resp. E^u) at the origin (corresponding to the steady state $g \equiv 1$). We refer to [12] for the functional framework. In the special case $n = 0$, it is straightforward to check that $W^u = E^u$.

Besides the existence of the stable/unstable manifolds, we are interested in a numerical approximation of the stable manifold W^s involving the Hessien of $\mathcal{F}$.

We consider the equation (21) together with the initial condition ($\varepsilon > 0$ is small) :

$$g(0) = 1 + \varepsilon\ \xi^s + \varepsilon^2\ \xi^u \tag{25}$$

where $\xi^s\ \varepsilon\ W^s$ is given and $\xi^u(\varepsilon)$ is to be determined in order to get a vanishing perturbation of the steady state as $t \to +\infty$, i.e., the solution belongs to W^s.

From (23), the perturbation v satisfies

$$\begin{cases} v_t = Lv + \dfrac{\varepsilon}{2} H(v,v) + \varepsilon^2 R(\varepsilon; v) \\ v(0) = \xi^s + \varepsilon\ \xi^u \end{cases} \tag{26}$$

to which apply the projection operators P and Q. Denoting Qv by φ^s and Pv by $\varepsilon\gamma^u$, we get :

$$\begin{cases} \varphi_t^s = L_s\ \varphi^s + \dfrac{\varepsilon}{2}\ QH(v,v) + \varepsilon^2 QR(\varepsilon; v) \\ \varphi^s(0) = \xi^s \end{cases} \tag{27}$$

$$\begin{cases} \gamma_t^u = L_u \gamma^u + \dfrac{1}{2}\ PH(v,v) + \varepsilon PR(\varepsilon; v) \\ \gamma^u(0) = \xi^u. \end{cases} \tag{28}$$

In (27) and (28), L_s and L_u stand for the restriction of L to E^s and E^u, and $v = \varphi^s + \varepsilon\gamma^u$.

The solution γ^u of (28) may be expressed as

$$\gamma^u(t) = e^{L_u t} \left(\xi^u + \int_0^t e^{-L_u \sigma} \{ \frac{1}{2}\ PH(v,v) + \varepsilon PR(\varepsilon; v)\} d\sigma \right). \tag{29}$$

As we want to follow an orbit entering the equilibrium point, $\gamma^u(t) \to 0$, as $t \to 0$; therefore

$$\xi^u = - \int_0^{+\infty} e^{-L_u \sigma} \{ \frac{1}{2}\ PH(v,v) + \varepsilon PR(\varepsilon; v)\} d\sigma. \tag{30}$$

Since we are concerned with the numerical approximation of the stable manifold, we will take $\varepsilon = 0$ in the formulas (27)-(30). The validity of this approximation may be proved by means of the implicit function theorem, the demonstration being deferred to a forthcoming article.

We adopt the ansatz

$$\begin{cases} \varphi^s = \varphi_0^s + \varepsilon \varphi_1^s + \cdots \\ \gamma^u = \gamma_0^u + \varepsilon\ \gamma_1^u + \cdots \\ \xi^u = \xi_0^u + \varepsilon\ \xi_1^u + \cdots, \end{cases} \tag{31}$$

and it turns out that φ_0^s, γ_0^u and ξ_0^u satisfy the *decoupled* system :

$$(32)\qquad \begin{cases} \varphi_{0t}^s = L_s\varphi_0^s, \ \varphi_0^s(0) = \xi^s \\ \gamma_{0t}^u = L_u\gamma_0^u + \frac{1}{2}\, PH(\varphi_0^s, \varphi_0^s) \\ \xi_0^u = -\frac{1}{2} \int\limits_0^{+\infty} e^{-L_u\sigma}\, PH(\varphi_0^s, \varphi_0^s) d\sigma. \end{cases}$$

It is convenient to expand $\xi^s = \sum\limits_{j=n+1}^{+\infty} \alpha_j^s w_j$, $\xi_0^u = \sum\limits_{j=0}^{n} \beta_j^u w_j$, so that

$$(33)\qquad \varphi_0^s(t) = \sum_{j=n+1}^{\infty} \alpha_j^s w_j e^{\lambda_j t}.$$

Denoting by P_j the projection onto the eigenspace spanned by w_j, we get explicity :

$$(34)\qquad \beta_j^u = -\frac{1}{2} \int\limits_0^{+\infty} e^{-\lambda_j t} P_j H(\varphi_0^s, \varphi_0^s)\, dt,\ j = 0, 1, \cdots n\,.$$

Since $P_j H(\varphi_0^s, \varphi_0^s) = \sum\limits_{k=n+1}^{\infty} \sum\limits_{k'=n+1}^{\infty} \alpha_k^s \alpha_k^s\, e^{(\lambda_k + \lambda_{k'})t} P_j H(w_k, w_{k'})$,

$$(35)\qquad \beta_j^u = -\frac{1}{2} \sum_{k=n+1}^{\infty} \sum_{k'=n+1}^{\infty} \int\limits_0^{+\infty} e^{(\lambda_k + \lambda_{k'} - \lambda_j)t}\, \alpha_k^s \alpha_{k'}^s\, P_j H(w_k, w_{k'}) dt$$

there follows

$$(36)\qquad \beta_j^u = \frac{1}{2} \sum_{k=n+1}^{\infty} \sum_{k'=n+1}^{\infty} \frac{1}{\lambda_k + \lambda_{k'} - \lambda_j}\, \alpha_k^s \alpha_{k'}^s\, P_j H(u_k, w_{k'}).$$

For example, since $P_0 f = \frac{1}{l} \int\limits_0^l f dy$,

$$(37)\qquad \beta_j^u = \frac{1}{2} \sum_{k=n+1}^{+\infty} \sum_{k'=n+1}^{+\infty} \frac{1}{\lambda_k + \lambda_{k'} - 1}\, \alpha_k^s \alpha_{k'}^s\, \frac{1}{l} \int\limits_0^l H(w_z, w_{k'}) dy.$$

We finish this section with some numerical computations in the special case : $0 < l < l_c$, namely $n = 0$.

For the sake of simplicity we take $\alpha_1^s = 1$, $\alpha_k^s = 0$, $k > 1$. Formula (37) becomes

$$\beta_j^u = \frac{1}{2l(2\lambda_1 - 1)} \int_0^l H(w_1, w_1) dy \tag{38}$$

and since

$$H(w_1, w_1) = -(w_{1y}^2 - (w_1)^2 + c w_1 w_{1yy} + \frac{c^2}{12} (w_{1yy})^2, \tag{39}$$

it turns out that

$$\frac{1}{2} \int_0^l H(w_1, w_1) dy = - \int_0^l \frac{\pi^2}{2l^2} \sin^2 \frac{\pi y}{l} dy$$

$$+ \int_0^l \cos^2 \frac{\pi y}{l} \left(-\frac{1}{2} - \frac{c\pi^2}{2l^2} + \frac{\pi^4 c^2}{24\, l^4} \right) dy = -\frac{\pi^2}{4\, l} - \frac{l}{4} - \frac{c\pi^2}{4\, l} + \frac{\pi^4 c^2}{48\, l^3}.$$

Finally

$$\beta_0^u = \frac{1}{l^2 - c\pi^2} \left(\frac{\pi^2}{4}(1+c) + \frac{l^2}{4} - \frac{\pi^4 c^2}{48\, l^2} \right). \tag{40}$$

Introducing the parameter $a = l/l_c$, $0 < a < 1$,

$$\beta_0^u = \frac{1}{2 - a^2} \left(-\frac{1}{12a^2} + \frac{1+c}{2c} + \frac{a^2}{4} \right). \tag{41}$$

This asymptotic value has been compared to numerical values obtained for several ε, for $c = 0.268$

a	$\varepsilon = 0.13$	$\varepsilon = 0.03$	$\varepsilon = 0.01$	β_0^u (Eqn. 41)
0.9	1.942	2.058	2.070	2.0717
0.8	1.686	1.754	1.761	1.7614
0.6	1.324	1.353	1.356	1.3562
0.4	1.010	1.023	1.024	1.0244

5. THE CRITICAL CASE $l = l_c$

In the special case $l = l_c$, i.e. $a = 1$, numerical computation is still in good agreement with the theoretical value of β_0^u calculated via (41), namely :

$$\varepsilon = 0.01,\ \widetilde{\beta}_0^u = 2.529,\ \beta_{0_{th}}^u = 2.532.$$

However in this case $\lambda_1 = 0$, and a center manifold analysis has to be performed in order to describe the dynamics.

We consider again the expansion (23), now up to the 3rd order :

$$v_t = Lv + G(v) \tag{42}$$

with

$$G(v) = \frac{1}{2}\ H(v,v) + \frac{1}{6}\ N(v,v,v) + \widetilde{\mathcal{R}}(v), \tag{43}$$

where L and H are respectively the linear operator defined by (24) and H the Hessien defined by (39).

The spectrum of L is here split into three parts : $\sigma_u = \{\lambda_0 = 1\}$, $\sigma_c = \{\lambda_1 = 0\}$, $\sigma_s = \{\lambda_2, \cdots, \lambda_n, \cdots\}$ corresponding respectively to the unstable eigenspace E^u generated by $\{1\}$, the center eigenspace E^c generated by $\{w_1 = \cos \frac{\pi y}{l_c}\}$ and the codimension 2 stable eigenspace E^s generated by $\{w_2, \cdots, w_n, \cdots\}$. We still denote by Q the projection operator onto E^s, while P is now split into P_u (onto E^u) and P_c (onto E^c). Denoting $P_u v$ by γ^u, $P_c v$ by z^c and Qv by φ^s, there follows (since $P_c L = 0$) :

$$\begin{cases} \varphi_t^s = L_s \varphi^s + Q\ G(v) \\ \gamma_t^u = \gamma^u + P_u\ G(v) \\ z_t^c = P_c\ G(v) \end{cases} \tag{44}$$

where $v = \varphi^s + \gamma^u + z^c$.

The existence of a center-unstable manifold $W^{c,u}$ which is *locally* the graph of a mapping h from $E^c \times E^u$ into E^s, $h(0,0) = h_1'(0,0) = h_2'(0,0) = 0$ follows from [12], th.3.1 and 3.2,; and therefore the question of stability of (44) is reduced to a 2-dimensional problem :

$$\begin{cases} z_t^c = P_c\ G(z^c + \gamma^u + h(z^c, \gamma^u)) \\ \gamma_t^u = \gamma^u + P^u\ G(z^c + \gamma^u + h(z^c, \gamma^u)). \end{cases} \tag{45}$$

Next, we consider a center-manifold $\gamma = \tau(z)$ for (45), tangent to E^c at the origin, defined by [13] :

$$\gamma^u|_{t=0} = -\int_0^{+\infty} e^{-\sigma}\ P_u G(z^c + \tau(z^c) + h(z^c,\ \tau(z^c)))\ d\sigma. \tag{46}$$

Finally wa are led to study the stability of the scalar problem

$$z_t^c = P_c\, G(z^c + \tau(z^c) + h(z^c, \tau(z^c))) \tag{47}$$

where $z = sw_1$. This equation describes the dynamics on a center manifold Σ of (44) for small values of s. For this matter we develop $\tau(sw_1)$ and $h(sw_1, \tau(sw_1))$ about the origin. The first order derivatives being zero, it turns out that :

$$\begin{aligned} \tau(sw_1) &= \frac{s^2}{2}\tau''(0)(w_1, w_1) + O(s^3) \\ h(sw_1, \tau(sw_1)) &= \frac{s^2}{2}\ h_1''(0,0) + 0(s^3). \end{aligned} \tag{48}$$

Then

$$\begin{aligned} G(z+\tau(z) + h(z, \tau(z))) &= \frac{s^2}{2}H(w_1, w_1) + \frac{s^3}{2}\ H(w_1, h_1''(0,0)(w_1, w_1)) \\ &+ \frac{s^3}{2}\ H(w_1, \tau''(0)\ (w_1, w_1)) + \frac{s^3}{6}\ N(w_1, w_1, w_1) + o(s^3). \end{aligned} \tag{49}$$

From (47) and (49) there follows :

$$\begin{aligned} s_t\, w_1 &= \frac{s^3}{2}\ P_cH(w_1, h_1''(0,0)(w_1, w_1)) + \frac{s^3}{2}\ P_cH(w_1, \tau''(0)\ (w_1, w_1)) \\ &+ \frac{s^3}{6}\ P_cN(w_1, w_1, w_1) + o(s^3) \end{aligned} \tag{50}$$

since $P_cH(w_1, w_1) = 0$, as $\int_0^{l_c} H(w_1, w_1).w_1 dy = 0$.

From (46) and (48), one has

$$\tau''(0)\ (w_1, w_1) = -\int_0^{+\infty} e^{-\sigma} P_u H(w_1, w_1) d\sigma = P_u H(w_1, w_1). \tag{51}$$

Similarly, according to the definition of the mapping h, [12] formula (3.8) :

$$h_1''(0,0)(w_1, w_1) = \int_{-\infty}^{0} e^{-\sigma L_s} QH(w_1, w_1)\ d\sigma. \tag{52}$$

Finally, (50) may be rewritten as :

$$\langle w_1, w_1 \rangle . s_t = As^3 + o(s^3) \tag{53}$$

where

$$A = \frac{1}{6}\langle N(w_1,w_1,w_1),w_1\rangle + \frac{1}{2}\langle H(w_1, \int_{-\infty}^{0} e^{-\sigma L_s} QH(w_1,w_1)d\sigma), w_1\rangle \tag{54}$$
$$+ \frac{1}{2}\langle H(w_1, P_u\, H(w_1,w_1)), w_1\rangle$$

where the brackets stand for the scalar product in $L^2(0,l_c)$. We write $A = A_1 + A_2 + A_3$.

Clearly the sign of A determines the stability on Σ : if $A < 0$ (resp. $A > 0$), the origin is stable (resp. unstable). From (39) it follows

$$\frac{1}{2}\, H(w_1,w_1) = (-\frac{1}{2c} - \frac{2}{3}) + (\frac{1}{2c} - \frac{2}{3})w_2, \tag{55}$$

$w_2 = \cos\frac{2\pi y}{l_c}$ being the eigenvector associated with the eigenvalue $\lambda_2 = -1$. Therefore, since $e^{-\sigma L_s} QH(w_1,w_1) = e^{-\sigma}\left(\frac{1}{2c} - \frac{2}{3}\right) w_2$, $A_2 = -\left(\frac{1}{2c} - \frac{2}{3}\right)\frac{l_c}{c}$.
Again with (55),

$$A_3 = \left(-\frac{1}{2c} - \frac{2}{3}\right)\int_0^{l_c} H(w_1,w_0)w_1\, dy = \left(\frac{1}{2c} + \frac{2}{3}\right) l_c.$$

Finally since $N(w_1,w_1,w_1) = \frac{c^2}{2} w_{1yy}^2 w_1 + 2w_1^3$, we have

$$A_1 = \int_0^{l_c} \left(\frac{c^2}{12}\, w_{1yy}^2\, w_1^2 + \frac{w_1^4}{3}\right) dy = \frac{l_c}{4}.$$

Therefore

$$A = l_c\left(-\frac{1}{2c^2} + \frac{7}{6c} + \frac{11}{12}\right). \tag{56}$$

For $c = 0.268$, $A = -1.69\ l_c$ is negative. The exchange of stability occurs for $c^* > 0$ corresponding to the positive root of

$$11c^2 + 14c - 6 = 0$$

namely $c^* \simeq 0.338$.

Finally, the stability of the origin (on the center-unstable manifold) is completely determined by the stability of the origin on Σ :

- if $0 < c < c^*$, the origin $v = 0$ (i.e. the steady state $g = 1$) is stable w.r.t. the unstable-manifold $W^{c,u}$.

- if $c^* < c$, it is unstable.

REMARK : The above formulae may be extended to the general center manifold case, namely the jth-eigenvalue λ_j becoming 0 for $l = \sqrt{j}\ l_c$.

ACKNOWLEDGMENT

Work partly supported by the Air Force Office of Scientific Research, and a NATO travel grant.

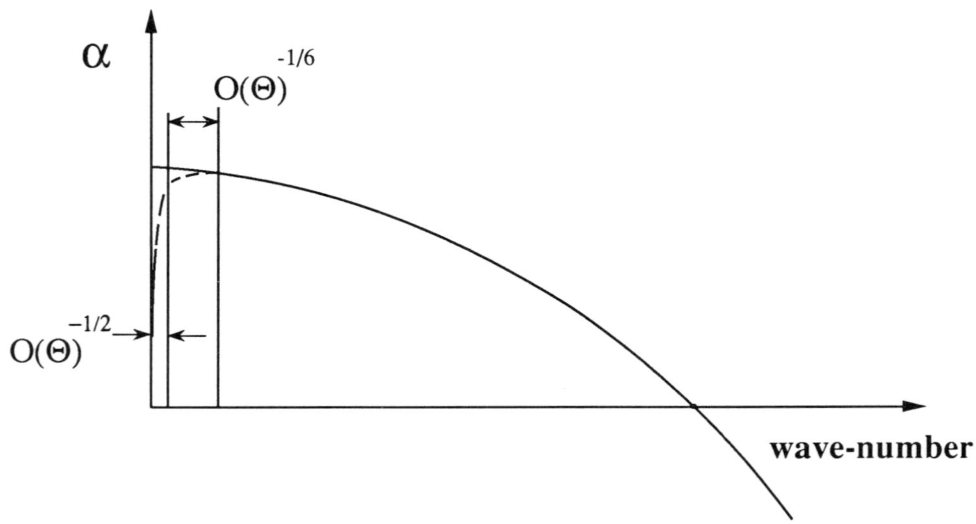

Fig. 1. Growth rate vs. wave-number. Within the boundary layer the dotted line is correct, the solid line is generated by the linearized form of eqn. (15).

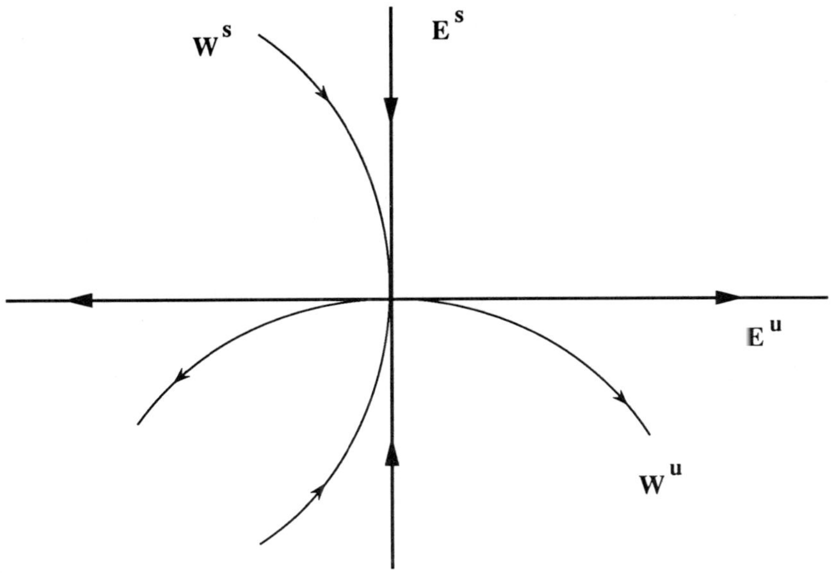

Fig. 2. Saddle-Point Configuration.

REFERENCES

[1] W. FICKETT and W.C. DAVIS, Detonation, University of California Press, 1976.

[2] J. BUCKMASTER and G.S.S. LUDFORD, Theory of Laminar Flames, Cambridge University Press, 1982.

[3] J. BUCKMASTER and G.S.S. LUDFORD, Twenty-First Symposium (International) on Combustion, The Combustion Institute, 1986, pp. 1669-1676.

[4] J. BUCKMASTER, Combust. Sci. and Tech., 61 (1988), pp. 1-20.

[5] J. BUCKMASTER, Combustion and Flame, 77 (1989), pp. 219-228.

[6] J. BUCKMASTER and J. NEVES, Phys. Fluids, 31 (1988), pp. 3571-3576.

[7] Cl. SCHMIDT-LAINÉ, J.W. DOLD, and J. BUCKMASTER, In Numerical Combustion, Lect. Notes in Physics 351, Springer, 1989, pp. 428-439.

[8] G. GARRIER, M. KROOK, and C. PEARSON, Functions of a complex Variable, McGraw-Hill, 1966.

[9] G. DA PRATO and P. GRISVARD, Ann. Mat. Pura Appli., 120 (1979), pp. 329-396.

[10] D. HENRY, Geometric Theory of Semilinear Parabolic Equations, Lectures Notes in Mathematics 840, Springer, 1981.

[11] J. CARR, Appl. Math. Sci., 35, Springer.

[12] G. DA PRATO and A. LUNARDI, Arch. Rat. Mech. Anal., 101 (1988), pp. 115-141.

[13] A. KELLEY, J. Differential Equations, 3 (1967), pp. 546-570.

C.-M. BRAUNER
Centre de Recherche en Mathématiques de Bordeaux
Université Bordeaux I
351, Cours de la Libération
33405 Talence Cedex
France

J. D. BUCKMASTER
Department of Aeronautical
and Astronautical Sciences
University of Illinois
104 S. Mathews Av.
Urbana, IL 61801 - USA

J. W. DOLD
School of Mathematics University of Bristol
Bristol BS8 1TW
U. K.

Cl. SCHMIDT-LAINÉ
Laboratoire de Mathématiques, CNRS UMR 128,
Ecole Normale Supérieure de Lyon
46, Allée d'Italie
69364 Lyon Cedex 07
France

4 Chemical kinetics

J WARNATZ

Chemical kinetics problems in combustion

1. INTRODUCTION

Although hydrocarbon combustion at present is one of the most important technical processes (more than 80% of energy use is basing on fossile fuel combustion), there is a lack of information in the literature on the reaction mechanisms involved in this process. However, the detailed knowledge of combustion mechanisms is important, for example [1-4]:

- for the control of processes like pollutant formation (e.g. nitrogen oxides, unburnt hydrocarbons, soot)
- for the understanding of induced ignition and auto-ignition or engine knock processes, and
- for extrapolation towards important but experimentally not accessible conditions.

Since combustion processes are the result from a complex interaction of chemical reaction, transport, and convection, simulation needs much computational efforts and has to be restricted to relatively simple cases. Simple flame fronts can then be described in terms of time-dependent or of stationary one-dimensional partial differential equation systems with temperature, species concentrations, and flow velocity used as the dependent variables.

However, even in the case of combustion of simple hydrocarbons, about 50 species involved in 500 reactions have to be considered, resulting in very large non-linear stiff partial differential equation systems [5,6]. For ignition problems presented here, the systems consist of up to 1500 reactions of 150 species.

By suitable separation and elimination of unimportant reactions, a mechanism can be developed with the aid of the present kinetic data for the elementary reactions involved [5,7]. This mechanism explains at high temperature ($T > 1200$ K) the currently available experimental results for laminar premixed flames of alkanes, alkenes, and acetylene (flame velocity and structure of free flames, concentration and temperature profiles in burner-stabilized flames [5,8-14]) and at low temperature ($T<1200$ K) experiments on induced and auto-ignition (ignition limits, minimum ignition energies, shock tube induction time determinations, engine knock measurements [3,4,6,15-17]).

These experimental results are simulated by the solution of the corresponding conservation equations with suitable models describing transport in the multicomponent mixture considered and including a full set of elementary reactions to describe chemistry.

2. COLLECTION OF RATE DATA, DEVELOPMENT OF REACTION MECHANISMS

2.1 Reaction Mechanism

From literature work it is known that the similarity of hydrocarbon flames (similar flame velocities, similar normalized composition profiles of reactants and main products [5,9,10]) is mainly due to the predominant influence of the reactions

$$H + O_2 \rightarrow OH + O \qquad \text{and} \qquad CO + OH \rightarrow CO_2 + H$$

which are non-specific for the hydrocarbon considered. Furthermore, it can be shown [18] that flame velocities of alkane flames can be modelled within a factor of two using the pure H_2–O_2–CO mechanisms (see Tab. 1) and only the initial attack of H, O, and OH on the hydrocarbon.

						A	b	E	No.
1. H_2-O_2 Chain Reactions									
O_2	+ H		→	OH	+ O	$2.00 \cdot 10^{14}$	0.00	70.3	(1,2)
H_2	+ O		→	OH	+ H	$5.06 \cdot 10^{4}$	2.67	26.3	(3,4)
H_2	+ OH		→	H_2O	+ H	$1.00 \cdot 10^{8}$	1.60	13.8	(5,6)
OH	+ OH		→	H_2O	+ O	$1.50 \cdot 10^{9}$	1.14	0.4	(7,8)
2. Recombination Reactions									
H	+ H	+ M*	→	H_2	+ M*	$9.70 \cdot 10^{16}$	−0.60	0.0	(9,10)
H	+ OH	+ M*	→	H_2O	+ M*	$2.20 \cdot 10^{22}$	−2.00	0.0	(11,12)
O	+ O	+ M*	→	O_2	+ M*	$2.90 \cdot 10^{17}$	−1.00	0.0	(13,14)
3. HO_2 Formation/Consumption									
H	+ O_2	+ M*	→	HO_2	+ M*	$2.30 \cdot 10^{18}$	−0.80	0.0	(15,16)
HO_2	+ H		→	OH	+ OH	$1.70 \cdot 10^{14}$	0.00	3.7	(17,18)
HO_2	+ H		→	H_2	+ O_2	$4.30 \cdot 10^{13}$	0.00	5.9	(19,20)
HO_2	+ H		→	H_2O	+ O	$3.00 \cdot 10^{13}$	0.00	7.2	(21,22)
HO_2	+ O		→	OH	+ O_2	$3.20 \cdot 10^{13}$	0.00	0.0	(23,24)
HO_2	+ OH		→	H_2O	+ O_2	$6.00 \cdot 10^{13}$	0.00	0.0	(25,26)
4. H_2O_2 Formation/Consumption									
HO_2	+ HO_2		→	H_2O_2	+ O_2	$1.00 \cdot 10^{12}$	0.00	0.0	(27)
OH	+ OH	+ M*	→	H_2O_2	+ M*	$3.25 \cdot 10^{22}$	−2.00	0.0	(28,29)
H_2O_2	+ H		→	H_2	+ HO_2	$1.70 \cdot 10^{12}$	0.00	15.7	(30,31)
H_2O_2	+ H		→	H_2O	+ OH	$1.00 \cdot 10^{13}$	0.00	15.0	(32,33)
H_2O_2	+ O		→	OH	+ HO_2	$6.80 \cdot 10^{11}$	0.00	16.6	(34,35)
H_2O_2	+ OH		→	H_2O	+ HO_2	$1.80 \cdot 10^{12}$	0.00	1.4	(36,37)

A(cm,mol,s), E(kJ/mol), $k = A \cdot T^b \cdot \exp(-E/RT)$
$[M^*] = [H_2] + 6.5\,[H_2O] + 0.35\,[O_2] + 0.40\,[N_2]$

Tab. 1: Example of reaction mechanism: hydrogen-oxygen system

Nevertheless, a detailed reaction mechanism is necessary for realistic calculation of the structure of hydrocarbon flames, particularly of rich flames, pollutant formation, and ignition processes [1-5]: In lean and moderately rich flames the hydrocarbon is attacked by O, H, and OH, in the first step. These radicals are produced by the chain-branching steps of the oxyhydrogen reaction.

The alkyl radicals formed in this way decompose to smaller alkyl radicals by fast thermal elimination of alkenes (see Fig. 1a). Only the relatively slow thermal decomposition of the smallest alkyl radicals (CH_3 and C_2H_5) competes with recombination and with oxidation reaction by O atoms and O_2 which are important processes in ignition. This part of the mechanism (Fig. 1b) is rate-controlling in the combustion of alkanes and alkenes, and is therefore the reason for the similarity of all alkane and alkene flames.

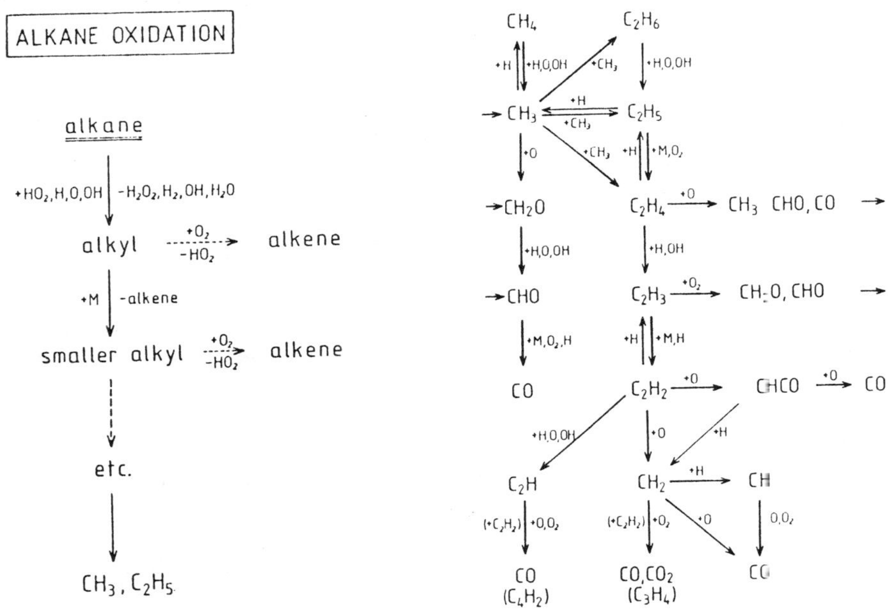

Fig. 1: Scheme for decomposition of primarily formed alkyl radicals of higher hydrocarbons (left picture, Fig. 1a) and detailed mechanism for reactions of the CH_3 and C_2H_5 radicals (right picture, Fig. 1b)

The reaction mechanism is based on a critical review on ≈ 200 elementary reactions (including several hundreds of literature references) which may occur in hydrocarbon combustion [7]. In eliminating unimportant reactions, the following two criteria for the insignificance of a reaction have been used [5]:

(1) Throughout the whole flame front thickness, the (molar scale) chemical formation or consumption rates of all species participating in this reaction must be negligible (less than e.g. 1 %) in comparison to the maximum of the corresponding rates of the other reactions.

(2) Furthermore, a reaction is assumed to be unimportant in any case, if the reaction term r in the mass conservation (see below) is small in comparison to the flow term.

It seems possible to simplify the complicated mechanism by proper use of partial equilibrium and by steady state assumptions [11,18,19]. This simplification and the correct use of the simplified mechanisms presuppose a detailed knowledge of the mechanism demonstrated here.

2.2 Sensitivity Analysis

Sensitivity analysis is a powerful method to identify the critical rate-determining steps in a reaction mechanism which should be object of further experimental consideration and are the key reactions to develop working reaction mechanisms for application in simulations of reactive flow. The method shall be described briefly and some illustrative examples be presented.

To perform local sensitivity analysis, the conservation equations given above are written for stationarity in the short form [6,20]

$$\frac{\partial \Phi}{\partial t} = F(\Phi) = O \quad \text{for stationarity}$$

with $\Phi = (w_{11}, \ldots, w_{s1}, T_1, v_1; w_{12}, \ldots, w_{s2}, T_2, v_2; \ldots; w_{1L}, \ldots, w_{SL}, T_L, v_L)$. To guarantee stationarity and proper performance of the sensitivity test, first a Newton iteration is carried out:

$$\frac{\partial f(\Phi)}{\partial \Phi}(\Phi_{n+1} - \Phi_n) = -F(\Phi_n) \quad ; \quad n = \text{ iteration parameter}$$

The equation $F(\Phi) = 0$ is written again, taking into account the explicit and implicit dependence of Φ in$F(\Phi(k), k) = 0$ from parameters $k = (k_1, ..., k_R)$, e.g. from preexponential factors of rate coefficients of R reactions. Differentiation gives the linear equation system

$$\frac{\partial f}{\partial \Phi}\frac{\partial \Phi}{\partial k} + \frac{\partial F}{\partial k} = 0$$

The Jacobian matrix $\partial F/\partial \Phi$ is known from the Newton iteration. The derivatives of the residuum F with respect to the parameters k can be determined numerically. Solution of this linear equation system leads to the sensitivities $\partial \Phi/\partial k$ or the relative ones $\partial(\ln \Phi)/\partial(\ln k)F$. These sensitivities are local parameters and dependent on the grid point number $l = 1, \ldots, L$. For this reason, maximum sensitivities are given in the plot presented.

Sensitivity tests with respect to the flame velocity of CH_4-air and C_3H_8-air flames at atmospheric pressure are given in Fig. 2 to demonstrate the benefits of this method.

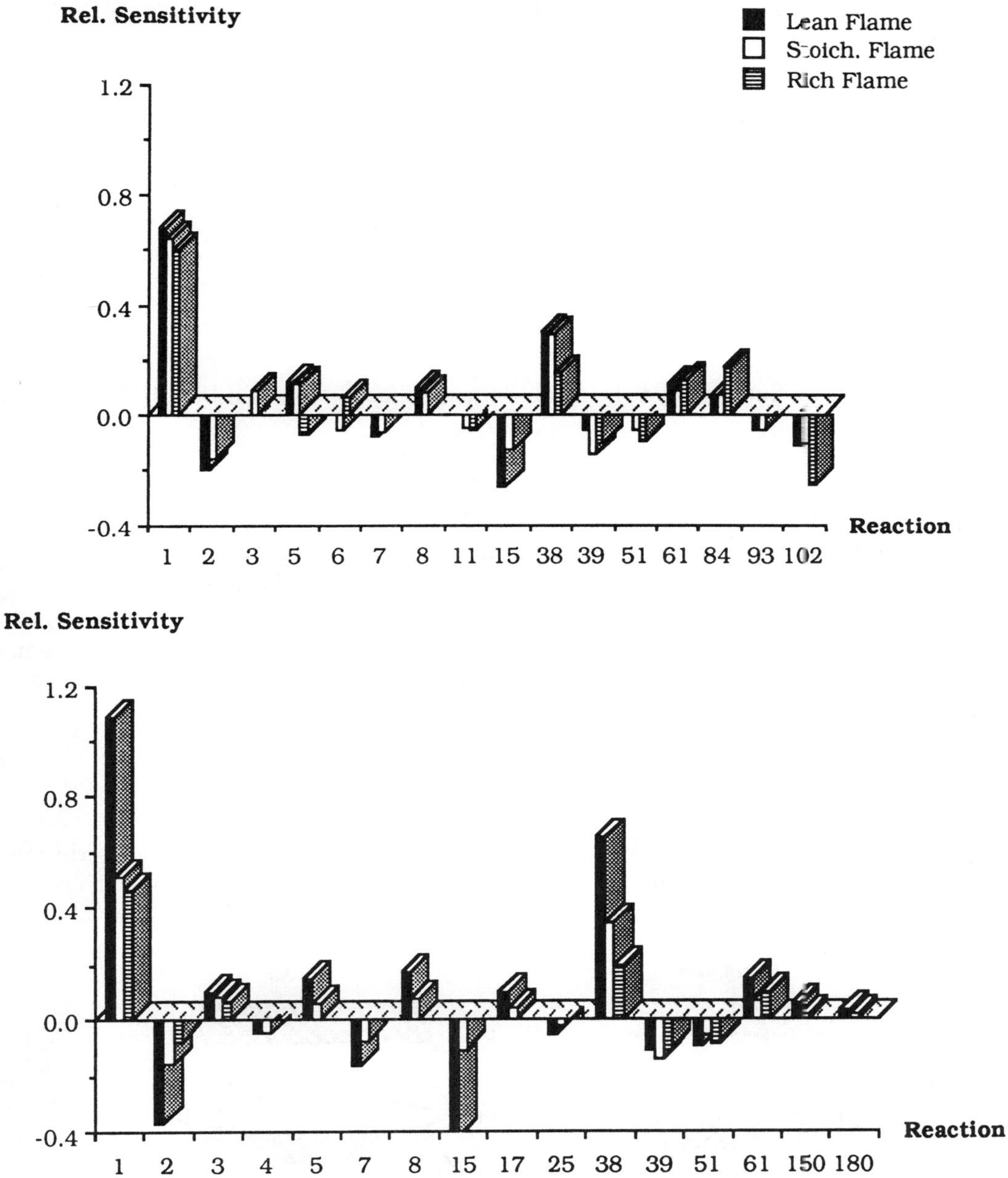

Fig. 2: Sensitivity tests in methane-air (upper drawing) and propane-air (lower drawing) flames at P = 1 bar, T_u = 298 K [20]

3. APPLICATION TO FLAME PROPAGATION AND FLAME STRUCTURE

3.1 Conservation Equations and Solution

The conservation equations for a laminar premixed one-dimensional flat flame are given by [5,21]

$$\rho\frac{\partial T}{\partial t} = -\rho v\frac{\partial T}{\partial z} - j_q\frac{\partial T}{\partial z} + \frac{1}{c_p}\frac{\partial}{\partial z}\left(\lambda\frac{\partial T}{\partial z}\right) - \frac{\sum r_i h_i}{c_p}$$

$$\rho\frac{\partial w_i}{\partial t} = -\rho v\frac{\partial w_i}{\partial z} - \frac{\partial j_i}{\partial z} + r_i$$

where the diffusion fluxes j_i and the mean diffusion flux j_q are given by [21-25]

$$j_q = \frac{\sum c_{p,i} j_i}{c_p} \quad ; \quad j_i = -D_i^M \rho\frac{\partial w_i}{\partial z} - D_i^T\frac{1}{T}\frac{\partial T}{\partial z}$$

and c_p = specific heat capacity, D_i^M = diffusion coefficient of species i into the mixture, D_i^T = thermal diffusion coefficient, h = enthalpy, λ = mixture heat conductivity, M = molar mass, ρ = mass density, r_i = (mass scale) rate of chemical reaction, T = absolute temperature, t = time, v = (center of mass) flow velocity, w_i = mass fraction, z = cartesian space co-ordinate, the index i denotes a single species in the mixture, and

$$D_i^M = \frac{1 - w_i}{\sum_{i\neq j} x_j / D_{ij}} \qquad \lambda = \frac{1}{2}\left[\sum_i x_i\lambda_i + \left(\sum\frac{x_i}{\lambda_i}\right)^{-1}\right]$$

Similar equations can be written down for different geometries (see Fig. 3), e.g. description of non-premixed laminar systems [13,26,27] and for strained flames [27,29,30].

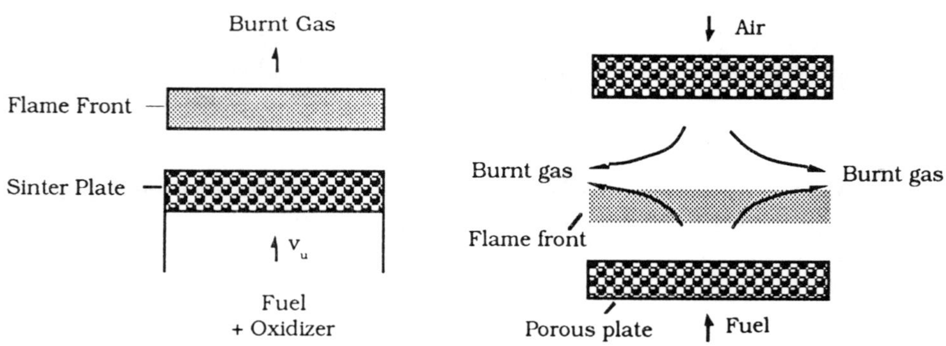

Fig. 3: Scheme of a laminar premixed one-dimensional flame (left picture, Fig. 1a) and of a laminar counterflow diffusion flame (right picture, Fig. 1b)

3.2 Results

Some representative results are presented in Fig. 4, showing a comparison of simulated concentration profiles with experimental ones [28] for reactants, intermediate species, and products.

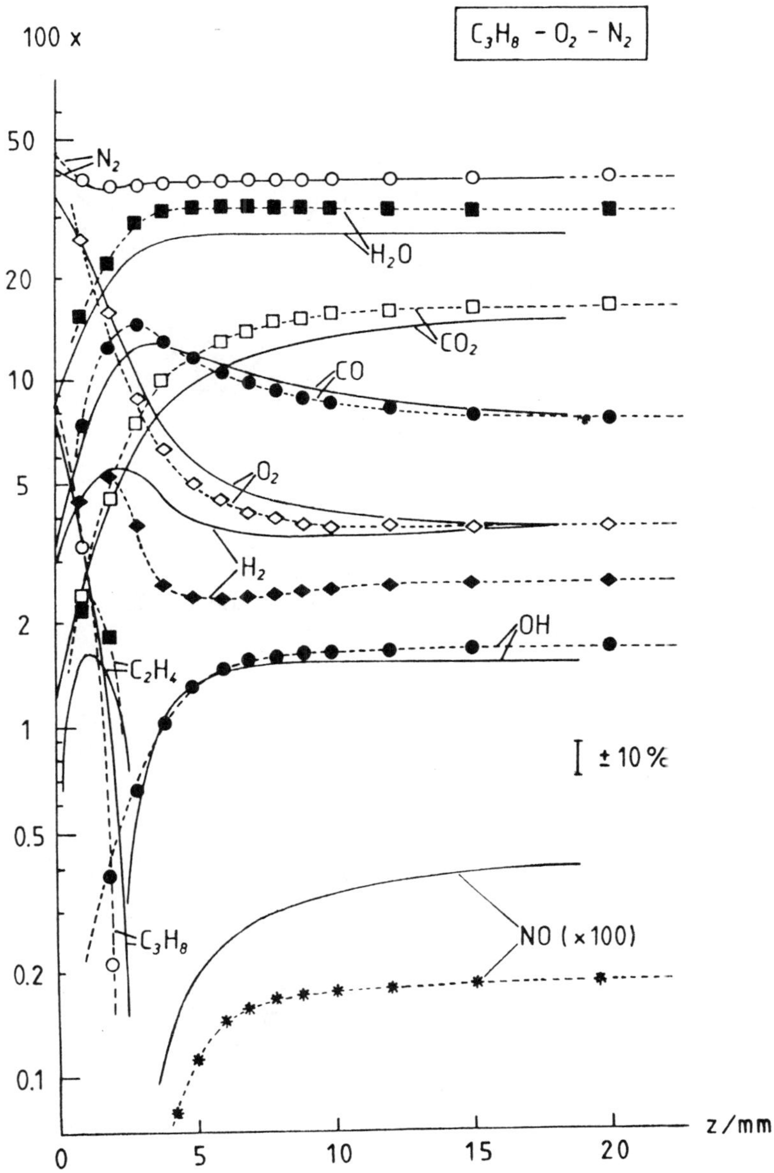

Fig. 4: Simulated and measured profiles of concentrations for a laminar premixed propane-air flame [27]

Furthermore, Fig. 5 shows a comparison of temperature profiles in a diffusion flame measured by non-intrusive coherent anti-Stokes Raman scattering (CARS) and simulated by the method given above [27]. Finally, Fig. 6 shows the strain rate aq at quenching for CH_4-air flames at $P = 1$ bar, $T_u = 298$ K as function of the equivalence ratio F, derived from a library of premixed flames at different F. Given is a comparison of measurements and calculations (see [29,30]).

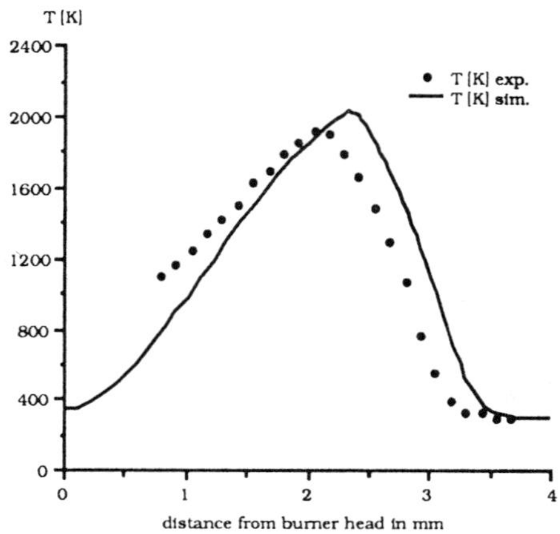

Fig. 5: Temperature profiles in a diffusion flame measured by anti-Stokes Raman scattering (CARS) and simulated by the method given above [27].

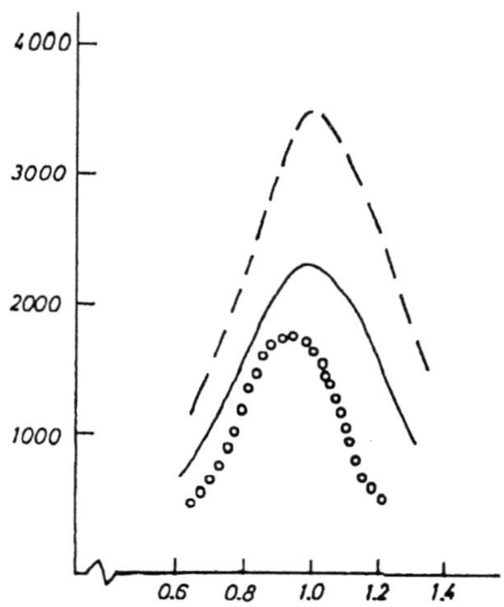

Fig. 6: Strain rate a_q at quenching for CH_4-air flames at $p = 1$ bar, $T_u = 298$ K as function of the equivalence ratio Φ, derived from a library of premixed flames at different Φ. Points: measurements (see [29,30]), broken line: calculation [29], full line: calculation [30].

4. APPLICATION IN IGNITION PROCESSES

A detailed knowledge of instationary combustion processes is required for the understanding of phenomena like ignition or extinction which are of importance in many practical applications (e.g. ignition and knock in Otto engines, safety considerations in industry [3,4,6]). Time-dependent 1D simulations have to be used to treat such combustion processes. They are performed by solving the corresponding conservation equations (i.e. of mass, energy, momentum and species mass) without restriction to systems with spatially uniform pressure, therefore allowing to study the interaction of pressure waves with the flame front and the influence of the "uniform pressure assumption".

Spatial discretization using finite differences and an adaptive grid point system leads to a differential/algebraic equation system which is solved numerically by both extrapolation and BDF codes [31,32]; an example is fiven in Fig. 7.

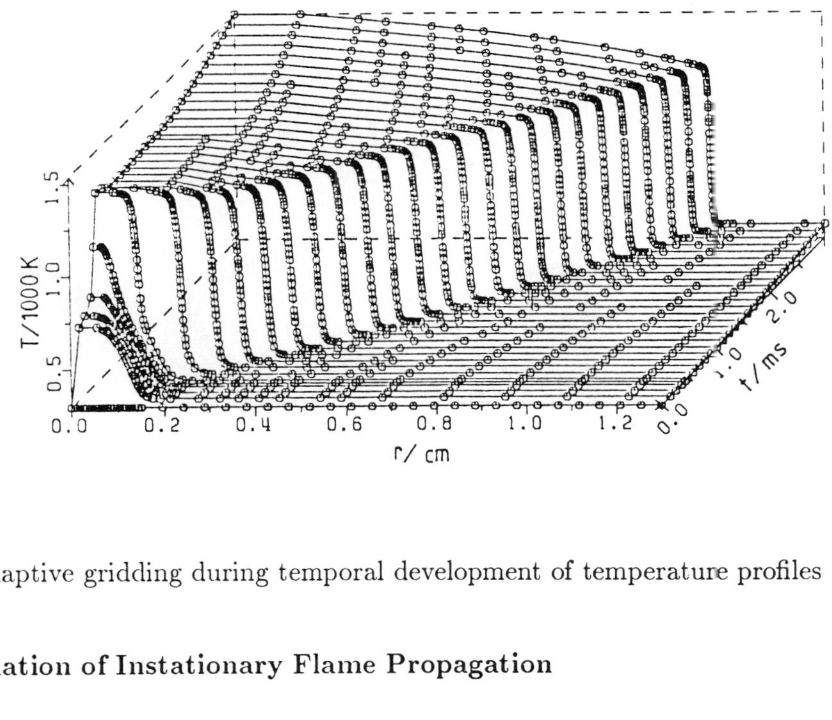

Fig. 7: Adaptive gridding during temporal development of temperature profiles

4.1 Simulation of Instationary Flame Propagation

Since the detection of $p-T$ explosion limits in H_2-O_2 mixtures in the Thirtieth and Fortieth, many efforts have been made to explain this phenomenon quantitatively. All these attempts included severe simplifications as steady state assumptions, truncation of the reaction mechanism or restriction to homogeneous (zero-dimensional) systems. The mathematical model described above is able to simulate all the explosion limits of the hydrogen-oxygen system without those simplifications, using a multi-species transport model, detailed chemistry, and realistic surface chemistry.

Ignition limits are determined by simulating the reaction in closed vessels for various initial pressures and temperatures. At the explosion limits, there is a quite sharp transition from slow reaction (moderate temperature rise over a long time, sometimes more than 100 s) to ignition (with an immediate temperature rise in the vessel center and propagation of the flame).

Calculated $p-T$ ignition limits for auto-ignition in hydrogen-oxygen mixtures are shown in Fig. 8. As can be seen, calculated and experimental values are in quite good agreement [6,33].

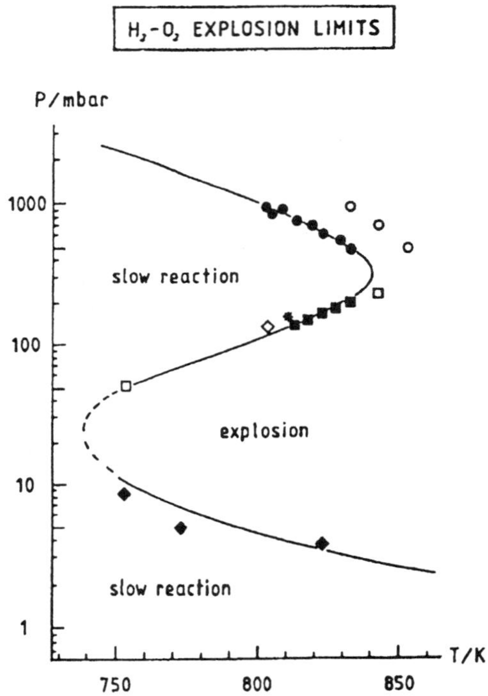

Fig. 8: Measured (points) and calculated (line) $p-T$ ignition limits for H_2-O_2 mixtures (see [6,30])

4.2 Simulation of the Auto-Ignition and Engine Knock Initialization

Engine knock is a many decades old problem connected with the operation of Otto engines, in special since the development of high performance versions (e.g. used for airplanes). Very early it became apparent, that knocking is kinetically controlled and closely connected with the chemistry of combustion, as can be seen for instance from the strong effect of addition of traces of inhibitors like lead tetraethyl (for reference see [3]).

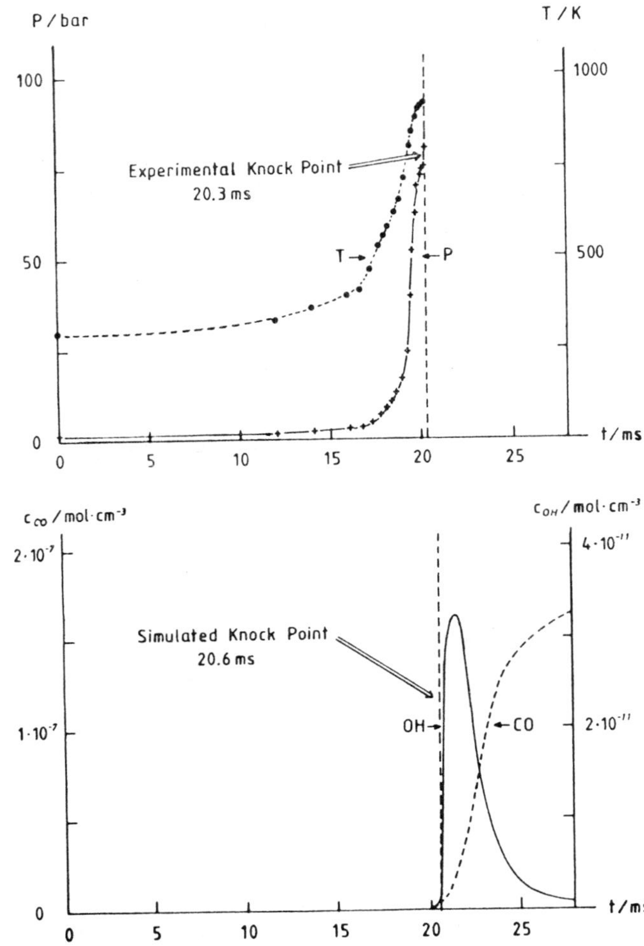

Fig. 9: Measured (upper drawing [32]) and calculated (lower drawing) knocking behaviour in an engine fueled with n-octane

Extension of detailed reaction schemes of the high temperature oxidation of hydrocarbons [5-7] to lower temperatures now allows the explanation of ignition delay up to octane using sophisticated mechanisms for the alkyl radical decomposition of higher hydrocarbons [3,4]. Calculation of ignition delay times is performed by solving the conservation equations of the reaction system, considering the respective experimental conditions of pressure, volume, temperature and volume which, in some cases (end-gas of Otto engines), are time dependent. Treating homogeneous mixtures, those equations represent a system of ordinary differential equations with the time t as independent variable and with the species concentrations ci as dependent variables. Solution of this initial value problem is obtained by use of the programs DASSL [31] or LIMEX [32], both of them designed to solve stiff differential/algebraic systems. Solution of the sensitivity equations is performed simultaneously with the solution of the model equations [20]].

The reaction mechanism used for the description of the oxidation of molecules up to C_8-species is very similar to that discussed in Section 2, with the exception that the reactions of the alkyl radicals with O_2 to form RO_2 and the corresponding alkene must be considered at lower temperatures. Isomerization reactions of the alkyl radicals by internal H atom transfer have been included into the mechanism. These reactions do not affect the overall reaction rate, but may be important for the main reaction channels. The resulting mechanisms of alkyl radical decomposition is given in [4] for octane and heptane.

The example considered here (Fig. 9) is knock in an engine driven with i-octane [34] as fuel. Both knocking and non-knocking case can be predicted with a mechanism taken from [4]. No global steps have to be included in the reaction mechanism for this relatively simple fuel.

5. TURBULENT COMBUSTION INCLUDING DETAILED CHEMISTRY

Two approaches to link complex chemistry to models of turbulent flow field predictions are applied to an enclosed turbulent jet diffusion flame. One is a conserved scalar approach employing a stretched laminar flamelet model, the other involves a direct closure of the chemical production terms by a pdf method.

The governing equations are solved in their full, elliptic form. In both approaches detailed chemistry is taken into account for a CO/air system (72 elementary reactions of 15 chemical species). The computational results of both approaches are compared with each other as well as with experimental data available in the literature.

5.1 Background

In the vast majority of industrial devices the characteristic dimensions and flow velocities are sufficiently large for the flow to be turbulent. Furthermore, for safety reasons most devices are operated such that combustion takes place in a diffusion flame. Therefore, turbulent diffusion flames are of central concern in practical equipment for combustion. Confident numerical predictions of such flames are desirable for design of equipment, diagnosis of malfunction problems, and development of modifications for improvement of performance.

As a consequence models of turbulent reacting flows are needed that are able not only to predict the flow field and global flame properties like the flame length or the overall heat release, but that allow equally well for an accurate and confident prediction of the temperature and the species concentrations throughout the flame. The detailed numerical prediction of the latter quantities requires detailed chemistry to be taken into account, which is an essential shortcoming of many of the models of turbulent combustion currently available.

Modeling of turbulent reacting flows is often based on either a laminar flamelet model or an approach that involves a direct closure of the chemical production terms. At least in principle both approaches allow complex chemistry to be included which, how ever, has seldom been practiced.

Stretched laminar flamelet models of non-premixed turbulent combustion are based on a concept suggested by *Williams* [36], viz., a turbulent diffusion flame is conceived as an ensemble of thin laminar stretched diffusion flames, the so-called laminar diffusion flamelets. The theory of these models has essentially been developed by *Peters*, (e.g. [37-40]), their practical application to turbulent diffusion flames with complex chemistry was first indicated by *Liew et al.* [41].

Recently the approach of the latter authors was extended by *Rogg, Behrendt, and Warnatz* [42,43]. In contrast to the conserved scalar approaches of laminar flamelet models, direct closure approaches imply the solution of the governing equations for the mean concentrations of the chemical species which entails the necessity of directly closing the chemical production terms by either a moment method or a pdf method. Direct closure approaches with simplified chemistry have been pursued by various authors (see [44,45]).

Here, a conserved scalar approach employing a stretched laminar flamelet model and a direct closure approach with closure of the chemical production terms by a pdf method are described with special regard to linking complex chemistry to models for turbulent flow field predictions. In particular, an enclosed turbulent jet carbon monoxide/air diffusion flame is numerically simulated, the reaction mechanism of this system comprising 72 elementary reaction steps of 15 chemical species. It is shown that for the flame considered here both approaches yield quantitatively comparable results that agree reasonably well with the experimental data available in the literature [46].

5.2 Conservation Equations and Models

In the present paper, the turbulent mixing of a fuel jet with air is described by the conservation equations for a swirl-free, axisymmetric flow where, because the main focus of interest is the chemical source term closure, a conventional eddy viscosity gradient model of turbulent transport and a $k-\epsilon$ turbulence model are adopted. Using of an elliptic turbulent transport operator L

$$L(\tilde{\Phi}) \equiv \frac{\partial(\bar{\rho}\tilde{u}\tilde{\Phi})}{\partial x} + \frac{1}{r}\frac{\partial(r\bar{\rho}\tilde{v}\tilde{\Phi})}{\partial r} - \frac{\partial}{\partial x}\left(\frac{\mu_t}{\sigma_\Phi}\frac{\partial\tilde{\Phi}}{\partial x}\right) - \frac{1}{r}\frac{\partial}{\partial r}\left(r\frac{\mu_t}{\sigma_\Phi}\frac{\partial\tilde{\Phi}}{\partial r}\right)$$

for any variable $\tilde{\Phi}$, these equations may be written as (Eqs. 1-5)

$$L(1) = 0$$

$$L(\tilde{u}) = -\frac{\partial\bar{p}}{\partial x} + \frac{\partial}{\partial x}\left(\mu_t\frac{\partial\tilde{u}}{\partial x}\right) + \frac{1}{r}\frac{\partial}{\partial r}\left(r\mu_t\frac{\partial\tilde{v}}{\partial x}\right) - \frac{2}{3}\frac{\partial}{\partial x}\left[\mu_t\left(\frac{\partial\tilde{u}}{\partial x} + \frac{1}{r}\frac{\partial(r\tilde{v})}{\partial r}\right) + \bar{\rho}k\right]$$

$$L(\tilde{v}) = -\frac{\partial\bar{p}}{\partial r} + \frac{\partial}{\partial x}\left(\mu_t\frac{\partial\tilde{u}}{\partial r}\right) + \frac{1}{r}\frac{\partial}{\partial r}\left(r\mu_t\frac{\partial\tilde{v}}{\partial r}\right) - \frac{2}{3}\frac{\partial}{\partial r}\left[\mu_t\left(\frac{\partial\tilde{u}}{\partial x} + \frac{1}{r}\frac{\partial(r\tilde{v})}{\partial r}\right) + \bar{\rho}k\right] - 2\frac{\mu_t\tilde{v}}{r^2}$$

$$L(k) = G_k - \bar{\rho}\epsilon$$

$$L(\epsilon) = \frac{\epsilon}{k}(C_{\epsilon 1}G_k - C_{\epsilon 2}\bar{\rho}\epsilon)$$

where in Eqs. (4) and (5), G_k is the rate of production of turbulence kinetic energy. In deriving Eqs. (1) to (5), the Favre averaging procedure has been used; tildes identify Favre averages, overbars time averages; x and r denote the longitudinal and radial coordinates, respectively, $\tilde{u}$ and $\tilde{v}$, the respective mean flow velocity components; $\bar{\rho}$ is the mean density, μ_t the turbulent diffusivity, k is the Favre turbulence kinetic energy, ϵ the Favre turbulence dissipation, $\bar{p}$ the mean pressure, and the σ_Φ's and $C_{\epsilon i}$'s are model constants with the numerical values $\sigma_u = \sigma_v = \sigma_k = 1.0$, $\sigma_\epsilon = 1.3$, $C_{\epsilon 1} = 1.40$, $C_{\epsilon 2} = 1.85$ [41,47]:

$$
\begin{aligned}
G_k = & \mu_t \left[2\left(\frac{\partial \tilde{u}}{\partial x}\right)^2 + 2\left(\frac{\partial \tilde{v}}{\partial r}\right)^2 + 2\left(\frac{\tilde{v}}{r}\right)^2 + \left(\frac{\partial \tilde{u}}{\partial r} + \frac{\partial \tilde{v}}{\partial x}\right)^2 \right] \\
& - \frac{3}{2}\left[\bar{\rho}k + \mu_t\left(\frac{\partial \tilde{u}}{\partial x} + \frac{\partial \tilde{v}}{\partial r} + \frac{\tilde{v}}{r}\right)\right]\left(\frac{\partial \tilde{u}}{\partial x} + \frac{\partial \tilde{v}}{\partial r} + \frac{\tilde{v}}{r}\right)
\end{aligned}
$$

For modeling details, the literature (e.g. [38-43]) should be consulted.

5.3 Detailed Chemistry

The same detailed elementary reaction mechanism of the CO/H_2/air system (see *Warnatz* [7]) consisting of 72 elementary reactions of 15 species (that are N_2, CO, CO_2, O_2, H_2, H_2O, H, O, OH, HO_2, H_2O_2, CHO, CH_2O, N, and NO) was employed in both approaches.

In the turbulent flame computations based on the conserved scalar approach, the profiles of burning laminar flamelets were drawn from a flamelet library cotaining 15 differently stretched flames. Extinguished flamelets were simulated by a cold, non-reactive mixing process with constant temperature, and carbon monoxide, hydrogen, oxygen, and nitrogen varying linearly with the conserved scalar. Some results of the laminar flamelet computations are given in Fig. 10. Here profiles of temperature and carbon dioxide mass fractions are plotted as functions of the conserved scalar for various values of the scalar dissipation rate. It may be gathered from Fig. 10 that temperature decreases with increasing stretch and, accordingly, carbon dioxide mass fractions decrease. For a comprehensive representation and a detailed discussion of these laminar flamelet computations, the literature should be consulted [48].

5.4 Numerical Results and Comparison with Experiments

In Fig. 11 axial profiles of the Favre averaged conserved scalar, i.e., the mixture fraction, and the mean temperature are compared with experimental data reported by *Razdan and Stevens* [46]. It is seen that the agreement between measured and computed mixture fraction profiles generally is good.

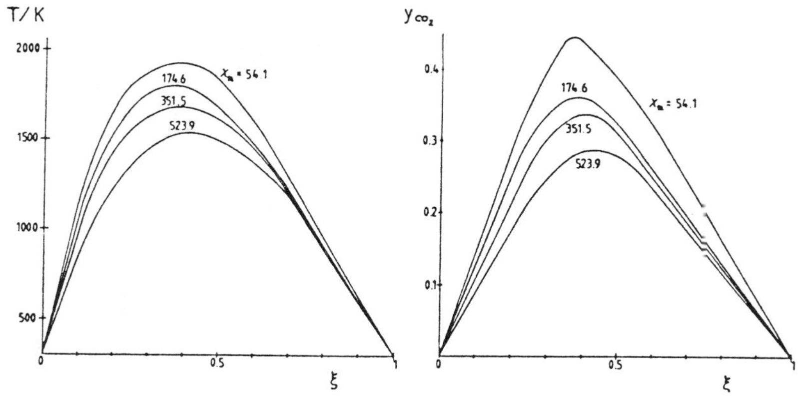

Fig. 10: Temperature (left hand side) and carbon dioxide mass fraction (right-hand side) as function of the conserved scalar for a laminar carbon monoxide-air counterflow diffusion flame with various strain rates

One source for the observed disagreement between the computed and measured temperatures are the boundary conditions that are not unambiguously defined by *Razdan and Stevens* [46] for the experimental system. Furthermore, temperatures were measured by unshielded thermocouples and, therefore, some of the discrepancies may be due to inconsistencies in calculating radiation corrected values for the measured temperature data.

In Fig. 12 are the Favre averaged mass fractions of carbon monoxide, carbon dioxide and oxygen on the centerline of the turbulent flame shown. As for the mixture fraction, with both methods reasonable agreement between measured and predicted profiles is obtained, and the differences between the models are moderate. Therefore, it seems to be worthwhile to apply these models to more complex combustion systems and to study systematically phenomena like pollutant formation in turbulent flames.

Finally, it should be noted that for the weakly turbulent CO/air diffusion flame investigated herein we found effects of stretch on the laminar flamelets being negligible small. In fact, for the flame under consideration from the stretched laminar flamelet approach practically the same results are obtained if the flamelet library consists only of one unstrained laminar diffusion flame.

For the turbulent flame considered herein this finding a posteriori justifies the one-variable approaches utilized in the numerical simulations by *Razdan and Stevens* [46] and *Gore, Jeng, and Faeth* [49].

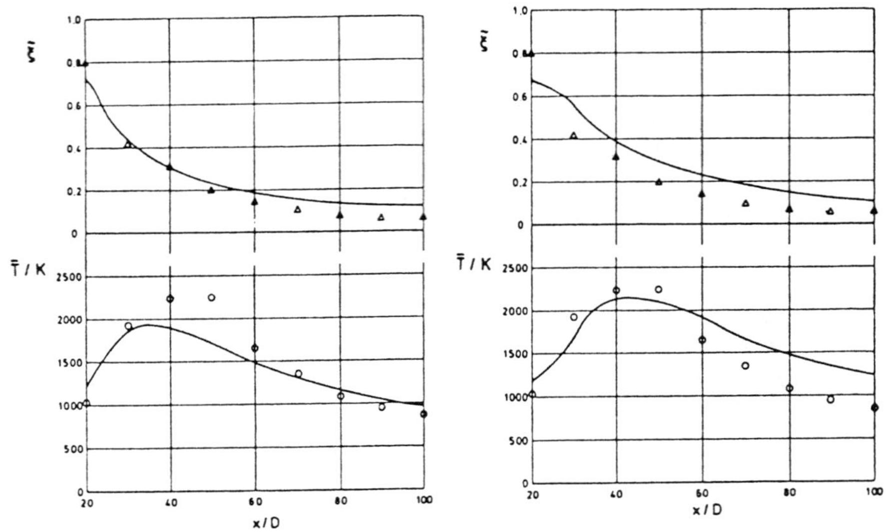

Fig. 10: Mixture fraction and temperature on the centerline of a turbulent carbon monoxide-air diffusion flame. Symbols: experimental data taken from Razdan and Stevens [46]; solid lines: numerical results obtained from flamelet model (left-hand side) and pdf method (right-hand side)

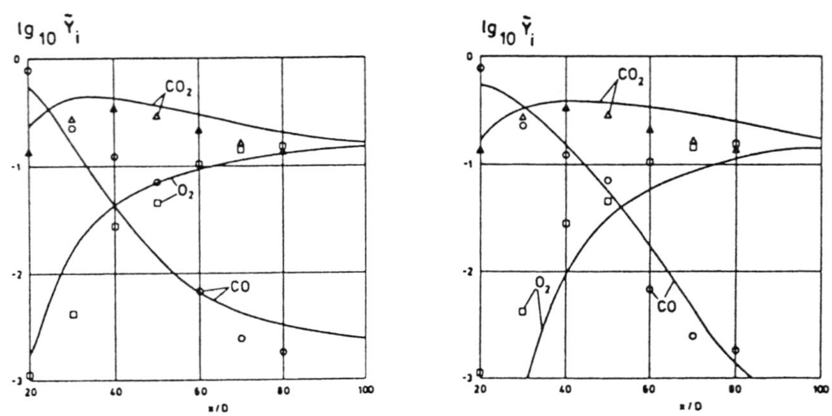

Fig. 10: Mass fractions of carbon monoxide, carbon dioxide and oxygen on the centerline of a turbulent carbon monoxide-air diffusion flame. Symbols and lines as in Fig. 11

6. ACKNOWLEDGEMENT

The author is very grateful to *"Bundesministerium für Forschung und Technologie"* (TecFlam-Project), *"Commission of the European Communities"* (JOULE project), *"Fonds der Chemischen Industrie"*, and *"Max-Buchner-Stiftung"* for financial support of this work and thanks *Dr. F. Behrendt, Dr. H. Bockhorn, Dr. U. Maas*, and *Dr. B Rogg* for valuable help.

ADDRESS: Prof. Dr. Jürgen Warnatz

Universität Stuttgart

Institut für Technische Verbrennung

Pfaffenwaldring 12

W-7000 Stuttgart 80

REFERENCES

[1] *J. Warnatz*, in: R. Zellner (ed.), Formation, Distribution, and Chemical Transformation of Air Pollutants, p. 21. DECHEMA, Frankfurt (1987)

[2] *M. Frenklach, J. Warnatz*: Comb. Sci. Technol. **51**, 265 (1987)

[3] *C. Esser, U. Maas, J. Warnatz*: Proc. Intl. Symp. on Diagnostics and Modeling of Combustion in Reciprocating Engines, p. 335. The Japan Society of Mechanical Engineering, Tokyo (1985)

[4] *W. J. Pitz, J. Warnatz, C. K. Westbrook*: 22nd Symposium (International) on Combustion, in press

[5] *J. Warnatz*: 18th Symposium (International) on Combustion, p. 369. The Combustion Institute, Pittsburgh (1981)

[6] *J. Warnatz, U. Maas*: Proc. 12th IMACS World Congress (1988), in press

[7] *J. Warnatz*, in: W. C. Gardiner (ed.), Combustion Chemistry. Springer, New York (1984)

[8] *J. Warnatz, H. Bockhorn, A. Möser, H.-W. Wenz*: 19th Symposium (International) on Combustion, p. 83. The Combustion Institute, Pittsburgh (1985)

[9] *J. Warnatz*: 20th Symposium (International) on Combustion, p. 845. The Combustion Institute, Pittsburgh (1985)

[10] *J. Warnatz*: Ber. Bunsenges. Phys. Chem. **87**, 1008 (1983)

[11] *J. Warnatz*: Comb. Sci. Technol. **34**, 177 (1983)

[12] *J. Warnatz*: Ber. Bunsenges. Phys. Chem. **83**, 950 (1979)

[13] *T. Dreier, B. Lange, J. Wolfrum, M. Zahn, F. Behrendt, J. Warnatz*: Ber. Bunsenges. Phys. Chem. **90**, 1010 (1986)

[14] *J. Warnatz*, in: K. Ebert, P. Deuflhard, W. Jäger (eds.), Modelling of Chemical Reaction Systems, p. 162. Springer, Heidelberg (1981)

[15] *B. Raffel, J. Warnatz, H. Wolff, J. Wolfrum, R. J. Kee*, in: J. R. Bowen, J.-C. Leyer, R. I. Soloukhin (eds.), Dynamics of Reactive Systems, Part II, p. 335. AIAA, New York (1986)

[16] *J. Warnatz*, in: R. Glowinski, B. Larrouturou, R. Temam, Numerical Simulation of Combustion Phenomena. Springer, Heidelberg (1985)

[17] *B. Raffel, J. Warnatz, J. Wolfrum*: Appl. Phys. **B37**, 189 (1985)

[18] *J. Warnatz*, in: J. R. Bowen, N. Manson, A. K. Oppenheim, R. I. Soloukhin (eds.), Combustion in Reactive Systems, p. 501. AIAA, New York (1981)

[19] *F. Behrendt, J. Warnatz*: Int. J. Hydrogen Energy **10**, 749 (1985)

[20] *U. Nowak, J. Warnatz*: Proc. 11th International Colloquium on Dynamics of Explosions and Reactive Systems (1988), in press

[21] *J. Warnatz*, in: N. Peters, J. Warnatz (eds.): Numerical Methods in Laminar Flame Propagation. Vieweg, Wiesbaden (1982)

[22] *R. J. Kee, J. Warnatz, J. A. Miller*: A Fortran Program Computer Code for the Evaluation of Gas-Phase Viscosities, Conductivities, and Diffusion Coefficients. SANDIA Report SAND83-8209 (1983)

[23] *R. J. Kee, G. Dixon-Lewis, J. Warnatz, M. E. Coltrin, J. A. Miller*: A Fortran Computer Code for the Evaluation of Gas-Phase Multicomponent Transport Properties. SANDIA Report SAND86- 8246 (1986)

[24] *J. Warnatz*: Ber. Bunsenges. Phys. Chem. **82**, 193 (1978)

[25] *J. M. Heimerl, T. P. Coffee*, in: N. Peters, J. Warnatz (eds.), Numerical Methods in Laminar Flame Propagation. Vieweg, Braunschweig (1982)

[26] *T. Dreier, B. Lange, J. Wolfrum, M. Zahn, F. Behrendt, J. Warnatz*: 21st Symposium (International) on Combustion. The Combustion Institute, Pittsburgh (1987)

[27] *E. Diessel, T. Dreier, B. Lange, J. Wolfrum, F. Behrendt, J. Warnatz*: 22nd Symposium (International) on Combustion, in press

[28] *W. Weyrauch, H. Bockhorn, J. Warnatz*: to be published

[29] *G. Stahl, B. Rogg, J. Warnatz*: Proc. 11th International Colloquium on Dynamics of Explosions and Reactive Systems (1988), in press

[30] *G. Stahl, J. Warnatz*: publication in preparation

[31] *P. Deuflhard, E. Hairer, J. Zugck*: One-Step and Extrapolation Methods for Differential/Algebraic Systems. Univ. Heidelberg, SFB 123: Tech. Rep. 318, (1985)

[32] *L. R. Petzold*: A Description of DASSL: A Differential/Algebraic System Solver. Sandia Report SAND82-8637, Sandia National Laboratories, Livermore (1982); IMACS World Congress, Montreal 1982

[33] *U. Maas, J. Warnatz*: Combust. and Flame, in press

[34] *P. Pinchon*: personal communication

[35] *J. Warnatz, W. Jäger* (eds.): Complex Chemical Reaction Systems: Mathematical Modelling and Simulation. Springer, Heidelberg/New York (1987)

[36] *F. A. Williams*, in: S. N. B. Murphy (Ed.), Turbulent Mixing in Nonreactive and Reactive Flows, p. 189. Plenum Press, New York (1975)

[37] *N. Peters, F. A. Williams*: AIAA J. **21**, 423 (1983)

[38] *N. Peters*: Combust. Sci. and Tech. **30**, 1 (1983)

[39] *N. Peters*: Prog. Energy Combust. Sci. **10**, 319 (1984)

[40] *N. Peters*: 20th Symposium (International) on Combustion, p. 353. The Combustion Institute, Pittsburgh (1984).

[41] *S. K. Liew, K. N. C. Bray, J. B. Moss*: Combust. Flame **56**, 199 (1984)

[42] *B. Rogg, F. Behrendt, J. Warnatz*: 21st Symposium (International) on Combustion, p. 1533. The Combustion Institute, Pittsburgh (1987)

[43] *B. Rogg, F. Behrendt, J. Warnatz*: Ber. Bunsenges. Phys. Chem. **90**, 1005 (1986)

[44] *H. T. Sommer, W. Adams*: Combust. Flame **49**, 1 (1983)

[45] *R. Borghi, E. Pourbais*: Physicochemical Hydrodynamics **2**, 65 (1981)

[46] *M. K. Razdan, J. G. Stevens*: Combust. Flame **59**, 289 (1985)

[47] *J. H. Kent, R. W. Bilger*: 16th Symposium (International) on Combustion, p. 1643. The Combustion Institute, Pittsburgh (1976)

[48] *F. Behrendt, J. Warnatz*:

[49] *J. P. Gore, S.-M. Jeng, G. M. Faeth*: AIAA Paper No. 86-0294 (1986)

N PETERS

Systematic reduction of flame kinetics: principles and details

1. INTRODUCTION

The general idea of reducing complex reaction schemes by the introduction of steady-state assumptions for intermediates has been known to chemists for a long time. For a homogeneous system, it is valid for those intermediate species that are produced by slow reactions and consumed by fast reactions, such that their maximum concentration remains small and essentially stationary. A mathematical justification of this assumption and thereby a better understanding was first given by Bowen et al. [1]. They showed for some typical reaction schemes that the steady-state assumption relies on a singular asymptotic expansion for large values of the rate coefficient of the dominant consumption reaction, large as compared to the other rate coefficients and the time derivative. Similarly, the partial equilibrium assumption may be justified by considering the asymptotic limit where the rate coefficients of both the forward and backward step of a single reaction are much larger than all the other rate coefficients. Both, steady-state and partial equilibrium assumptions reduce the number of differential equations for the reacting species to be solved. In general, steady-state assumptions, if applicable, are easier to handle than partial equilibrium assumptions and lead directly to a set of global reactions between the remaining species.

The first systematic examination of steady-state and partial equilibrium assumptions for hydrogen flames was completed by Dixon-Lewis [2]. His conclusion was that the steady-state assumption for all but one radical yields good results, whereas the results obtained with partial equilibrium assumptions deviate from these considerably.

In a series of recent papers on lean premixed methane flames (Peters [5]), lean-to-rich premixed methane, methanol, and propane flames (Paczko et al. [4]), and methane-air diffusion flames (Peters and Kee [6]), reduced reaction mechanisms have been derived and compared with numerical calculations based on full kinetic schemes. On the basis of these formulations, the detailed structure of stoichiometric methane-air premixed flames (Peters and Williams [7]) and methane-air diffusion flames (Seshadri and Peters [8]) has been analyzed using asymptotic expansions. The result of these papers will be summarized in the following sections.

2. REDUCTION STRATEGY

In this section, the principle steps of reducing a detailed kinetic mechanism to a small number of global reactions will be outlined. The principles to be observed are stated and augmented by detailed arguments and examples.

Principle Step 1

Define a problem-oriented starting mechanism.

Details

If one does not have a lot of experience in combustion kinetics, it is not easy to decide which one of the many reactions in a complete reaction scheme, as proposed, for example, by Warnatz [9] or Westbrook and Dryer [10], is important and which one could possibly be neglected. Some guidance about the important reactions is provided by sensitivity diagrams [9], but since the specific reactions that must be retained depend on the problem under consideration, additional considerations are necessary. Peters and Williams [7] in analyzing the asymptotic structure of premixed methane-air flames conceived a selection procedure for flame calculations. It is based on an order-of-magnitude estimate of the reaction partners, as well as an evaluation of the rate coefficient at an estimated reaction zone temperature. This temperature can be estimated for premixed methane flames as a function of pressure as

$$\frac{T}{T_r} = (0.4\,p)^{1/6.63} \tag{1}$$

where $T_r = 1600\,\mathrm{K}$ and p is in atmospheres. It ranges from $1400\,K$ for $p = 1\,\mathrm{atm}$ to $2200\,K$ for $p = 20\,\mathrm{atm}$. The same estimate should be valid for other hydrocarbon flames.

In order to identify the importance of a specific reaction, one must compare it to those that compete for the same species in the same reaction chain. For methane flames, it is sufficient to consider only the C_1 chain from CH_4 via CH_3, CH_2O and CHO to CO and CO_2, which corresponds to reactions 10–22 in Table 1. It is seen that the reaction partners of hydrocarbon species are the radicals H, OH, and O as well as H_2 and O_2 and an inert body M. (If the reverse reaction 12 was considered, CH_3 reacts also with H_2O.) The orders of magnitude of these species in the flame zone can be estimated as

$$\begin{aligned} x_{\mathrm{H}} &= \mathrm{O}(\mu^m)\,, & x_{\mathrm{OH}} &= \mathrm{O}(\mu^{1+m/2})\,, & x_{\mathrm{O}} &= \mathrm{O}(\mu^{1+m})\,, \\ x_{\mathrm{H_2}} &= \mathrm{O}(\mu^{m/2})\,, & x_{\mathrm{O_2}} &= \mathrm{O}(\mu^{m/2})\,, & x_{\mathrm{H_2O}} &= \mathrm{O}(\mu^0) \end{aligned} \tag{2}$$

where the x_i are normalized mole fractions $x_i = (X_i/X_{CH_4,0})$ and $X_{\mathrm{CH_4,0}}$ is the mole fraction of CH_4 in the unburnt mixture. The exponent m is related to the small number $\mu = 0.1$ and the thickness of the reaction zone ϵ and is estimated as $m = 1$ for $1\,\mathrm{atm}$ and $m = 2$ for $20\,\mathrm{atm}$. Now it suffices to realize that the relative concentration of the third body is $x_M = \mathrm{O}(\mu^{-1})$ and, starting from a full mechanism, we may calculate the competing rate coefficients for each hydrocarbon species in terms of powers of μ at the typical temperature. For example, for a stoichiometric CH_4 air flame with rate coefficients from Warnatz [9], we have at 1600 K for the ratios of forward rate coefficients

$$\frac{k_{f12}}{k_{f11}} = 0.68 \approx \mathrm{O}(\mu^0)$$

(where the indices correspond to the reaction numbers in Table 1). For the reaction (111) $CH_4 + O \rightarrow CH_3 + OH$, one obtains the ratio

$$\frac{k_{f111}}{k_{f11}} = 1.02 \approx O(\mu^0)\,.$$

This indicates in view of (2) that the reaction $CH_4 + O \rightarrow CH_3 + OH$ can be neglected at all pressures compared to $CH_4 + H \rightarrow CH_3 + H_2$, whereas the reaction $CH_4 + OH \rightarrow CH_3 + H_2O$ should be retained at least at high pressures. This procedure is performed for all reactions of the C_1 and C_2 chain. Then, reactions 11–22 in Table 1 turn out to be the most important reactions of the C_1 chain for methane flames.

The order of magnitude of the CH_3 radical may be estimated using the rate coefficients of reactions 11 and 13 as $x_{CH_3} = O(\mu^{1+m/2})$. Then at high pressures and in moderately fuel-rich flames, the C_2 chain also must be considered. Again, competing reactions must be compared. This has led (Paczko et al. [4]) to a starting mechanism of 50 reactions for methane, 26 reactions for methanol, and 64 reactions for propane flames covering the entire range of stoichiometry. For lean methane flames, a starting mechanism of only 18 reactions had been considered (Peters [5]).

Principle Step 2

Identify steady-state species.

Details

Steady-state assumptions are based on order-of-magnitude estimates of reaction coefficients. Those that produce an intermediate species must be smaller than those that consume it, such that the species remains small. Therefore, an alternative way to identify a steady-state species is to analyze the outcome of a numerical calculation based on the starting mechanism and to consider its relative order of magnitude. The general balance equation for a reactive species may be written in terms of $\overline{Y}_i = y_i/M_i$ as

$$L(\overline{Y}_i) = \sum_{k=1}^{r} \nu_{ik}\, w_k\,, \tag{3}$$

where the operator L is defined by

$$L(\overline{Y}_i) \equiv \rho\frac{\partial \overline{Y}_i}{\partial t} + \rho\, v_\alpha \frac{\partial \overline{Y}_i}{\partial x} + \frac{\partial}{\partial x}\left(\frac{j_{i\alpha}}{M_i}\right) \qquad (i = 1, 2, \ldots, n)\,. \tag{4}$$

Here, Y_i is the mass fraction of species i, M_i its molecular weight, ρ the density, t the time coordinate, v_α the velocity component in the x_α direction, $j_{i\alpha}$ the diffusion flux in the x_α direction, ν_{ik} the stoichiometric coefficient of species i in reaction k and w_k the rate of the k-th reaction. There are r reactions between n chemical species in the system. We use the variable $\overline{Y}_i$ rather than the usual mass fraction Y_i because then the reaction rates in (3) are no longer multiplied by the molecular weight and the convective and diffusive terms are directly comparable between the species equations. For order-of-magnitude estimates of different terms in (2), we write the diffusion flux of the i-th species $j_{i\alpha}$ using the binary diffusion coefficient with respect to nitrogen as

$$\frac{j_{i\alpha}}{M_i} = -\rho\, D_{i,N_2} \frac{\partial \overline{Y}_i}{\partial x_\alpha} \,. \tag{5}$$

For a steady-state species, the concentration remains very small and all terms in the operator $L(\overline{Y}_i)$ on the l.h.s. of (3) may be neglected compared to the reaction rates. The L operator is equal to the time derivative in homogeneous systems (which led to the notion of "steady state"). In the balance equations for steady flames, the L operator contains the convective and diffusive term; the assumption therefore requires that both of these terms are small. Since the binary diffusion coefficients are approximately proportional to $1/\sqrt{M_{i,N_2}}$, where

$$M_{i,N_2} = \frac{2M_i M_{N_2}}{M_i + M_{N_2}} \tag{6}$$

the species concentrations must be weighted with this factor for the order-of-magnitude estimate. The relevant concentrations are the mole fractions X_i, rather than the mass fractions Y_i, since X_i is proportional to Y_i/M_i. In Table 2, we list the maximum mole fractions of intermediate species obtained from the calculation with a large mechanism for a stoichiometric premixed methane-air flame at one atmosphere. We have also listed the mole fractions weighted with $\sqrt{M_{N_2}/M_{i,N_2}}$. The values fall essentially into two groups: those well below 1% and those well above it.

The first group includes the intermediate species OH, O, HO_2, CH_3, CH_2O, and CHO which in the following are assumed to be in a steady state. The second group contains CO, H_2, and H if the weighted mole fractions are considered. Those species will therefore not be assumed to be in steady state. The procedure adopted here is therefore an indirect one. Rather than performing an asymptotic analysis as in Bowen et al. [1] with the ratio of rate coefficients as a small parameter, the examination is based on the outcome of a numerical calculation using a large mechanism. This procedure is easier and less likely subject to errors, given the large number of rate coefficients. Also, by weighting the species with the diffusion coefficients, we use a result from the asymptotic analysis of flames that states within the reactive layers, the diffusive terms are dominant compared to the convective terms and are the only ones to balance the reaction terms.

The present choice of retaining H as a non-steady-state species finds a further justification, if we recall that according to Warnatz [9], $H + O_2 \rightleftharpoons O + OH$ (reaction 1) is the most important one for flame calculations and that it competes with $H + O_2 + M \rightarrow HO_2 + M$ (reaction 5) at the flammability limits. Therefore, it is important to calculate the H concentration more accurately than those of O and OH. However, since O and OH appear in the backward reaction of reaction 1 and this is almost as important as the forward reaction, particular considerations will be necessary for the product of the concentrations of O and OH. These will be presented in the details to principle step 5.

Principle Step 3

Define a main chain.

Details

For methane flames, it is now possible to use the six steady-state conditions in order to eliminate at least six reaction rates from the system. We want to eliminate the fastest reactions that consume each steady-state species and thereby construct a main chain. Using the same procedure as in principle step 1 to estimate the magnitude of rates, we find that for the oxidation of CH_4 via CH_3, CH_2O and CHO to CO, this main chain is

$$\begin{array}{ll} 11 & CH_4 + H \rightarrow CH_3 + H_2 \\ 13 & CH_3 + O \rightarrow CH_2O + H \\ 17 & CH_2O + H \rightarrow CHO + H \\ 22 & CHO + M \rightarrow CO + H + M\,. \end{array} \tag{7}$$

We will therefore use the steady-state relations for CH_3, CH_2O, and CHO in order to eliminate the rates w_{13}, w_{17}, and w_{22} from the balance equations. In addition, we will use the steady-state relations for O, OH, and HO_2 to eliminate the rates w_2, w_3, and w_7 and thereby define the main chain for the hydrogen-oxygen system

$$\begin{array}{ll} 1 & H + O_2 \rightarrow OH + O \\ 2 & O + H_2 \rightarrow H + OH \\ 3 & OH + H_2 \rightarrow H + H_2O \end{array} \tag{8}$$

as well as for the recombination reactions

$$\begin{array}{ll} 5 & H + O_2 + M \rightarrow HO_2 + M \\ 7 & H + HO_2 \rightarrow O_2\,. \end{array} \tag{9}$$

The elimination of the rates w_2, w_3, w_7, w_{13}, w_{17}, and w_{22} from the balance equations of the non-steady-state species is just a problem of linear algebra. By linear combinations of the equations, we find the following balance equations for the remaining species, if, as in Peters [5], only the reactions 1–13, 17–19, and 21–22 are considered.

$$
\begin{aligned}
&L(\overline{Y}_{\mathrm{H}}) + \Big[L(\overline{Y}_{\mathrm{OH}}) + 2L(\overline{Y}_{\mathrm{O}}) - L(\overline{Y}_{\mathrm{CH_3}}) + L(\overline{Y}_{\mathrm{CHO}}) - L(\overline{Y}_{\mathrm{HO_2}})\Big] \\
&\quad = 2w_1 - 2w_5 + 2w_6 - 2w_9 - 2w_{11} - 2w_{16} - 2w_{19} - 2w_{21} \\
&L(\overline{Y}_{\mathrm{H_2}}) + \Big[-L(\overline{Y}_{\mathrm{OH}}) - 2L(\overline{Y}_{\mathrm{O}}) + 3L(\overline{Y}_{\mathrm{CH_3}}) + L(\overline{Y}_{\mathrm{CH_2O}}) + L(\overline{Y}_{\mathrm{HO_2}})\Big] \\
&\quad = -3w_1 + w_5 - 3w_6 + w_9 + w_{10} + 4w_{11} + 4w_{12} + w_{19} + w_{21} \\
&L(\overline{Y}_{\mathrm{O_2}}) + \Big[L(\overline{Y}_{\mathrm{HO_2}})\Big] = -w_1 - w_6 \\
&L(\overline{Y}_{\mathrm{H_2O}}) + \Big[L(\overline{Y}_{\mathrm{OH}}) + L(\overline{Y}_{\mathrm{O}}) - L(\overline{Y}_{\mathrm{CH_3}})\Big] \\
&\quad = 2w_1 + 2w_6 - w_{10} - w_{11} - w_{12} \\
&L(\overline{Y}_{\mathrm{CO}}) + \Big[L(\overline{Y}_{\mathrm{CH_3}}) + L(\overline{Y}_{\mathrm{CH_2O}}) + L(\overline{Y}_{\mathrm{CHO}})\Big] \\
&\quad = -w_{10} + w_{11} + w_{12} \\
&L(\overline{Y}_{\mathrm{CO_2}}) = w_{10} \\
&L(\overline{Y}_{\mathrm{CH_4}}) = -w_{11} - w_{12}\,.
\end{aligned}
\tag{10}
$$

Here, the terms in square brackets on the l.h.s. can be neglected. It is interesting to note that the rates w_4, w_7 and w_{17} have appeared automatically, since they are linearly dependent on reactions that have been eliminated.

Principle Step 4

Identify global reactions and their principle rates.

Details

Based on the selection of steady-state species and the main chain, a set of global reactions may be identified by considering the stoichiometric coefficients in front of the reactive terms in the balance equations of the remaining species. Although all the physical information is already contained in the balance equations, these global reactions mainly serve to illustrate the overall reaction process. They only represent the stoichiometry and are not unique, since any linear combination of a set of global reactions represents an equally valid set. In the present example for methane, the stoichiometric coefficients on the r.h.s. of (10) lead to the following set of four global reactions:

$$
\begin{array}{ll}
\mathrm{I} & \mathrm{CH_4 + 2H + H_2O = CO + 4H_2} \\
\mathrm{II} & \mathrm{CO + H_2O = CO_2 + H_2} \\
\mathrm{III} & \mathrm{H + H} + M = \mathrm{H_2} + M \\
\mathrm{IV} & \mathrm{O_2 + 3H_2 = 2H + 2H_2O}\,.
\end{array}
\tag{11}
$$

The rates are given in terms of the rates of the elementary reactions by the following expressions:

$$\begin{aligned} w_{\mathrm{I}} &= w_{11} + w_{12} \\ w_{\mathrm{II}} &= w_{10} \\ w_{\mathrm{III}} &= w_5 + w_9 + w_{19} + w_{11} \\ w_{\mathrm{IV}} &= w_{11} + w_{16}\,. \end{aligned} \tag{12}$$

Between the seven reacting non-steady-state species CH_4, H, H_2, CO, H_2O, CO_2, and O_2, there are only four independent reactions possible, since they contain three elements C, H, and O. This has led to a four-step mechanism. On the other hand, in the case of propane, the existence of two parallel reaction paths is illustrated in Paczko et al. [4] by replacing reaction I by the two global reactions

$$\begin{aligned} &\mathrm{Ia} \qquad C_3H_8 + 2H + 2H_2O + O_2 = 2CO + 7H_2 + CO_2 \\ &\mathrm{Ib} \qquad C_3H_8 + 4H + 3H_2O = 3CO + 9H_2 \end{aligned} \tag{13}$$

which are, of course, linearly independent and could be converted into each other using the global steps II–IV.

For analytical investigations, in particular asymptotic analysis, it is useful to relate each one of the global reactions a single rate. These should be the rates of the most important reactions and will be called principle rates. For methane, by using the procedure outlined in principle step 1, one can identify the most important reactions as

$$\begin{aligned} \mathrm{I} \rightarrow 11 &\qquad CH_4 + H \rightarrow CH_3 + H_2 \\ \mathrm{II} \rightarrow 10 &\qquad CO + OH \rightleftharpoons CO_2 + H \\ \mathrm{III} \rightarrow 5 &\qquad H + O_2 + M \rightarrow HO_2 + M \\ \mathrm{IV} \rightarrow 1 &\qquad H + O_2 \rightleftharpoons O + OH\,. \end{aligned}$$

It is not surprising that 1, 5 and 10 correspond to the most important reactions identified by Warnatz [9], whereas reaction 11 is the most important fuel consumption step.

Principle Step 5

Introduce partial equilibrium assumptions.

Details

Some of the steady-state species appear in the reaction rates and must therefore be determined on the basis of their steady-state relations. This will, in general, lead to a set of nonlinear algebraic equations that are difficult to handle analytically. One of the main goals in the recent series of papers was to obtain explicit algebraic expressions for the steady-state species such that remaining rates can be formulated in computer codes in a similar way as for elementary reactions. One way to achieve this goal was to truncate the steady-state expression for OH and O, such that only the most important terms appear. It turns out that the forward and backward step of reaction 3 is very fast. In the steady-state relation for OH, all other rate expressions may be neglected as compared to those two, which then is identical to the partial equilibrium assumption

$$C_{OH} = C_H C_{H_2O}/(C_{H_2} K_{C3}) . \tag{14}$$

Here, the C_i's are molar densities $C_i = \rho Y_i / M_i$, and K_{C3} is the equilibrium constant of reaction 3. A similar approach was originally used in Peters [5] for the steady-state relation of O, where the forward and backward steps of reaction 2 were assumed to be the fastest, such that the partial equilibrium of this reaction determines the concentration of O. However, it has turned out that by using both partial equilibrium assumptions, the concentrations of both OH and O are underestimated in the reaction zone. This leads to a large error in the backward rate of reaction 1, where they appear as a product, and this is responsible for the large burning velocities computed in Peters [5]. A better choice is to use only reaction 1 and 2 in the steady-state relation of O, which still leads to an explicit algebraic expression

$$C_O = \frac{k_{1f} C_{O_2} C_H + k_{2b} C_{OH} + C_H}{k_{1b} C_{OH} + k_{2f} C_{H_2}} . \tag{15}$$

This expression was used in Paczko et al. [4] and also gave good results for the hydrogen flames to be presented next.

Whereas the combination of steady-state and partial equilibrium assumptions, if applicable, leads to an important simplification, the partial equilibrium assumption for non-steady-state species is more difficult to handle. In Peters and Williams [7], partial equilibrium between CO and H_2 was assumed for the downstream oxidation layer (in addition to the steady state of H), which led to the two-step mechanism

$$\begin{aligned} CH_4 + O_2 &= \frac{2}{1+\alpha} X + \frac{2\alpha}{1+\alpha} H_2O + \frac{1-\alpha}{1+\alpha} CO_2 \\ \frac{2\alpha}{1+\alpha} X + O_2 &= \frac{2}{1+\alpha} H_2O + \frac{2}{1+\alpha} CO_2 . \end{aligned} \tag{16}$$

where the intermediate species X is a combination of H_2 and CO

$$X = H_2 + \alpha CO \tag{17}$$

and

$$\alpha = X_{CO}/(X_{H_2O} K_{II}) \tag{18}$$

relates H_2 and CO to each other as $X_{CO} = \alpha X_{H_2}$. Here, the X_i's are again normalized mole fractions and K_{II} is the equilibrium constant of the second global reaction in (11). Since both H_2 and CO are not steady-state species, their balance equations must be combined and an equation for the combined species X solved. If the diffusivities are not assumed equal, this can lead to complications if both the convective and diffusive terms in the balance equations must be considered.

Principle Step 6

Justify the assumptions by asymptotic analysis.

Details

Even if satisfactory results are obtained for the comparison of a reduced mechanism with the full mechanism (as shown in Fig. 1 for the flame velocity of methane, methanol, and propane flames, reproduced from Paczko et al. [4]), the success may appear fortuitous. The question is whether a different choice for the steady-state species would have given equally good results or if one could introduce even further steady-state assumptions, for instance, for the H atom. This becomes clearer if one considers the asymptotic structure of methane flames derived by Peters and Williams [7] and shown in Fig. 2. Here, in fact, the additional assumption of the steady state of H was originally introduced, which leads to the three-step mechanism

$$\begin{array}{ll} \mathrm{I} & CH_4 + O_2 = CO + H_2 + H_2O \\ \mathrm{II} & CO + H_2O = CO_2 + H_2 \\ \mathrm{III} & O_2 + 2H_2 = 2H_2O\,. \end{array} \qquad (19)$$

Four different layers could be identified in the structure of the premixed stoichiometric methane flame (and similarly in a methane-air diffusion flame as analiyzed by Seshadri and Peters [8]). In terms a coordinate nondimensionalized with ℓ_F, the flame structure consists of a chemically inert preheat zone of thickness O(1), a thin fuel consumption layer of thickness δ, where CH_4 is consumed and H_2 and CO are formed due to reaction I, and a downstream CO/H_2-oxidation layer of thickness ϵ governed by the rate of the reaction $H + O_2 + M \rightarrow HO_2 + M$. At the leading edge of the oxidation layer, there is a nonequilibrium layer of reaction II with thickness ν that tends to equilibrium downstream. The ordering of the relative width of the layers is $\delta < \nu < \epsilon < 1$. A blowup of the fuel consumption layer in the lower part of Fig. 2 shows the existence of an even thinner radical consumption layer between the fuel consumption layer and the chemically inert preheat zone upstream. In this layer, where the radicals are depleted by the fuel due to the global reaction $CH_4 + 2H + H_2 = CO + 4H_2$, the steady-state assumption for H is no longer valid. However, although the existence of the radical consumption layer is essential for the existence of a thin chemically reacting part of the flame, its detailled structure does not need to be known for the analysis. This explains why the original assumption of a steady state of H is acceptable, but requires additional care in solving the steady-state equations. Since H cannot coexist with a large amount of CH_4, one would obtain nonpositive values for H and also the other radicals in this part of a flame, which must be set equal to zero for physical reasons. In addition to these considerations, asymptotic analysis can justify the original choice of reactions within the starting mechanism as discussed in principle step 1.

Principle Step 7

Identify limitations of a reduced reaction mechanism.

Details

It is evident that the approximations introduced by the steady-state and partial equilibrium assumptions will reduce the accuracy of the result as compared to the complete set of equations and rates. Differences are expected in particular for those species that were assumed to be in steady state, whereas global properties such as flame speed and heat release rates will be much less affected. As a result, the prediction of properties that depend essentially on the concentrations of minor species remains a challenge. Such properties of a flame are, for instance, the formation of NO that depends on the O atom concentration and of soot that depends on higher hydrocarbon radicals. One may argue that the prediction of these species concentrations remains inaccurate even with large kinetic schemes, as long as the kinetic mechanism is not well enough established and as long as a certain inaccuracy of the rate coefficients persists. Nevertheless, the investigation of the approximations with respect to pollution formation needs to be pursued.

3. HYDROGEN AIR-FLAMES

In a similar way as for hydrocarbon-air flames, hydrogen-air flames have been calculated with the mechanism 1–11 of Table 3 as well with a reduced mechanism. The reduction assumes that only O and HO_2 are in steady state such that there results after elimination of w_2 and w_7 the three-step mechanism

$$\begin{aligned} &\mathrm{I} && \mathrm{H_2 + O_2 = 2OH} \\ &\mathrm{II} && \mathrm{OH + H_2 = H_2O + H} \\ &\mathrm{III} && \mathrm{H + H} + M = \mathrm{H_2} + M \end{aligned} \tag{20}$$

with the rates

$$\begin{aligned} w_{\mathrm{I}} &= w_1 + w_6 + w_9 \\ w_{\mathrm{II}} &= w_3 + w_4 + w_8 + w_9 + w_{11} \\ w_{\mathrm{III}} &= w_5 + w_{10} + w_{11} \,. \end{aligned} \tag{21}$$

Here again, the truncated steady-state relation given by (15) was used for O, whereas HO_2 was determined from its steady-state relation. Calculated flame velocities of H_2-air flames at 1 atm obtained with a computer program provided by Warnatz and his rates (Warnatz [9]) are plotted versus the mole fraction of H_2 in Fig. 3. With the reduced mechanism, numerical convergence could only be obtained for mole fractions up to 45%, whereas there was no such difficulty with the full mechanism. It is seen that the flame velocities from the three-step mechanism differ by less than 10% from the results obtained with the full mechanism. A further reduction to a two-step mechanism that corresponds to III and IV in (11) was also investigated. It showed satisfactory results near stoichiometry, but deviated from the results of the full mechanism close to the flammability limits. Therefore, for hydrogen flames, a three-step mechanism seems to be the best choice.

For stoichiometric H_2-air flames at 1 atm, the profiles of the species and the temperature that are not in steady state are compared for both calculations in Fig. 4. Here again, there is hardly any difference between the two cases.

4. SELF-IGNITION OF HYDROGEN-OXYGEN MIXTURES

One of the most simple combustion problems and therefore a challenging application for reduced mechanisms is the self-ignition of a homogeneous mixture. It was generally believed that different from flame calculations, the homogeneous explosion problem is more sensitive to approximations of the governing equations. It is true, for instance, that the partial equilibrium assumption of reaction 3, (14), breaks down at low temperatures. Differently from flames, reactions that involve H_2O_2 now become important. Therefore, starting from the entire mechanism of Table 3 and by assuming O, HO_2, and H_2O_2 in steady state, by which w_2, w_7 and w_{17} are eliminated, one obtains again the three-step mechanism (20) with the rates given by

$$\begin{aligned} w_{\mathrm{I}} &= w_1 + w_6 + w_9 - w_{12} + w_{14} + w_{16} - w_{18} - w_{19} \\ w_{\mathrm{II}} &= w_3 + w_4 + w_8 + w_9 + w_{11} + w_{13} + w_{14} + w_{15} - w_{16} - w_{18} \\ w_{\mathrm{III}} &= w_5 + w_{10} + w_{11} + w_{12} + w_{15} \,. \end{aligned} \tag{22}$$

For a homogeneous stoichiometric hydrogen-oxygen mixture at a pressure of 0.1 mbar and different initial temperatures, the evolution of the concentrations and the temperature were calculated. For an initial temperature of 1000 K, this is shown in Fig. 5. The ignition delay time t_i is defined by the time where the fuel has decreased by one-third. It is seen in Fig. 5 that it is a time where most of the chemical activity happens. Any appropiate alternative definition of the ignition delay time would lead to very similar values. In Fig. 6, the logarithm of t_i is plotted versus the inverse of the initial temperature T_0 in an Arrhenius diagram. From the slope, an overall activation energy of 13.6 kJ/kmol can be deduced at $T_0 = 1000$ K. However, at initial temperatures lower than $T_0 = 800$ K, the curve no longer follows an Arrhenius behavior. The ignition delay times increase very rapidly, indicating the vincinity of the second flammability limit that is due to the competition of reactions 1 and 5 in Table 3. The second flammability limit taken from Lewis and von Elbe [3] is also shown in Fig. 6. It should be noted, however, that the corresponding experiments were performed in a vessel of finite size and that reactions that occur at the walls may have slightly influenced the experimental results. The difference between the ignition delay times obtained from the full mechanism and the reduced mechanism still varies by about 20%, but the global behavior, as shown in Fig. 6, is qualitatively similar. Therefore, we believed that the reduced mechanism can also profitably be derived for self-ignition problems for hydrocarbon mixtures.

5. CONCLUSIONS

A systematic reduction strategy for flame kinetics has been outlined and results are discussed for premixed hydrocarbon and hydrogen flames but also for the self-ignition of hydrogen-oxygen mixtures. Calculated flame velocities agree well with those obtained from the full mechanism. The agreement is better for hydrocarbon than for hydrogen flames. But even for the ignition of H_2-O_2 mixtures at low pressures, a good overall agreement of global properties can be obtained.

Table 1 Starting mechanism for methane flames

No.	Reaction		
1	O_2+H	⇌	OH+O
2	O+H_2	⇌	H+OH
3	OH+H_2	⇌	H+H_2O
4	OH+OH	⇌	H_2O+O
5	H+O_2+M	→	HO_2+M
6	HO_2+H	→	H_2+O_2
7	HO_2+H	→	H_2+O_2
8	HO_2+OH	→	H_2O+O_2
9	H+OH+M	→	H_2O+M
10	CO+OH	⇌	CO_2+OH
11	CH_4+H	→	H_2+CH_3
12	CH_4+OH	→	H_2O+CH_3
13	CH_3+O	→	CH_2O+H
14	CH_3+OH	→	CH_2O+H+H
15	CH_3+OH	→	CH_2O+H_2
16	CH_3+H	→	CH_4
17	CH_2O+H	→	CHO+H_2
18	CH_2O+H	→	CHO+H_2
19	CHO+H	→	CO+H_2
20	CHO+OH	→	CO+H_2O
21	CHO+O_2	→	CO+HO_2
22	CHO+M	→	CO+H+M

Underlined reaction numbers indicate reactions of the main chain.

Table 2 Maximum percent mole fractions in a premixed stoichiometric methane-air flame

	H_2	CO	H	OH	O	CH_3	CH_2O	CHO	HO_2
Original	1.9	4.3	0.64	0.57	0.21	0.3	0.13	0.001	0.022
Weighted	5.185	4.3	2.428	0.656	0.246	0.359	0.128	0.001	0.021

Table 3 A starting mechanism for homogeneous explosion of H_2-O_2-mixtures

No.	Reaction		
1	O_2+H	$\rightleftharpoons$	OH+O
2	O+H_2	$\rightleftharpoons$	H+OH
3	OH+h_2	$\rightleftharpoons$	H_2O+H
4	OH+OH	$\rightleftharpoons$	H_2O+H
5	H+O_2+M′	$\rightleftharpoons$	H_2O+M′
6	H+HO_2	$\rightleftharpoons$	OH+OH
7	H+HO_2	$\rightleftharpoons$	H_2+O_2
8	OH+HO_2	$\rightleftharpoons$	H_2O+H
9	H+HO_2	$\rightleftharpoons$	H_2O+O
10	H+H+M′	$\rightleftharpoons$	H_2+M′
11	H+OH+M′	$\rightleftharpoons$	H_2O+M′
12	O+O+M′	$\rightleftharpoons$	O_2+M′
13	O+HO_2	$\rightleftharpoons$	OH+O_2
14	HO_2+HO_2	$\rightleftharpoons$	H_2O_2+O_2
15	OH+OH+M′	$\rightleftharpoons$	H_2O_2+M′
16	H_2O_2+H	$\rightleftharpoons$	H_2+HO_2
17	H_2O_2+H	$\rightleftharpoons$	H_2O+OH
18	H_2O_2+O	$\rightleftharpoons$	OH+HO_2
19	H_2O_2+OH	$\rightleftharpoons$	H_2O+HO_2

Underlined reaction numbers indicate reactions of the main chain.

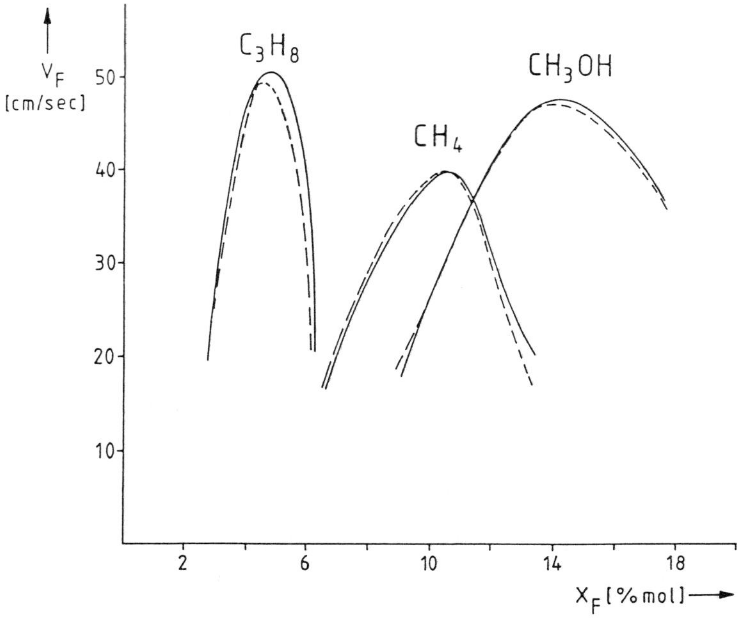

Fig. 1 Flame velocities at $p = 1\,\text{atm}$ as a function of the fuel mole fraction (– – – starting mechanism, ——— reduced four-step mechanism).

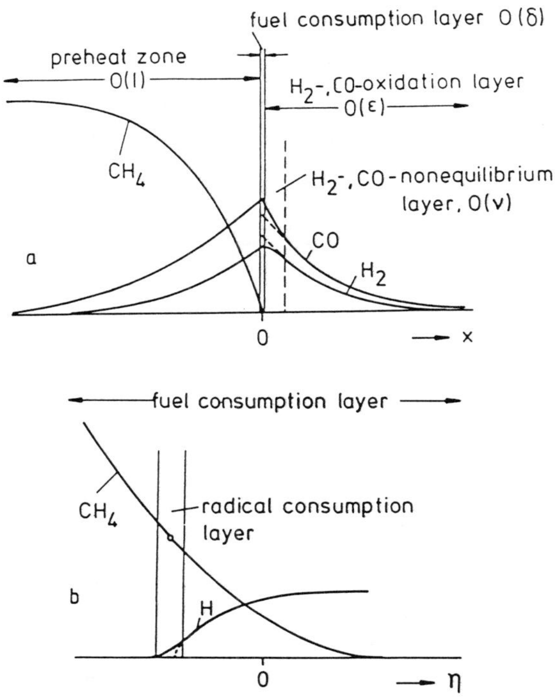

Fig. 2 The asymptotic structure of a premixed CH_4-air flame.

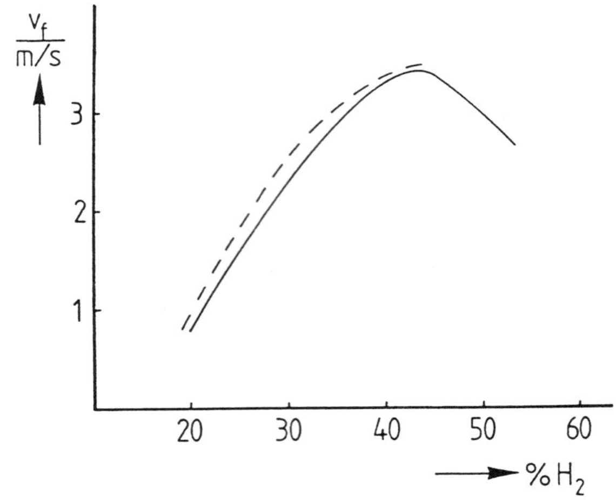

Fig. 3 Flame velocities of H_2-air flames at $p = 1\,\text{atm}$ as a function of the H_2 mole fraction (——— starting mechanism, – – – reduced three-step mechanism).

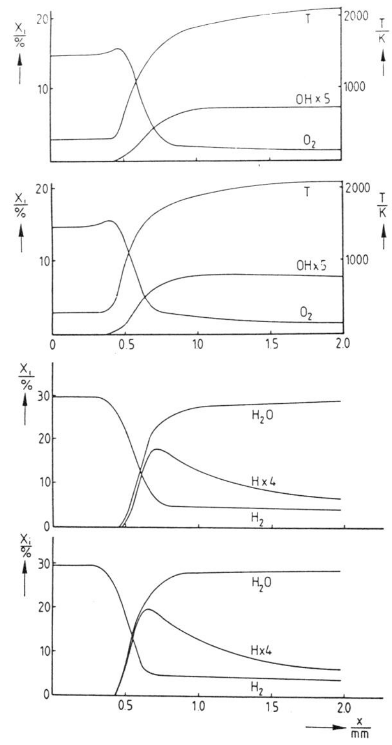

Fig. 4 Profiles of mole fractions and temperature for stoichiometric H_2-air flames at 1 atm. Upper part in each diagram: starting mechanism. Lower part: reduced three-step mechanism.

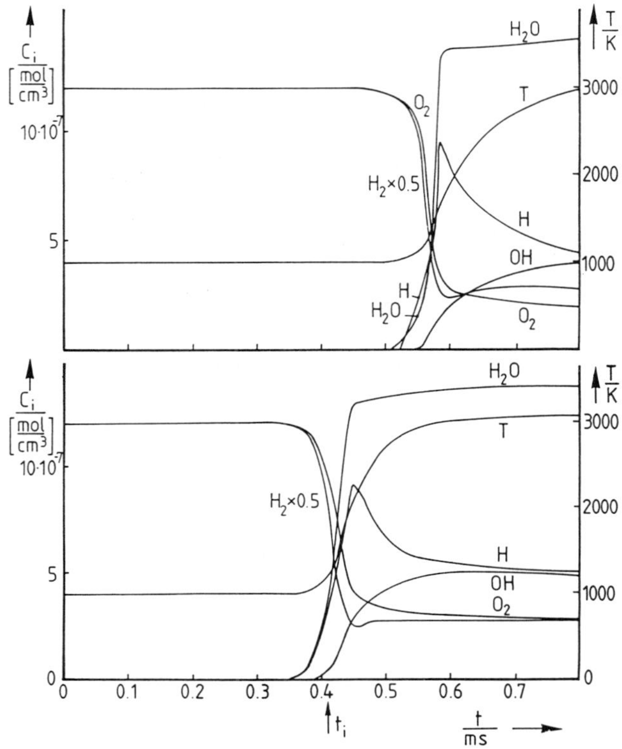

Fig. 5 Evolutions of concentrations and temperature during self-ignition of a stoichiometric H_2-O_2 mixture at 0.1 mbar and initial temperature $T_0 = 1000\,K$. Upper part: starting mechanism. Lower part: reduced three-step mechanism.

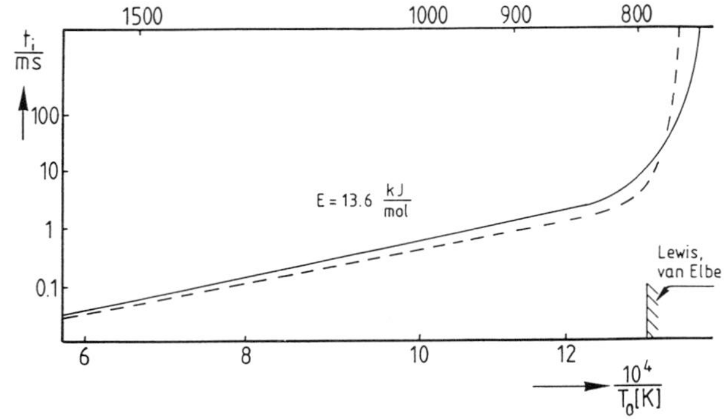

Fig. 6 Ignition delay times for stoichiometric H_2-O_2 mixtures at 0.1 mbar as a function of the initial temperature.

REFERENCES

[1] J.R. Bowen, A. Acrivos, A.K. Oppenheim, *Singular perturbation refinement to quasi-steady-state approximation in chemical kinetics*, Chem. Eng. Sci. **18**, 177–188, 1963.

schemes for hydrocarbon flames, Lecture Notes in Physics, 90–109, 1985.

[6] N. Peters, R.J. Kee, *The computation of stretched laminar methane-air diffusion flames using a reduced four-step mechanism*, Combust. Flame **68**, 17–30, 1987.

[7] N. Peters, F.A. Williams, *The asymptotic structure of stoichiometric methane-air flames*, Combust. Flame **68**, 185–207, 1987.

[8] K. Seshadri, N. Peters, *Asymptotic structure and extinction of methane-air diffusion flames*, Combust. Flame **73**, 23–44, 1988.

[9] J. Warnatz, *Rate coefficients in the C/H/O-system*, Combustion Chemistry (edited by W.C. Gardiner Jr.), Springer-Verlag, New York, 197–360, 1984.

[10] C.K. Westbrook, F.L. Dryer, *Chemical kinetic modeling of hydrocarbon combustion*, Progr. Energy Combust. Sci. **10**, 1–57, 1984.

N. Peters
Institut für Technische Mechanik, RWTH Aachen
Templergraben 64
D-5100 Aachen
Federal Republik of Germany

5 Numerical combustion

B LARROUTUROU

Mathematics contribute to the progress of combustion science

1. INTRODUCTION

The interest of the numerical simulations for the progress of combustion theory is now widely accepted by all concerned scientists, and the interaction between the formal investigations of the physicists and the numerical simulation of combustion phenomena can now be illustrated by numerous examples. But such is not the case for the mathematical studies in combustion, which are still sometimes considered are purely academic exercises.

It is true that some of these studies rigorously prove some results which the physicists had already shown using formal methods, and moreover that the rigorous proofs of these results by the mathematicians are more lengthy and technical than the corresponding arguments of the physicists. But this is far from being always the case, and we believe that mathematical studies now actually contribute to increasing our knowledge on several basic combustion phenomena: mathematics both contribute to our understanding of combustion and help for the numerical simulation of combustion.

This is what we want to show in this paper. We present in Section 2 two examples where the rigorous mathematical studies of some academic models of flame propagation theory either concern a problem which has not been or has been only partially investigated using formal asymptotics, or give a new light on a physically interesting situation. In Section 3, we describe three examples where the mathematical results are helping the numerical analysts for designing adequate numerical methods for the simulation of various combustion phenomena.

2. CONTRIBUTING TO UNDERSTAND COMBUSTION

2.1. Analysis of the multidimensional thermo-diffusive model

The so-called thermo-diffusive model, which is derived in the framework of the constant density approximation (see Barenblatt *et al.* [3], Larrouturou [20], Sivashinsky [29], [30]), is a well-known simplified model which decouples the reaction-diffusion and the hydrodynamical effects in the flame phenomenon. This model is essentially valid for situations where the hydrodynamical effects play a secondary role with respect to the reactive and diffusive effects, in other words, when the gas flow is not too far from being uniform. But it has allowed numerous improvements in the understanding of flame propagation phenomena, because it retains many features of flame propagation phenomena, including cellular flame instabilities; let us finally mention that this model formally leads (using high activation energy asymptotics) to the Kuramoto-Sivashinsky equation for the evolution of the flame front (see [20], [29], [30]).

We will consider the classical academic situation of a flame propagating in a mixture of two species $\mathcal{R}$ and $\mathcal{P}$, with the single one-step reaction $\mathcal{R} \to \mathcal{P}$, in an infinite tube $S = \{(x, \bar{y}) \in I\!R^3,\ x \in I\!R,\ \bar{y} \in \mathcal{O}\}$, where $\mathcal{O}$ is a bounded domain in $I\!R$ or $I\!R^2$. In the framework of the constant-density approximation, the velocity field is considered as given and is divergence free: in the case under consideration, it is natural to choose a velocity field parallel to the tube walls: $\vec{v} = [u(\bar{y}), 0]$, and the normalized thermo-diffusive model then takes the form (we refer to e.g. [8], [20] for the details):

$$\begin{cases} T_t + u(\bar{y})T_x = Y f_{0\epsilon}(T) + \Delta T \ , \\ \\ Y_t + u(\bar{y})Y_x = -Y f_{0\epsilon}(T) + \dfrac{1}{\mathcal{L}}\Delta Y \ . \end{cases} \tag{2.1}$$

Here, T is the normalized temperature of the mixture, Y is the normalized mass fraction of the reactant $\mathcal{R}$, $\mathcal{L}$ is the Lewis number (a positive constant), and Δ is the Laplace operator in S. The term $Y f_{0\epsilon}(T)$ in the right-hand side of (2.1) is the reaction term; the subscript ϵ is indicated here to emphasize that the non linear term $f_{0\epsilon}$ depends on the activation energy: ϵ is a (small) positive parameter proportional to the inverse of the activation energy. The subscript 0 indicates that we consider here a model with an ignition temperature (see (2.8) below).

We will consider the steady, or more precisely the travelling-wave solutions $T(x + ct, \bar{y})$, $Y(x + ct, \bar{y})$ of (2.1). They are given by the system:

$$\begin{cases} [c + u(\bar{y})]T_x = Y f_{0\epsilon}(T) + \Delta T \ , \\ \\ [c + u(\bar{y})]Y_x = -Y f_{0\epsilon}(T) + \dfrac{1}{\mathcal{L}}\Delta Y \ \text{ in } S = I\!R \times \mathcal{O} \ , \end{cases} \tag{2.2}$$

with the normalized boundary conditions:

$$\begin{cases} T(-\infty, \bar{y}) = 0 \ , \quad Y(-\infty, \bar{y}) = 1 \ , \\ T(+\infty, \bar{y}) = 1 \ , \quad Y(+\infty, \bar{y}) = 0 \ \text{ for } \bar{y} \in \mathcal{O} \ , \end{cases} \tag{2.3}$$

at the tube ends, and:

$$\frac{\partial T}{\partial n} = 0 \ , \quad \frac{\partial Y}{\partial n} = 0 \ \text{ on } I\!R \times \partial\mathcal{O} \ , \tag{2.4}$$

at the tube walls; we have used the notation n for the outward unit normal on $\partial S = I\!R \times \partial\mathcal{O}$.

The unknowns in (2.2)-(2.4) are the functions $T(x, \bar{y})$, $Y(x, \bar{y})$ ($\mathcal{C}^2$ functions in S) and the real number c. We recall that the velocity $u(\bar{y})$ is given.

When $\mathcal{L} = 1$, it is easy to check that any solution of (2.2)-(2.4) satisfies the identity $T + Y \equiv 1$. The problem then reduces to a single partial differential equation:

$$[c + u(\bar{y})]T_x = g_{0\epsilon}(T) + \Delta T \ \text{ in } S = I\!R \times \mathcal{O} \ , \tag{2.5}$$

$$T(-\infty, \bar{y}) = 0 \ , \quad T(+\infty, \bar{y}) = 1 \ \text{ for } \bar{y} \in \mathcal{O} \ , \tag{2.6}$$

$$\frac{\partial T}{\partial n} = 0 \ \text{ on } I\!R \times \partial\mathcal{O} \ ; \tag{2.7}$$

we have set $g_{0\epsilon}(s) = (1 - s) f_{0\epsilon}(s)$.

One of the main interests of models (2.2)-(2.4) and (2.5)-(2.7) lies in the fact that they are the simplest models where we can investigate the influence of a non uniform velocity field on flame propagation (of course, the opposite influence of the flame on the flow has been removed in these simplified models). The general question of the existence of a travelling-wave solution for any given velocity field u and any Lewis number $\mathcal{L}$ is far from being solved, either from a formal point of view or from a rigorous mathematical point of view, and one may wonder if, in the framework of these models, a highly non uniform shear flow would have the effect of extinguishing the flame or not (see Berestycki-Sivashinsky [17]).

The system (2.2)-(2.4) for a non unit Lewis number has only been studied in [17] using formal asymptotics; a first and recent mathematical result on this problem can be found in [10]. On the opposite, the model (2.5)-(2.7) with $\mathcal{L} = 1$ has been the subject of numerous mathematical investigations, whose conclusions are of interest from both mathematical and physical standpoints. We now present the main results of these works:

Theorem 1: [9], [16]

Assume that:

$$\begin{cases} g_{0\epsilon} \text{ is Lipschitz} - \text{continuous on } [0,1] \ , \\ \exists \theta \in]0,1[\ , \quad g_{0\epsilon} \equiv 0 \text{ on } [0,\theta] \ , \quad g_{0\epsilon} > 0 \text{ on }]\theta, 1[\ , \quad g_{0\epsilon}(1) = 0 \ , \end{cases} \tag{2.8}$$

and that the given velocity field $u(\bar{y})$ is continuous on $\mathcal{O}$, and set: $<u> = \frac{1}{|\mathcal{O}|}\int_{\mathcal{O}} u(\bar{y})d\bar{y}$. Then the problem (2.4)-(2.7) has a unique solution (c, T), which satisfies:

$$<u> + c > 0 \ , \tag{2.9}$$

$$0 \leq T(x, \bar{y}) \leq 1 \quad \text{for all } (x, \bar{y}) \in S \ , \tag{2.10}$$

$$T_x > 0 \quad \text{in } S \ . \ \bullet \tag{2.11}$$

The proof of this result is naturally much more complex that the proof of the analogous result for planar flames (see Berestycki *et al.* [14]). It uses a priori estimates obtained from Schauder estimates or from the maximum principle for elliptic partial differential equations, a topological degree argument, and in particular a monotonicity result shown in Berestycki-Nirenberg [15] for semi-linear elliptic equations in cylindrical domains. The existence of the solution (c, T) has been proved in Berestycki *et al.* [9], and its uniqueness (which is to be understood as uniqueness up to a translation of the origin) has been shown by Berestycki-Nirenberg [16] (see also [11]).

The existence of a solution (c, T) had first been shown using somewhat simpler mathematical arguments in Berestycki-Larrouturou [7] under the additional hypothesis that the velocity field u was "not too far from being uniform". More precisely, the following result was shown in [7]:

Proposition 2: [7]

Assume that the hypotheses of Theorem 1 hold, and that:

$$\left(\max_{\bar{y}\in\mathcal{O}} u(\bar{y}) - \min_{\bar{y}\in\mathcal{O}} u(\bar{y})\right) < \sqrt{2\int_0^1 g_{0\epsilon}(s)ds} \ . \tag{2.12}$$

Then the solution (c, T) of (2.5)-(2.7) satisfies:

$$c + \min_{\bar{y}\in\mathcal{O}} u(\bar{y}) > 0 \ . \ \bullet \tag{2.13}$$

In contrast with this situation, the solutions obtained for a general velocity field do not necessarily satisfy (2.13): $c + u(\bar{y})$ may in general change sign in the domain $\mathcal{O}$. This phenomenon may be interpreted as an "inversion of the velocity field". Indeed, (2.5) shows that $c + u(\bar{y})$ is the mixture velocity in the reference frame R_f in which the solution is stationary (R_f is a reference frame attached to the flame and moves with the velocity $-c$ with respect to the original reference frame R_0). Then, (2.13) means that, at every point, the velocity in the reference frame R_f points from left to right, i.e. from the fresh mixture towards the burnt gases, a physically natural situation. But, for solutions satisfying:

$$c + \min_{\bar{y}\in\mathcal{O}} u(\bar{y}) < 0 \ , \tag{2.14}$$

there are regions of the tube (where $c + u(\bar{y}) < 0$) where the velocity is directed in the opposite way, from the burnt gases towards the fresh mixture! It is important to realize here that this non classical situation is by no means non physical: it has indeed been known for a long time (not for a flame in a tube, but in other geometrical configurations, such as for a counterflow diffusion flame; see Williams [31, p. 418]) that the *mixture* velocity in the neighbourhood of the flame may be pointing from the burnt gases towards the fresh mixture. This simply means that the convective effects are locally dominated by the diffusive effects. Moreover, in these conditions, (2.9) says that the average velocity in the reference frame R_f is necessarily positive.

Moreover, one proves in [9] that solutions of (2.5)-(2.7) which exhibit this inversion of the velocity field (i.e. which satisfy (2.14)) actually exist. In fact, one shows that condition (2.12) which insures that the inversion of the velocity field does not occur is optimal in the following sense:

Proposition 3: [9]

For any $\delta > 0$, there exist functions u and $g_{0\epsilon}$ satisfying the assumptions of Theorem 1 and the inequalities:

$$\sqrt{2\int_0^1 g_{0\epsilon}(s)ds} < \left(\max_{\bar{y}\in\mathcal{O}} u(\bar{y}) - \min_{\bar{y}\in\mathcal{O}} u(\bar{y})\right) < \sqrt{2\int_0^1 g_{0\epsilon}(s)ds} + \delta \ , \tag{2.15}$$

and such that the corresponding solution (c, T) of (2.4)-(2.7) has the inversion of the velocity field (i.e. satisfies (2.14)). $\bullet$

The existence result obtained in [9] for any given velocity field had in fact been conjectured in [7] in view of some numerical results obtained by Benkhaldoun-Larrouturou [5].

In these numerical experiments, one considers a tube of width L (the domain $\mathcal{O}$ is reduced to the interval $]0, L[$), with a velocity field of the form $u(y) = A \sin\left(\frac{\pi y}{L}\right)$, and one observes that the inversion of the velocity in the flame reference frame does occur as soon as the amplitude A is large enough.

These results even enlight in a new and interesting way the conclusions of the formal analysis of Berestycki-Sivashinsky [17]: in [17], one examines the same problem as in [5], again with the velocity $u(y) = A \sin\left(\frac{\pi y}{L}\right)$, but now for a non unit Lewis number, and one investigates, in the limit of high activation energies, the formal asymptotic expansions of the solutions (c, T, Y) of (2.2)-(2.4) (which are assumed to exist), *assuming that the inversion of the velocity does not occur* ($c + u(y) > 0$ for all y). Then, one observes that the asymptotic expansion of these solutions no longer exists if the amplitude A of the velocity field becomes too large, and the authors conclude from this fact that the extinction of the flame occurs when the amplitude A is increased above some critical value. In view of the previously recalled mathematical and numerical analyses (which concern the simplest case where $\mathcal{L} = 1$), it seems that increasing A does not produce flame quenching, but more simply produces the inversion of the velocity.

To conclude this paragraph, we mention two other physically interesting mathematical results on problem (2.5)-(2.7). First, the rigorous analysis of the travelling-wave solutions of (2.5)-(2.7) has been carried out by Berestycki *et al.* [13] and Roquejoffre [27]: for any given velocity field u (that is, even if the inversion of the velocity field occurs), the steady solution of (2.5)-(2.7) is stable. The linearized stability analysis is presented in [13], while Roquejoffre [27] investigates the nonlinear stability by considering the unsteady problem (2.1) with an initial data which is close enough to the steady solution.

Lastly, we turn to the asymptotic analysis of problem (2.5)-(2.7) for high activation energies. We will again assume here that $\mathcal{O}$ is the interval $]0, L[$, and that $g_{0\epsilon}$ satisfies (2.8) and the following hypotheses:

$$g_{0\epsilon} \in \mathcal{C}^1([0,1], I\!R_+) \ , \quad {g_{0\epsilon}}'(1) < 0 \ , \tag{2.16}$$

$$\forall\, \delta > 0, \ \lim_{\epsilon \to 0} \left[\max_{0 \le s \le 1-\delta} g_{0\epsilon}(s) \right] = 0 \ , \tag{2.17}$$

$$\lim_{\epsilon \to 0} \int_0^1 g_{0\epsilon}(s) ds = m > 0 \ . \tag{2.18}$$

We will call now $(c_{0\epsilon}, T_{0\epsilon})$ the unique solution of (2.5)-(2.7) which moreover satisfies the condition:

$$\max_{y \in \mathcal{O}} T_{0\epsilon}(0, y) = \theta \ , \tag{2.19}$$

and we will simply quote here a result of Berestycki-Nirenberg [16]:

Theorem 4: [16]

Let (ϵ_n) be a sequence of positive real numbers with $\lim_{n \to +\infty} \epsilon_n = 0$. Under the assumptions (2.8) and (2.16)-(2.18), on can extact a subsequence (still denoted ϵ_n) such that the corresponding solutions $(c_{0\epsilon_n}, T_{0\epsilon_n})$ of (2.5)-(2.7) and (5.19) satisfy the following:

$$c_{0\epsilon_n} \to c \quad \text{with } 0 < c < +\infty \ , \tag{2.20}$$

$$T_{0\epsilon_n} \to T \quad \text{in } \mathcal{C}(\bar{S}) \cap H^1_{loc}(S) \ . \tag{2.21}$$

The limit T is continuous in $\bar{S}$, it is non decreasing in the x-direction, and satisfies (2.6)-(2.7). Furthermore:

$$\exists\ \hat{x} > 0 \ , \quad T(x,y) = 1 \quad \forall x \geq \hat{x} \ , \quad \forall y \in \mathcal{O} \ . \tag{2.22}$$

Denoting $\mathcal{S}^- = \{(x,y) \in S,\ T(x,y) < 1\}$, we also have:

$$-\Delta T + cu(y)T_x = 0 \quad \text{in } \mathcal{S}^- \ . \tag{2.23}$$

Lastly, if one assumes that the boundary $\Gamma = \partial\mathcal{S}^- \cap S$ is regular (say, piecewise $\mathcal{C}^1$), and denoting T^- the restriction of T in $\mathcal{S}^-$ and ν the outward unit normal to $\mathcal{S}^-$ on Γ, we have:

$$\frac{\partial T^-}{\partial \nu} = \sqrt{2m} \quad \text{on } \Gamma \ . \ \bullet \tag{2.24}$$

A can be seen from this result, the limit (c, T) is a solution of a free-boundary problem, which consists in finding the boundary Γ and (c, T) such that :

$$-\Delta T + cu(y)T_x = 0 \quad \text{in } \mathcal{S}^- \ , \tag{2.25}$$

$$T(-\infty, y) = 0 \quad \text{for } y \in \mathcal{O} \ , \tag{2.26}$$

$$\frac{\partial T}{\partial n} = 0 \quad \text{on } \partial S \cap \partial\mathcal{S}^- \ , \tag{2.27}$$

$$T = 1 \quad \text{and} \quad \frac{\partial T}{\partial \nu} = \sqrt{2m} \quad \text{on } \Gamma \ . \tag{2.28}$$

This rigorous asymptotic analysis for curved thermo-diffusive flames therefore recovers the same conclusions as the classical formal asymptotic analysis (see e.g. [20], [30]). The questions of the uniqueness of the solution (c, T, Γ) of problem (2.25)-(2.28) and consequently of the global convergence of the solutions $(c_{0\epsilon}, T_{0\epsilon})$ towards the limit (c, T) are studied in Berestycki *et al.* [6].

2.2. Analysis of the cold boundary difficulty

Let us now consider the well-known cold boundary difficulty, which arises when investigating the simplest problem of flame theory, namely the propagation of a steady planar flame in the infinite tube S. We will see here how some mathematical results of Roquejoffre [26] and Roquejoffre *et al.* [28] provide a mathematical answer to this classical paradox of the combustion theory. The analysis of the cold boundary difficulty which we now present is also described in detail in Berestycki *et al.* [12].

To state the problem, we again assume a one-step chemical mechanism, and we use the classical isobaric approximation. Referring to [12], [20] or [26] for the details, we simply write (using normalized variables and a Lagrangian space coordinate) the equations for an unsteady planar flame in the form:

$$\begin{cases} T_t = Y f_\epsilon(T) + T_{xx} + \dfrac{\gamma - 1}{\gamma} \dfrac{T_u'(t)}{T_u(t)} T \ , \\ \\ Y_t = -Y f_\epsilon(T) + \dfrac{1}{\mathcal{L}} Y_{xx} \ , \end{cases} \tag{2.29}$$

$$T(x,0) = T^0(x) \ , \quad Y(x,0) = Y^0(x) \ , \tag{2.30}$$

$$\begin{cases} T(-\infty,t) = T_u(t) \ , \quad Y(-\infty,t) = Y_u(t) \ , \\ T(+\infty,t) = T_b(t) \ , \quad Y(+\infty,t) = 0 \ , \end{cases} \tag{2.31}$$

$$\begin{cases} T_u'(t) = \gamma Y_u(t) f_\epsilon[T_u(t)] \ , \quad T_u(0) = T_u^0 = T^0(-\infty) \ , \\ \\ T_b'(t) = \dfrac{\gamma - 1}{\gamma} \dfrac{T_u'(t)}{T_u(t)} T_b(t) \ , \quad T_b(0) = T_b^0 = T^0(+\infty) \ , \\ \\ Y_u'(t) = -Y_u(t) f_\epsilon[T_u(t)] \ , \quad Y_u(0) = Y_u^0 = Y^0(-\infty) \end{cases} \tag{2.32}$$

In these equations, γ is the specific heat ratio of the mixture (the same for both species).

The travelling-wave solutions $T(x + ct)$, $Y(x + ct)$ of (2.29)-(2.32) are given by:

$$\begin{cases} cT_x = Y f_\epsilon(T) + T_{xx} \ , \\ \\ cY_x = -Y f_\epsilon(T) + \dfrac{1}{\mathcal{L}} Y_{xx} \ , \end{cases} \tag{2.33}$$

$$\begin{cases} T(-\infty) = T_u^0 \ , \quad Y(-\infty) = Y_u^0 \ , \\ T(+\infty) = T_b^0 \ , \quad Y(+\infty) = 0 \ . \end{cases} \tag{2.34}$$

We will henceforth assume that the boundary values T_u^0, T_b^0 and Y_u^0 satisfy:

$$T_b^0 = T_u^0 + Y_u^0 \ , \quad Y_u^0 > 0 \ ; \tag{2.35}$$

the first relation in (2.35) is a balance equation obtained by integrating over $I\!R$ the sum of the two differential equations (2.33), and the inequality (2.35) says that the fresh gases at the left end of the tube actually contain some reactant (and that the burnt gases at the right end of the tube are hotter than the fresh mixture).

We assume in this section that the non linear reaction term f_ϵ is the true Arrhenius term (i.e., that this term has not been modified by introducing an ignition temperature, as in (2.8)): f_ϵ is proportional to the Arrhenius exponential term $\exp\left(-\dfrac{\mathcal{E}}{RT}\right)$, where $\mathcal{E}$ is the activation energy. This is the reason why the boundary conditions $T_u(t)$, $T_b(t)$ and $Y_u(t)$ depend on time in (2.29)-(2.32).

Now, problem (2.33)-(2.34) raises the well-known cold boundary difficulty. The difficulty is the following: the value of the reaction rate in the fresh mixture does not vanish: $Y_u^0 f_\epsilon(T_u^0) \neq 0$. It is then easy to see that (2.33)-(2.34) has no solution (the fact that $\lim_{x \to -\infty} [cT_x - T_{xx}] \neq 0$ contradicts the fact that T is bounded in the neighbourhood of $-\infty$).

From the physical point of view, the origin of the difficulty is clear: the state (Y_u^0, T_u^0) prescribed at $-\infty$ is not an equilibrium state, and the problem is therefore ill-posed.

On the other hand, there is a well-established experimental evidence that steady planar premixed flames do exist, and solutions of (2.33)-(2.34) are therefore expected to exist.

In fact, the difficulty is essentially mathematical: the actual value of the reaction rate inside the fresh mixture is non zero, but is extremely small compared to the reaction term value inside the flame: this is due to the fact that the activation energy $\mathcal{E}$ is large, i.e. that the ratio $\dfrac{\mathcal{E}}{RT}$ is large for temperatures T in the considered range of temperature $[T_u^0, T_b^0]$. In practice, the ratio $\dfrac{e^{-\mathcal{E}/RT_u^0}}{e^{-\mathcal{E}/RT_b^0}}$ may well be of the order of e^{-50} . In other words, the characteristic time τ_u of the chemical reaction in the fresh mixture (at temperature T_u^0) is extremely large (as large as a big number of billions of years !). One may therefore think, if one believes in the Arrhenius expression of the chemical reaction term, that the experimentally observed "steady premixed flames" are not really steady; but they are evolving over a characteristic time which is of the same order of magnitude as τ_u. Therefore, although there exists no steady solution in the mathematical sense to the above systems, one may expect that these systems have an unsteady solution, *which remains "during a long time" "very close" to a steady flame.* This is exactly what has been proved in [12], [28], which we now describe.

Let us begin by stating the existence and uniqueness result for unsteady solutions. Introducing a function Γ satisfying:

$$\begin{cases} \Gamma \in \mathcal{C}^\infty(I\!R, I\!R) , \\ \Gamma \equiv 0 \text{ on } (-\infty, -1) , \quad \Gamma \equiv 1 \text{ on } (1, +\infty) , \end{cases} \tag{2.36}$$

and setting:

$$\begin{cases} T^0(x) = T_u^0 + (T_b^0 - T_u^0)\Gamma(x) + \phi^0(x) , \\ Y^0(x) = Y_u^0 - Y_u^0\Gamma(x) + \psi^0(x) , \end{cases} \tag{2.37}$$

we will assume that:

$$\phi^0 \in H^4(I\!R) \cap L^1(I\!R) , \quad \psi^0 \in H^4(I\!R) \cap L^1(I\!R) , \tag{2.38}$$

$$T_{min} = \inf_{x \in I\!R} T^0(x) > 0 , \quad Y^0(x) \geq 0 . \tag{2.39}$$

Moreover, we assume that:

$$\begin{cases} f_\epsilon \in \mathcal{C}^3(I\!R_+, I\!R_+) , \\ f_\epsilon \text{ is bounded and Lipschitz} - \text{continuous on } I\!R_+ . \end{cases} \tag{2.40}$$

We then have the two following results:

Proposition 5: [26]

Assume that the hypotheses (2.38)-(2.40) hold. Then, there exists a unique solution $[T_u(t), T_b(t), Y_u(t)]$ to system (2.32). This solution is bounded and exists for all time, and satisfies:

$$\begin{cases} T_u(t) > 0 \ , \quad T_u'(t) \geq 0 \ , \\ T_b(t) > 0 \ , \quad T_b'(t) \geq 0 \ , \\ Y_u(t) \geq 0 \ , \quad Y_u'(t) \leq 0 \ , \quad \lim_{t\to+\infty} Y_u(t) = 0 \ . \ \bullet \end{cases} \tag{2.41}$$

Theorem 6: [26]

Assume that the hypotheses (2.38)-(2.40) hold. Then, there exists a unique solution $[T(x,t), Y(x,t)]$ on $I\!R \times I\!R_+$ to system (2.29)-(2.32). This solution satisfies:

$$\begin{cases} T,\ Y \in L^{\infty}_{loc}(I\!R_+, L^{\infty}(I\!R)) \ , \\ 0 \leq Y(x,t) \leq \sup_{x\in I\!R} Y^0(x) \ , \\ T(x,t) \geq T_{min} \ . \ \bullet \end{cases} \tag{2.42}$$

We will consider two different applications of these results. On one hand, we consider problem (2.29)-(2.30), where the nonlinear term f_ϵ corresponds to the actual Arrhenius reaction term; we will call this problem $(\mathcal{P}_\epsilon)$. On the other hand, we will also consider the similar problem where we replace f_ϵ by a function $f_{0\epsilon}$ having an ignition temperature $T_i^0 \in]T_u^0, T_b^0[$; we will call this second problem $(\mathcal{P}_{0\epsilon})$. We will therefore assume that:

$$\begin{cases} f_{0\epsilon} \in \mathcal{C}^3(I\!R_+, I\!R_+) \ , \\ f_{0\epsilon} \text{ is bounded and Lipschitz} - \text{continuous on } I\!R_+ \ . \\ \exists T_i^0 \in (T_u^0, T_b^0) \ , \quad f_{0\epsilon} \equiv 0 \text{ on } [T_u^0, T_i^0] \ , \quad f_{0\epsilon} > 0 \text{ on } (T_i^0, +\infty) \ , \end{cases} \tag{2.43}$$

and we will set $g_{0\epsilon}(T) = (T_b^0 - T) f_{0\epsilon}(T)$ (we use here the same notation as in Section 2.1, but the assumptions on $g_{0\epsilon}$ are different; any confusion should be avoided). Moreover, we will assume for problem $(\mathcal{P}_{0\epsilon})$ that the Lewis number is unity, and that the initial data (2.30) satisfy:

$$T^0 + Y^0 \equiv T_b^0 \quad \text{on } I\!R \ . \tag{2.44}$$

Problem $(\mathcal{P}_{0\epsilon})$ then reduces to (2.45)-(2.46) below:

$$\begin{cases} T_t = g_{0\epsilon}(T) + T_{xx} \ , \\ T(x,0) = T^0(x) \ , \\ T(-\infty,t) = T_u^0 = T^0(-\infty) \ , \quad T(+\infty,t) = T_b^0 = T^0(+\infty) \ , \end{cases} \tag{2.45}$$

$$T(.,t) + Y(.,t) \equiv T_b^0 \ . \tag{2.46}$$

Lastly, we will call $(\hat{c}_{0\epsilon}, \hat{T}_{0\epsilon})$ the travelling-wave solution of (2.45):

$$\begin{cases} \hat{c}_{0\epsilon}(\hat{T}_{0\epsilon})_x = g_{0\epsilon}(\hat{T}_{0\epsilon}) + (\hat{T}_{0\epsilon})_{xx} \ , \\ \hat{T}_{0\epsilon}(-\infty) = T_u^0 \ , \quad \hat{T}_{0\epsilon}(+\infty) = T_b^0 \ . \end{cases} \tag{2.47}$$

About problems (5.45)-(5.46) and (5.47), we have the two following results:

Theorem 7: [14]

Under the hypotheses (2.43), there exists a solution $(\hat{c}_{0\epsilon}, \hat{T}_{0\epsilon})$ of (2.47), with $\hat{c}_{0\epsilon} > 0$. Moreover, this solution is unique up to a translation of the origin. •

The next result now shows that this travelling-wave solution is stable:

Theorem 8: [28]

Assume that the hypotheses (2.38)-(2.39) and (2.43)-(2.44) hold. Let $(T_{0\epsilon}, Y_{0\epsilon})$ be the unique solution of problem $(\mathcal{P}_{0\epsilon})$ (i.e. (2.45)-(2.46)), and let $(\hat{c}_{0\epsilon}, \hat{T}_{0\epsilon})$ be the solution of (2.47). Assume moreover that $T_u^0 \leq T^0(x) \leq T_b^0$ for any $x \in I\!R$, and that $\lim_{x \to -\infty} \dfrac{T^0(x)}{\exp(\hat{c}_{0\epsilon} x)}$ exists in $I\!R_+^*$. Then there exist a real number x_0 and two positive constants K and r such that, for any x and t:

$$|T_{0\epsilon}(x,t) - \hat{T}_{0\epsilon}(x + x_0 + ct)| \leq K e^{-rt} \ . \bullet \tag{2.48}$$

These results can be used to perform a mathematical large activation energy analysis. Keeping in mind that f_ϵ is the actual Arrhenius non linear reaction term and that ϵ is proportional to the inverse of the activation energy, we now use the fact that ϵ is small (but finite) to choose a function $f_{0\epsilon}$ satisfying (2.43) (i.e. having an ignition temperature) such that $\delta_\epsilon = \sup_{s>0} |f_{0\epsilon}(s) - f_\epsilon(s)|$ is also a small number. In fact, we can easily choose $f_{0\epsilon}$ such that:

$$\lim_{\epsilon \to 0} \delta_\epsilon = 0 \ . \tag{2.49}$$

Comparing the solutions of problems $(\mathcal{P}_\epsilon)$ and $(\mathcal{P}_{0\epsilon})$ is then the subject of the next lemma:

Lemma 9:

Assume that the hypotheses of Theorems 6 and 8 hold, and that (2.49) hold. Let (T_ϵ, Y_ϵ) be the solution of $(\mathcal{P}_\epsilon)$, and let $(T_{0\epsilon}, Y_{0\epsilon})$ be the solution of $(\mathcal{P}_{0\epsilon})$.

Then there exist two sequences (δ'_ϵ) and (t_ϵ) with $\lim_{\epsilon \to 0} \delta'_\epsilon = 0$ and $\lim_{\epsilon \to 0} t_\epsilon = +\infty$ such that:

$$\|T_\epsilon(.,t) - T_{0\epsilon}(.,t)\|_\infty + \|Y_\epsilon(.,t) - Y_{0\epsilon}(.,t)\|_\infty \leq \delta'_\epsilon \ , \quad \forall t \leq t_\epsilon \ . \bullet \tag{2.50}$$

Now, this result provides a mathematical answer to the paradox of the cold boundary difficulty. Under the preceding assumptions, Lemma 9 shows that, if ϵ is small enough (that is, if the activation energy is large enough), the time-dependent solution of the Arrhenius model $(\mathcal{P}_\epsilon)$ remains close during a long time to the time-dependent solution of the model with an ignition temperature $(\mathcal{P}_{0\epsilon})$; but Theorem 8 then says that the latter solution converges exponentially to a travelling wave. Therefore, we have proved that *the solution of the Arrhenius model behaves like a travelling wave during a period of time which is larger and larger as the activation energy increases.* This is the mathematical answer to the cold boundary difficulty.

3. CONTRIBUTING TO SIMULATE COMBUSTION

We now present in this section three examples of mathematical examples which are useful for the numerical simulation of flames and reacting flows.

3.1. Numerical analysis of the unsteady isobaric one-dimensional model

The first example concerns the convergence of a class of schemes for the numerical approximation of the unsteady isobaric one-dimensional model. To simplify the analysis, we will consider here instead of (2.29)-(2.32) the simpler problem (2.45).

To give an example of convergence result, let us consider the numerical solution of (2.45) with a computational grid which is equally spaced in both space and time on a bounded interval $[-a, a]$, and with an integration scheme which is implicit for the diffusion term and explicit for the reaction term. Moreover, we will use Neumann conditions at the boundaries $-a$ and $+a$ of the computational domain. The resulting scheme then writes:

$$\frac{T_j^{n+1} - T_j^n}{\Delta t} = \frac{T_{j-1}^{n+1} - 2T_j^{n+1} + T_{j+1}^{n+1}}{\Delta x^2} + g_{0\epsilon}(T_j^n) \quad \text{pour } 1 \leq j \leq N \, , \tag{3.1}$$

$$\frac{T_0^{n+1} - T_0^n}{\Delta t} = \frac{T_1^{n+1} - T_0^{n+1}}{\Delta x^2} + g_{0\epsilon}(T_0^n) \quad \text{pour } j = 0 \, , \tag{3.2}$$

$$\frac{T_{N+1}^{n+1} - T_{N+1}^n}{\Delta t} = \frac{T_N^{n+1} - T_{N+1}^{n+1}}{\Delta x^2} + g_{0\epsilon}(T_{N+1}^n) \quad \text{pour } j = N+1 \, , \tag{3.3}$$

where Δt is the time step, N is the number of interior nodes, and $\Delta x = \dfrac{2a}{N+1}$. At time $t = 0$, we set of course $T_j^0 = T^0(x_j)$, with $x_j = j\Delta x - a$ for $0 \leq j \leq N+1$.

The analysis of the scheme (3.1)-(3.3) leads to the following result:

Theorem 10: [26]

Assume that the hypotheses (2.38)-(2.39) and (2.43) hold, and that there exist two positive constants K and δ such that the initial temperature T^0 satisfies:

$$\begin{cases} |T^0(x) - T_u^0| + |T^{0\prime}(x)| \leq K \exp(\delta x) \ \ \forall x < 0 \, , \\ |T^0(x) - T_b^0| + |T^{0\prime}(x)| \leq K \exp(-\delta x) \ \ \forall x > 0 \, . \end{cases} \tag{3.4}$$

Let T be the unique solution of (2.45), and (T_j^n, Y_j^n) be the numerical solution given by the scheme (3.1)-(3.3), and assume that the time step Δt satisfies:

$$\Delta t \leq \|f_{0\epsilon}\|_\infty^{-1} \, . \tag{3.5}$$

Then for any $t_0 > 0$, there exists a positive constant $M = M(t_0)$ such that, for all j and n satisfying $0 \leq j \leq N+1$, $0 \leq n\Delta t \leq t_0$, we have:

$$|T(x_j, n\Delta t) - T_j^n| \leq M \left(\Delta t + \Delta x + \frac{\exp(-\delta a)}{\Delta x} \right) \, . \bullet \tag{3.6}$$

Therefore, this result shows the convergence of the scheme (3.1)-(3.3): for all j and n satisfying $0 \leq j \leq N+1$ and $0 \leq n\Delta t \leq t_0$, the error $|T(x_j, n\Delta t) - T_j^n|$ can be made less than any positive constant $\hat{\epsilon}$ provided that Δx and Δt are chosen small enough, and a large enough.

Remark 1: The exponential term in (3.6) comes from the fact that, under the assumption (3.4), the quantities $T'(x,t)$ and $T(x,t) - T_u^0$ (resp. $T(x,t) - T_b^0$) vanish exponentially as x tends to $-\infty$ (resp. $+\infty$), uniformly for $t \leq t_0$. •

Remark 2: Many similar convergence results can be shown, for instance for approximations which use an adaptive grid, or for various time integration schemes, or for the multi-dimensional thermo-diffusive model (see Benkhaldoun-Larrouturou [4] and [20]). All these results rely on the fact that, under some technical conditions, the considered schemes preserve the "discrete maximum principle", that is, they preserve the inequalities $0 \leq Y \leq 1$, $T_{min} \leq T$. •

3.2. Truncated approximations of travelling-wave solutions

For our second example, we again consider the basic problem of the steady planar flame with a unit Lewis number. Using here for simplicity the same normalisation as in Section 2.1, we will call (c, T) the only solution of:

$$\begin{cases} cT_x = g_{0\epsilon}(T) + T_{xx} \ , \\ T(-\infty) = 0 \ , \quad T(+\infty) = 1 \ , \end{cases} \tag{3.7}$$

which moreover satisfies $T(0) = \theta$.

In addition to the unsteady approach considered in the preceding section, the methods used for the numerical approximation of an unsteady planar flame consist in solving a problem which is analogous to (3.7) but posed in a truncated domain $[-a, a]$: for instance, one looks for c_a and T_a satisfying:

$$\begin{cases} c_a(T_a)_x = g_{0\epsilon}(T_a) + (T_a)_{xx} \text{ in } [-a, a] \ , \\ T_a(-a) = 0 \ , \quad T_a(0) = \theta \ , \quad T_a(a) = 1 \ . \end{cases} \tag{3.8}$$

The following result show the interest of this approach for the approximation of (3.7): the truncated solution (c_a, T_a) converges exponentially towards the steady solution (c, T):

Theorem 11: [18]

Assume that the hypotheses (2.8) hold, and that $g_{0\epsilon}$ is locally non increasing in the neighbourhood of 1, and is differentiable at $s = 1$ with $g'_{0\epsilon}(1) < 0$.

Then problem (3.8) as a unique solution (c_a, T_a), for all $a > 0$. When a tends to $+\infty$, c_a converges to c, and T_a converges locally towards T. Moreover, there exist positive constants a_0, K and k such that, for all $a \geq a_0$, the following estimates hold:

$$|c_a - c| \leq Ke^{-ka} \ , \quad \|\Theta_a - \Theta\|_{C^2(-a,a)} \leq Ke^{-ka} \ . \bullet \tag{3.9}$$

Remark 3: Other similar problems posed in the truncated domain $[-a, a]$ are also used for the numerical approximation of steady flames: one substitutes in (3.8) one of the boundary conditions at $-a$ or a by a Neumann boundary condition. But the preceding results about the existence, uniqueness and exponential convergence of the truncated solution does not hold for all for all these truncated solutions irrespective of whether Dirichlet or Neumann conditions are employed (see [18]). •

3.3. Maximum principle for multi-species compressible flows

Our last example concerns the more general problem of the numerical approximation of multi-species compressible gaseous flows. In this context, preserving the maximum principle for the mass fractions (that is, preserving the inequalities $0 \leq Y \leq 1$ for all species in the mixture) requires some particular care, and many of the commonly used methods fail to preserve it. We will see that, for the very classical upwind methods which are based on an "approximate Riemann solver", the preservation of the maximum principle for all mass fractions is based on some mathematical property of system (3.10) below.

As a first step, let us consider the one-dimensional "multi-species Euler equations", which we write as:

$$\begin{cases} \rho_t + (\rho u)_x = 0 \ , \\ (\rho u)_t + (\rho u^2 + p)_x = 0 \ , \\ E_t + [u(E+p)]_x = 0 \ , \\ (\rho Y)_t + (\rho u Y)_x = 0 \ , \end{cases} \tag{3.10}$$

with the equation of state:

$$p = (\gamma - 1)\left(E - \frac{1}{2}\rho u^2 - Q\rho Y\right) \ . \tag{3.11}$$

Here, ρ is the mixture density, E is the total energy of the mixture per unit volume, Q is the heat released by the reaction, and γ, which may in general depend on the mixture composition, is the specific heat ratio of the mixture. We refer to e.g. Larrouturou-Fezoui [25] and [20] for the details.

We introduce the notations:

$$W = \begin{pmatrix} \rho \\ \rho u \\ E \\ \rho Y \end{pmatrix} = \begin{pmatrix} W^1 \\ W^2 \\ W^3 \\ W^4 \end{pmatrix} \ , \quad F = \begin{pmatrix} \rho u \\ \rho u^2 + p \\ u(E+p) \\ \rho u Y \end{pmatrix} = \begin{pmatrix} F^1 \\ F^2 \\ F^3 \\ F^4 \end{pmatrix} \tag{3.12}$$

For the numerical approximation of (3.10) (which writes $W_t + F_x = 0$), we will consider three-points explicit conservative schemes of the form:

$$\frac{W_j^{n+1} - W_j^n}{\Delta t} + \frac{\phi_{j+1/2}^n - \phi_{j-1/2}^n}{\Delta x} = 0 \ , \tag{3.13}$$

where the numerical flux $\phi_{j+1/2}^n$ is evaluated using a "numerical flux function" Φ which defines the scheme under consideration:

$$\phi_{j+1/2}^n = \Phi(W_j^n, W_{j+1}^n) \ . \tag{3.14}$$

We will write:

$$\phi_{j+1/2}^n = \begin{pmatrix} \phi_{j+1/2}^{1,n} \\ \phi_{j+1/2}^{2,n} \\ \phi_{j+1/2}^{3,n} \\ \phi_{j+1/2}^{4,n} \end{pmatrix} \ . \tag{3.15}$$

Most of the existing schemes which have been developed for the classical (single-component) Euler equations can be extended to these multi-component Euler equations. We refer to Abgrall-Montagné [2] and Larrouturou-Fezoui [25] for the extension of the Godunov scheme, to Abgrall [1] and Fernandez-Larrouturou [19] for the generalisation of Roe's scheme, and to [2] and [25] for the extensions of the schemes of Osher and Van Leer respectively. One can show that the multi-species schemes of Godunov and Van Leer preserve the maximum principle for the mass fractions, but that the extended Roe and Osher schemes do not. Why do these schemes based on approximate Riemann solvers fail to preserve the maximum principle that is preserved by the scheme of Godunov, which uses an exact Riemann solver ?

The answer to this question is given in [22], [23], and relies on some mathematical property of the Riemann problem for (3.10). In order to present it, let us first recall some basic facts about problem (3.10): it is easy to show that the fluxes F^k are functions of the conservative variables W^k, and that the system $W_t + F(W)_x = 0$ is hyperbolic: the Jacobian matrix $\left(\dfrac{\partial F_i}{\partial W_j}\right)$ has always four real eigenvalues $u-c$, u, u, et $u+c$, where $c = \sqrt{\dfrac{\gamma p}{\rho}}$ is the sound speed, and four real eigenvectors. Moreover, for any states W_L and W_R, one can calculate the exact solution $W^{\mathcal{R}}(x,t)$ of the Riemann problem:

$$\begin{cases} W_t + F_x = 0 \quad \text{for } x \in I\!R\,,\quad t \geq 0\,, \\ W(x, t=0) = \begin{cases} W_L & \text{if } x < 0\,, \\ W_R & \text{if } x > 0\,. \end{cases} \end{cases} \tag{3.16}$$

As in the single-component case, this solution consists of four constant states separated by shocks or rarefaction waves and a contact discontinuity (see Figure 1). An essential point here is the fact that the mass fraction Y remains constant across the shock and rarefaction waves and only varies accross the contact discontinuity (see [25] for the proof of these properties).

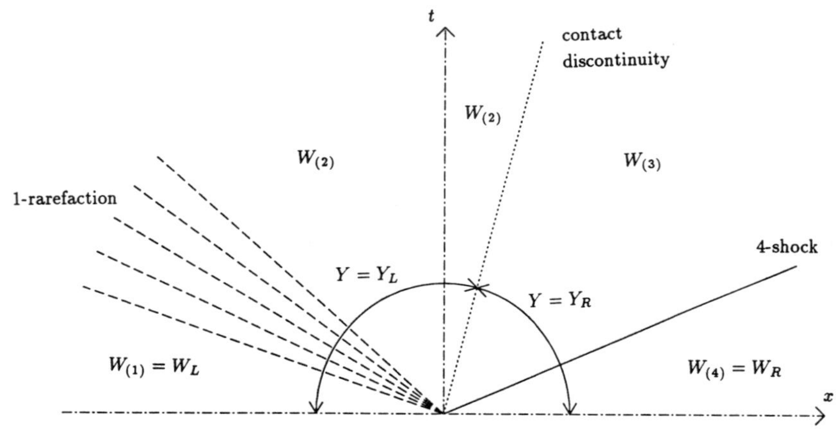

Figure 1: Exact solution of the multi-component Riemann problem.

For $\sigma \in I\!\!R$ and $t > 0$, call $\mathcal{W}(\sigma; W_L, W_R)$ the value of $W^{\mathcal{R}}(\sigma t, t)$, which does not depend on t. The main result of [23] is the following:

Proposition 12: [23]

For any states W_L and W_R, the following identity holds:

$$F^4[\mathcal{W}(0; W_L, W_R)] = F^1[\mathcal{W}(0; W_L, W_R)] \times \begin{cases} Y_L & \text{if } F^1[\mathcal{W}(0; W_L, W_R)] > 0 \,, \\ Y_R & \text{if } F^1[\mathcal{W}(0; W_L, W_R)] < 0 \,. \bullet \end{cases} \tag{3.17}$$

It appears in [23] that (3.17) is precisely the basic reason why Godunov scheme (where $\phi^n_{j+1/2} = F[\mathcal{W}(0; W^n_j, W^n_{j+1})]$) preserves the maximum principle for the mass fractions. Moreover, if one modifies the fourth component $\phi^{4,n}_{j+1/2}$ of the numerical flux for Roe and Osher schemes by imposing the relation:

$$\phi^{4,n}_{j+1/2} = \phi^{1,n}_{j+1/2} \times \begin{cases} Y^n_j & \text{if } \phi^{1,n}_{j+1/2} > 0 \,, \\ Y^n_{j+1} & \text{if } \phi^{1,n}_{j+1/2} < 0 \,, \end{cases} \tag{3.18}$$

which mimics (3.17), then the modified multi-component Roe and Osher schemes also preserve the maximum principle for the mass fractions.

Therefore, the property (3.17) is essential for the construction of maximum principle preserving schemes for multi-component compressible flows. It is shown in [23] that these schemes can be extended to reactive multi-dimensional flows, while still preserving the maximum principle for all mass fractions. Many numerical experiments show that the construction of these schemes has been an important progress for the numerical simulation of several types of reactive flows, including flames and transonic combustion (see e.g. [24]).

REFERENCES

[1] R. ABGRALL, "Généralisation du schéma de Roe pour le calcul d'écoulements de mélanges de gaz à concentrations variables", La Recherche Aérospatiale, pp. 31-43, **6**, (1988).

[2] R. ABGRALL & J. L. MONTAGNE, "Généralisation du schéma d'Osher pour le calcul d'écoulements de mélanges de gaz à concentrations variables et de gaz réels", La Recherche Aérospatiale, **4**, pp. 1-13, (1989).

[3] G. I. BARENBLATT, Ya. B. ZELDOVICH & A. G. ISTRATOV, Prikl. Mekh. Tekh. Fiz., **2**, p. 21, (1962).

[4] F. BENKHALDOUN & B. LARROUTUROU, "Numerical analysis of the two-dimensional thermo-diffusive model for flame propagation", Mod. Math. et Anal. Num., **22**, (4), pp. 535-560, (1988).

[5] F. BENKHALDOUN & B. LARROUTUROU, "A finite-element adaptive investigation of curved stable and unstable flame front", Comp. Meth. Appl. Mech. Eng., **76**, pp. 119-134, (1989).

[6] H. BERESTYCKI, L. CAFFARELLI & L. NIRENBERG, to appear.

[7] H. BERESTYCKI & B. LARROUTUROU, "A semilinear elliptic equation in a strip arising in a two-dimensional flame propagation model", J. für Reine und Angewandte Mathematik, **396**, pp. 14-40, (1989).

[8] H. BERESTYCKI & B. LARROUTUROU, "Quelques aspects mathématiques de la propagation des flammes prémélangées", Nonlinear partial differential equations and their applications, Collège de France seminar, Brezis Lions eds., **10**, Pitman Research Notes in Mathematics, Longman Scient. Techn., Harlow (1990).

[9] H. BERESTYCKI, B. LARROUTUROU & P. L. LIONS, "Multi-dimensional traveling wave solutions of an equation describing premixed flame propagation", Arch. Rat. Mech. Anal., **111**, pp. 33-49, (1990).

[10] H. BERESTYCKI, B. LARROUTUROU & P. L. LIONS, "Stationary solutions of the thermo-diffusive model of flame propagation for non unit Lewis number", to appear.

[11] H. BERESTYCKI, B. LARROUTUROU & L. NIRENBERG, "A nonlinear elliptic problem describing the propagation of a curved premixed flame", Mathematical modelling in combustion and related topics, Brauner & Schmidt-Lainé eds., pp. 11-28, NATO ASI Series E, Nijhoff, Doordrecht, (1988).

[12] H. BERESTYCKI, B. LARROUTUROU & J. M. ROQUEJOFFRE, "Mathematical investigation of the cold boundary difficulty in flame propagation theory", Dynamical issues in combustion, IMA Volume, Springer Verlag, to appear, (1991).

[13] H. BERESTYCKI, B. LARROUTUROU & J. M. ROQUEJOFFRE, "Stability analysis of curved thermo-diffusive flames - Part I: Linear analysis", Arch. Rat. Mech. Anal., to appear.

[14] H. BERESTYCKI, B. NICOLAENKO & B. SCHEURER, "Traveling wave solutions to combustion models and their singular limits", SIAM J. Math. Anal., **16**, (6), pp. 1207-1242, (1985).

[15] H. BERESTYCKI & L. NIRENBERG, "Monotonicity, symmetry and antisymmetry of solutions of semilinear elliptic equations", J. Geom. and Phys., (special issue dedicated to I. M. Gelfand), to appear.

[16] H. BERESTYCKI & L. NIRENBERG, "Some qualitative properties of solutions of semilinear elliptic equations in cylindrical domains", Analysis et Cetera, Rabinowitz Zehnder eds., pp. 115-164, Academic Press, New-York, (1990).

[17] H. BERESTYCKI & G. I. SIVASHINSKY, "Flame extinction by periodic flow field", to appear.

[18] R. FERRETTI & B. LARROUTUROU, "Error estimates for truncated domain approximations of travelling-front solutions of a combustion model", J. Nonlinear Anal - Theor. Meth. Appl., to appear.

[19] G. FERNANDEZ & B. LARROUTUROU, "Hyperbolic schemes for multi-component Euler equations", Nonlinear hyperbolic equations - Theory, numerical methods and applications, Ballmann & Jeltsch eds., pp. 128-138, Notes on Numerical Fluid Mechanics, **24**, Vieweg, Braunschweig, (1989).

[20] B. LARROUTUROU, "Introduction to combustion modelling", Springer series in computational physics, to appear.

[22] B. LARROUTUROU, "Problème de Riemann et approximation numérique pour les é-

quations d'Euler multi-espèces", C. R. Acad. Sci. Paris, **309**, Série I, pp. 893-896, (1989).

[23] B. LARROUTUROU, "How to preserve the mass fraction positivity when computing compressible multi-component flows", J. Comp. Phys., (1991), to appear.

[24] B. LARROUTUROU, "Recent progress in reactive flow computations", Computing methods in applied sciences and engineering, Glowinski Lichnewsky éds., pp. 249-272, SIAM, Philadelphia, (1990).

[25] B. LARROUTUROU & L. FEZOUI, "On the equations of multi-component perfect or real gas flow", Nonlinear hyperbolic problems, Carasso Charrier Hanouzet & Joly eds., pp. 69-98, Lecture notes in mathematics, **1402**, Springer Verlag, Heidelberg, (1989).

[26] J. M. ROQUEJOFFRE, "Etude mathématique et numérique de la propagation d'une flamme plame instationnaire", Rapport INRIA, (1991), to appear.

[27] J. M. ROQUEJOFFRE, "Stability analysis of curved thermo-diffusive flames - Part II: Non linear analysis", Arch. Rat. Mech. Anal., to appear.

[28] J. M. ROQUEJOFFRE, L. SAINSAULIEU & D. TERMAN, "On the stability of steady planar premixed flames", to appear.

[29] G. I. SIVASHINSKY, "Instabilities, pattern formation and turbulence in flames", Ann. Rev. Fluid Mech., **15**, pp. 179-199, (1983).

[30] G. I. SIVASHINSKY, "Instabilities, pattern formation and turbulence in flames", Modélisation des phénomènes de combustion, Direction des Etudes et Recherches d'Electricité de France, Eyrolles, Paris, (1985).

[31] F. A. WILLIAMS, "Combustion theory", second edition, Benjamin Cummings, Menlo Park, (1985).

B. LARROUTUROU
CERMICS, INRIA, Sophia-Antipolis, 06560 VALBONNE, FRANCE.

E S ORAN AND J P BORIS

Numerical simulation of reacting flows

1. Introduction

This paper summarizes a series of eight presentations at the Instituto per le Applicazioni del Calcolo "Mauro Picone" in which we described the results of computations of chemically reacting flows. Specifically, we

- presented an general review of reactive-flow modeling at NRL,
- described the numerical methods developed at NRL specifically aimed at solving problems of reacting flows,
- described selected computations of flames and detonations, with an emphasis on the new physical understandings gained from them,
- described selected simulations of complex transitional and turbulent flows, again emphasizing on new insights gained into the physical processes and the numerical methods used, and
- discussed advanced numerical methods and computer architectures that would greatly enhance our ability to perform these calculations.

These seminars described the approach to numerical simulations we have developed over the last ten years. Some of the material presented was a summary and update of the numerical algorithms and approach presented previously by Oran and Boris [1]. However, the presentations also showed the application of these methods to a range of problems in complex and reacting flows.

This paper highlights some of the major issues in the broad range of topics discussed in these presentations. The examples used draw heavily on work done in collaboration with T.R. Young, K. Kailasanath, and G. Patnaik.

2. The Equations of Gas-Phase Reacting Flows

The equations used to model neutral gas-phase reactive flows are the continuum time-dependent equations for conservation of mass density ρ, individual chemical species number densities, $\{n_i\}$, momentum density $\rho\mathbf{v}$, and total energy E,

$$\frac{\partial \rho}{\partial t} = -\nabla \cdot (\rho \mathbf{v}), \tag{1}$$

$$\frac{\partial n_i}{\partial t} = -\nabla \cdot (n_i \mathbf{v}) - \nabla \cdot (n_i \mathbf{v}_{di}) + Q_i - L_i n_i, \quad i = 1, \ldots N_s, \tag{2}$$

$$\frac{\partial \rho \mathbf{v}}{\partial t} = -\nabla \cdot (\rho \mathbf{v}\mathbf{v}) - \nabla \cdot \mathbf{P} + \sum_i \rho_i \mathbf{a}_i, \tag{3}$$

and

$$\frac{\partial E}{\partial t} = -\nabla\cdot(E\mathbf{v}) - \nabla\cdot(\mathbf{v}\cdot\mathbf{P}) - \nabla\cdot(\mathbf{q}) + \mathbf{v}\cdot\sum_i m_i\mathbf{a}_i + \sum_i \mathbf{v}_{di}\cdot m_i\mathbf{a}_i\ . \tag{4}$$

The first term on the right hand side of each of these equations describes the convective fluid dynamics effects. The remaining terms contain the source, sink, coupling, external force, and diffusive transport terms that drive the fluid dynamics. Pressure is a tensor,

$$\mathbf{P} \equiv P(N,T)\mathbf{I} + \left(\frac{2}{3}\mu_m - \kappa\right)(\nabla\cdot\mathbf{v})\mathbf{I} - \mu_m[(\nabla\mathbf{v}) + (\nabla\mathbf{v})^T], \tag{5}$$

where μ_m and κ_m are the shear and bulk viscosity coefficients and the superscript T means matrix transpose. The heat flux, $\mathbf{q}$, total number density, N, and internal energy density, ϵ, can be found from

$$\mathbf{q}(N,T) \equiv -\lambda_m\nabla T + \sum_i n_i h_i\mathbf{v}_{di} + P\sum_i K_i^T\mathbf{v}_{di} + \mathbf{q}_r, \tag{6}$$

$$N \equiv \sum_i n_i, \tag{7}$$

and

$$E \equiv \frac{1}{2}\rho\mathbf{v}\cdot\mathbf{v} + \rho\epsilon. \tag{8}$$

Here K_i^T is the thermal diffusion coefficient of species i, λ_m is the mixture thermal conductivity, $\mathbf{v}_{di}$ is the diffusion velocity for species i, $\mathbf{q}_r$ is the heat flux due to radiation, and h_i is the specific enthalpy.

In addition, we need an equation of state for gases, usually taken as

$$P = Nk_BT = \rho RT\ , \tag{9}$$

and the relation

$$h_i = h_{io} + \int_{T^o}^{T} c_{pi}\, dT\ , \tag{10}$$

where h_{io} and c_{pi} are the heat of formation and specific heat of species i.

In the reactive flows we will be discussing, the external force is gravity, so that the accelerations are the same for the individual species present, $\mathbf{a}_i = \mathbf{g}$. A great deat of physics is hidden in the chemical production terms and loss rates, $\{Q_i\}$ and $\{L_i\}$, which can be complicated functions of of temperature, pressure, and the other species densities. The set of species diffusion velocities $\{\mathbf{v}_{di}\}$ are found by inverting the matrix equation

$$\mathbf{W}_i = \sum_k \frac{n_i n_k}{N^2 D_{ik}}(\mathbf{v}_{dk} - \mathbf{v}_{di}), \tag{11}$$

Table 1. Terms in the Reactive Flow Equations

$\frac{\partial \rho}{\partial t} = \gamma\rho + S$	Local Processes: Source, Sink, Coupling, Chemical Reactions
$\frac{\partial \rho}{\partial t} = \nu\nabla^2\rho$	Diffusive Processes: Molecular Diffusion, Thermal Conduction, Thermal Diffusion
$\frac{\partial \rho}{\partial t} = -\nabla \cdot (\rho\mathbf{v})$	Convective Processes: Advection, Compression, Rotation
$\frac{\partial^2 \rho}{\partial t^2} = c_w^2\,\nabla^2\rho$	Waves and Oscillations: Sound Waves, Gravity Waves, Oscillations

where the source terms $\{\mathbf{W}_i\}$ in Eq. (11) are defined as

$$\mathbf{W}_i \equiv \nabla(\frac{n_i}{N}) - \left(\frac{\rho_i}{\rho} - \frac{n_i}{N}\right)\frac{\nabla P}{P} - K_i^T\frac{\nabla T}{T} \,. \tag{12}$$

The diffusion velocities are subject to the constraint

$$\sum_i \rho_i \mathbf{v}_{di} = 0. \tag{13}$$

These reactive-flow equations are an extremely rich set. Combined with appropriate initial and boundary conditions, they describe flames, detonations, and turbulence. There are basically four types of physical processes in this set of equations, and these are summarized in Table 1. The first two, chemical reactions and diffusive transport, originate in the atomic and molecular nature of matter. The third and fourth, convection and wavelike properties, are collective phenomena.

Chemical reactions (or chemical kinetics), represented by the production and loss terms Q_i and $L_i n_i$ in Eq. (2), are "local" phenomena and do not depend on spatial gradients. Other examples of local phenomena are phase changes, external source terms such as laser or spark heating, and sinks such as optically thin radiation loss. The macroscopic models of these processes used in the continuum equations arise from averages over microscopic effects.

The diffusion (or diffusive transport) terms have the general form $\nabla \cdot Y$ where Y may be $n_i \mathbf{v}_{di}$, $\mu\nabla\mathbf{v}$, or $\mathbf{q}$. The last two terms in Eq. (5) describe the diffusive effects of viscosity. The last terms in Eq. (6) describe the change in energy due to molecular diffusion, chemical reactions, and radiation transport.

The third equation in Table 1 is a continuity equation describing convection. Convection effects are represented in the equations by fluxes of conserved quantities through volumes, e.g., $\nabla \cdot \mathbf{v}X$ where X is ρ, n_i, E, $\rho\mathbf{v}$ or P. Convection is a continuum concept, which assumes that quantities such as density, velocity, and energy are smoothly varying functions of position.

Wavelike and oscillatory behavior are described implicitly in the reactive flow equations by coupled continuity equations. The important point about wave motion is that energy can be transported throughout the system without equivalent convective fluid or particle motions. Energy is transferred from one element of the fluid to others by waves that can travel much faster than the fluid. The main type of waves considered are shock waves, which move as discontinuities through the system, and sound waves, in which there are alternating compressions and rarefactions in density and pressure of the fluid. Other types of waves included in these reactive flow equations are gravity waves and chemical reaction waves.

The complete set of reactive flow equations is so difficult to solve that entire disciplines and scientific communities flourish solving subsets of these equations for particular applications. As complicated as this set of equations looks, it becomes even more complicated for high-temperature ionized flows and for multiphase flows. Several things make it difficult to solve the equations: lack of knowledge of the input parameters such as chemical reaction rates, physical diffusion coefficients, or boundary conditions; inadequate numerical methods to resolve the physical phenomena; and inadequate computer time and memory.

3. Numerical Methods

We have concentrated most numerical development effort on the solution of continuity equations, on ordinary differential equations, methods for coupling algorithms for various types of processes, and on methods for vector and parallel processing. The guiding principle for ordinary differential equations is that the method must be particularly efficient when coupled to convection calculations. The guiding principle for solutions of the continuity equation is that the method must be accurate, efficient computationally, and allow variably spaced and moving grids.

The method we use for solving ordinary differential equations is the algorithm CHEMEQ [2]. This is a hybrid method combining a second-order Adams-Moulton method and an asymptotic method. The method used for each equation depends on whether the equation is mathematically stiff or not. The method is self starting, crucial for multiprocess numerical simulations, and fully optimized for vector and parallel processing with respect to species and computational cells. It is extremely efficient when coupled to algorithms for convection.

The Flux-Corrected Transport (FCT) method was developed for solving continuity equations and extended in a straightforward way to the solution of coupled continuity equations [3]. FCT is a compressible, one-dimensional, fourth-order, nonlinear monotone convection algorithm that is particularly good at resolving and maintaining gradients. It is also straightforward to combine the FCT modules and construct multidimensional programs using directional-splitting methods. FCT was the first higher-order monotone method correctly enforcing the positivity in the solution.

A history of its development from 1971 until recently is summarized in [4].

Several generic approaches have evolved for solving sets of equations such as Eqs. (1)–(10) in which a number of different physical processes, represented by different types of mathematical terms, interact. The major methods are the global-implicit method, also called the block-implicit method, and the fractional-step method, also called timestep splitting. Other approaches to coupling, such as the method-of-lines and general finite-element methods, are also used. The various approaches are often combined into "hybrid" algorithms. Our method of choice has been timestep splitting. The reasons for this are its relatively straightforward applicability and its flexibility in incorporating the best available numerical algorithm for each particular physical process described in Table 1. For example, we have found that small errors in calculating convection can cause large errors in the overall reactive-flow simulation. This means that very accurate calculations of convection are generally required. However, the solution of ordinary differential equations representing chemical kinetics does not have to be as accurate when coupled to convection as it does when it is solved alone at constant volume or constant pressure. With timestep splitting, the order of the algorithms can be adjusted accordingly. Specific tricks for using timestep splitting for fast and slow reactive flows are discussed in [1].

The latest algorithmic developments focus on more sophisticated adaptive gridding to maintain high resolution at gradients, applications of monotone finite-element methods with unstructured grids, very fast near-neighbors and constraint algorithms for particle dynamics, more sophisticated methods for coupling physical processes with disparate time scales, and software especially optimized for the new computer architectures. Here we describe three new algorithms developed at NRL that directly address these problems: Barely Implicit Flux-Corrected Transport (BIC-FCT), for flows in the subsonic to nearly incompressible range, the finite-element Flux-Corrected Transport (FEM-FCT) for flows with irregular obstacles, flow structures, or boundary conditions, and the Monotonic Lagrangian Grid (MLG) algorithm for tracking particles in space. The first two algorithms are clearly for computational fluid dynamics. The last algorithm has been applied to molecular dynamics, which will be briefly mentioned, and is now the focus of attempts to develop a better Lagrangian fluid-dynamics algorithm.

Barely Implicit Flux-Corrected Transport

The key to being able to simulate the properties of time-dependent compressible reactive flows is to use a nonlinear, monotone, positivity-preserving method such as FCT. A monotone method does not add unphysical maxima or minima to a computed profile and does not allow originally positive quantities to become negative unphysically. The algorithm is nonlinear because the numerical coefficients associated with the finite-differences at each timestep at each location depend on the value of the quantity being convected at that location. Other monotone approaches include MUSCL [5], PPM [6], and the various TVD schemes (see, for example, [7]) The explicit FCT algorithms have been used extensively on quadrilaterial finite-difference grids and the extension to nonorthogonal grids is straightforward.

In cases where the flow velocities are very low, it is not practical to use a criterion based on the sound speed to select the timestep for integrating fluid equations. In such cases, a treatment which is implicit in pressure is useful and allows longer timesteps based on slower properties of the system. At the same time, it is extremely useful to keep the advantages of a nonlinear, monotone convection method. Most implicit methods, however, are based on linearizing the convective terms, thus obviating the advantages of a monotone method such as FCT.

The Barely Implicit Correction to FCT, *BIC-FCT*, pioneered by Patnaik [8], removes the timestep limit imposed by the sound speed by adding one implicit elliptic equation. The method is based on an analysis that shows that it is not necessary to treat all terms in the fluid dynamic equations implicitly to use longer timesteps than allowed by explicit stability limits. Only the two terms which force the numerical stability limit need to be coupled and solved implicitly.

The BIC-FCT method solves the compressible, gas dynamic conservation equations for density ρ, momentum density $\rho\mathbf{v}$, and total energy density E, by a two-stage algorithm. The first stage is an explicit predictor method that determines the estimated new values $\tilde{\rho}$ and $\tilde{\mathbf{v}}$. The second step is an implicit correction step which produces δP, the additional pressure required to accelerate the fluid momentum from the explicit predictor values, $\tilde{\rho}\tilde{\mathbf{v}}$, to the final implicitly corrected values, $\rho^n\mathbf{v}^n$. Because this correction deals with the sound waves (which were improperly computed by the explicit monotone predictor step in this algorithm) and not with the convection, there is no interference with the bulk convection in the FCT algorithm. Only one elliptic equation must be solved to make the fluid equations implicit.

One particularly surprising aspect of this method is its relatively low cost: *One implicit timestep using BIC-FCT is about the same as one explicit FCT timestep on a CRAY-XMP.* Normally we would have expected the implicit method to cost five to ten times more per timestep. This is primarily because the standard explicit implementation of FCT uses both a half-step prediction and a whole-step correction procedure each timestep. BIC-FCT does not have to use a half-step prediction, but instead requires solving one finite-difference elliptic equation.

Finite-Element Flux-Corrected Transport

Finite-element techniques have been used for years to solve difficult, practical problems in structural engineering. Later they became interesting to mathematicians who formulated their properties in terms of broad classes of approximations. Triangles and tetrahedra are commonly used elements for finite-element grids in two and three dimensions, respectively. They are particularly useful because they can be used to describe irregular objects or irregularly shaped internal boundaries and interfaces.

More recently, finite-element methods have been modified for a variety of problems in fluid dynamics and heat transfer, and most recently for time-dependent fluid dynamics. Their application to solutions of systems of continuity equations are of interest here. In particular, it is advantageous to combine the triangle-based, unstructured grid approach of a finite-element method with the accuracy of a nonlinear, monotone finite-volume fluid dynamics algorithm.

A major drawback of finite-element methods is their cost. If time is advanced by an explicit formula, it is necessary to solve a linear matrix problem at each timestep. When the time is advanced implicitly, there are additional problems. The cost in computer time for large matrix inversions at each timestep is often prohibitive. In multidimensions, finite-element methods require a great deal of computer storage. The cost of recomputing geometric and shape functions for each physical variable transported is often prohibitive. Thus these variables are usually stored rather than being recomputed, requiring dozens of quantities per node in two dimensions and many more in three dimensions.

There are compensating advantages to hybrid finite-element finite-volume methods. They are relatively straightforward to generalize for solving multidimensional problems in several dependent variables. In particular, the shapes of the domains do not have to be regular, so the method is well suited to problems with irregular boundaries. The domain can also change in the course of time to allow improvements in resolution where needed or to allow for deformable structures in the flow.

In the method developed by Löhner et al. [9], time is advanced by an explicit second-order Lax-Wendroff discretization and has been extended to meshes that adapt in time. The degrees of freedom are the values of the dependent variables. The nodes are subdivided in front of advancing gradients and flow structures and recombined behind them according to rules based on gradients in the known properties of the evolving system. This is a rather straightforward method which combines the accuracy and flexible grid of finite elements with the convenient features and high accuracy of monotone finite-volume methods.

Finite-element methods have not been exploited for fluid calculations as fully as finite-difference and spectral methods. In general the number of operations per finite-element node substantially exceeds that per mesh point in finite-difference or finite-volume approaches, so finite-element methods must be more accurate if they are to be competitive. The recent algorithms which incorporate the monotone FCT algorithm into a finite-element framework, are competitive with the best finite-difference methods. One drawback of this approach is the amount of computer memory per mesh point that it requires, a serious issue when memory is at a premium. Another current problem is inherent in implementing boundary conditions on triangle-based grids. However, these methods are accurate, relatively fast, and allow very flexible, adaptive gridding. Extensions to three dimensions are underway and potentially affordable with the introduction of supercomputers with very large high-speed memories.

Monotonic Lagrangian Grid

A major part of a particle dynamics or Lagrangian fluid-dynamics calculation is the continual reevaluation of which nodes are near neighbors in a large set of nodes that often seem to be moving randomly. The expense arises because each node can potentially interact with any of $N-1$ other nodes in the system. This is called the N^2 problem because only a few of the N^2 interactions possible between pairs of nodes are potentially important.

The Monotonic Lagrangian Grid (MLG) is a new method designed to beat the N^2 problem [10]. The MLG is really a data structure for storing the positions and other data needed to describe N moving nodes. These N nodes could, for example, represent fluid elements, atomic particles, molecules, droplets, or dust particles that must be tracked in a multiphase detonation calculation. A node with three spatial coordinates has three indices in the MLG data arrays. The data relating to each node are stored in the memory locations indicated by these indices and ensure that nodes which are close to each other in real space are near neighbors in the MLG data arrays. This simplifies and speeds most interaction and correlation calculations. A computer program based on the MLG data structure does not need to check $N-1$ possible distances to find which nodes are close to a particular node. The indices of the neighboring nodes are automatically known because the MLG indices vary monotonically in all directions with the Lagrangian coordinates. The cost of the algorithm in practical calculations is dominated by the calculation of the interactions of nodes with their near neighbors, and the timing thus scales as N. There are significant advantages to N or $N \log N$ versus N^2 or N^3 timing for computing interactions among many nodes.

Using an MLG to index the positions and physical properties describing a node in computer memory, a Lagrangian neighborhood in the MLG can be determined by a maximum index offset, N_c. Computations for the interactions affecting a particular node are only considered for nodes whose indices are offset from the index of the particular node by an amount less than or equal to N_c. Such computations can be vectorized efficiently because nodes that are close to each other are indexed through contiguous memory. From an arbitrary data set, an MLG can be built from randomly located nodes by an algorithm that also scales as $N \log N$. When node motions in real space destroy some of the monotonicity conditions, another faster $N \log N$ algorithm exists to swap the node data between adjacent cells to restore the MLG order.

Using the MLG for molecular dynamics offers large potential improvlements in the time it takes to do such calculations. In addition, Lambrakos et al. [10] have recently derived a new algorithm for use with the MLG that maintains constraints in a very efficient way, and so expands the types of calculations that can be done with the method.

One potential application of the MLG is to Lagrangian fluid dynamics. There are two approaches here. One is a macroparticle approach, which in spirit is somewhat like molecular dynamics. Each Lagrangian node has a potential associated with it, and attempts are made to choose potentials that simulate fluid dynamics. Choosing these potentials is now an area of active research. The other approach is a finite-difference or finite-volume approach on a distorted quadrilateral grid. Here it is important to ensure that the particular discretization maintains the physical conservation properties of the fluid equations. Of the two approaches, the finite-difference approach has the most promise for producing a quantitatively accurate algorithm. The macroparticle approach is the most straightforward to implement if the potentials are known, but statistical fluctuations limit the applicability of the method.

There is a great deal of current activity now in using molecular-dynamics methods to study microscopic processes. In this approach, a series of particles are defined which can be atoms or molecules, or molecular fragments, and potential energies are defined that describe their interaction. The interactions are usually classical because of the complexity and number of the particles. The computations use Newton's Law to evaluate how the system evolves in time. Until now, there have been some enlightening calculations, but very little has been really quantitative in the area of chemically reactive flows. With the advent of new types of computers and new algorithms such as the MLG, use of molecular dynamics to determine fundamenntal aspects of reactive flows is becoming possible. Examples of this type of computaton are the work of Peyrard et al. and Lambrakos et al. [11], studying propagating reaction and detonation waves in crystal lattices.

4. New Computer Architectures

In the last several years, a number of different supercomputer architectures have emerged. These machines are categorized in several ways: 1) by the maximum floating point operation count in megaflops (million floating-point operations per second), 2) by the number of processors available, that is, the "grain" or degree of parallelism possible, 3) by the power of the individual processors, that is, whether pipelined operations or even hardware floating-point operations are possible in each processor, 4) by the communication and memory distribution strategy connecting the processors, and 5) by the type of program coordination and control adopted.

1. *Supercomputers.* These are the very fast general-purpose machines available today with operation counts of a gigaflop or more. They include the CRAY X-MP, CRAY 2, Fujitsu, Hitachi, NEC, and ETA 10 systems. They have vector capabilities implemented by arithmetic pipelines and some parallelism. The degree of parallelism is not usually very high, often two, four, or eight processors, and is often called "coarse-grained."
2. *Mini-supericomputers.* These have speeds of 10 to 100 megaflops and some modest level of parallel processing is possible. They are designed in a more or less standard way and are less expensive and somewhat more cost effective than the true supercomputers. These mini-supercomputers include the CDC CYBER computers, the Convex C1 and C2, the SCS 40, the Multiflow, the Alliant, and the Elexi machines. The performance is comparable to current technology array processors with much better software and with interactive operating systems. The parallelism in these system is also quite coarse.
3. *Highly Parallel Processors.* These have operation counts of 50 to 2500 megaflops and usually have more radical designs specifically to exploit parallelism. They are, in principle, expandable. They are competitive with conventional supercomputers, although perhaps not as versatile for some time to come. These include, for example, the TMC Connection Machine, the BBN Butterfly Machine, the Navier-Stokes Computer, the various hypercube machines such as the NCUBE system, and systolic arrays. The grain of the parallelism for this class of supercomputer is either moderate, consisting of 16 to 128 processors, or "fine" consisting of more than 256 processors.
4. *Hybrid Systems.* These are combinations of the above. One example is the GAPS (Graphical and Array Processing System), assembled in our laboratory at NRL and described briefly below. These are heterogeneous assemblies of computing components with overall performance in the supercomputer range, currently starting around 200 megaflops. It is highly unlikely that any one highly parallel architecture will be optimal or even efficient on all classes of CFD problems. Structured and unstructured grids, for example, have very different computational constraints as do implicit and explicit time advancement strategies for CFD models. A Connection Machine, with approximately 64,000 processors, could be used effectively as an array processor with a Cray. Fluid calculations can be executed efficiently on the Cray as each node of the Connection Machine simultaneously integrates a set of ordinary differential equations describing the chemistry in each of the fluid elements in the Cray. A heterogenous element supercomputer system with optically networked components could be the most flexible system in the near future, provided the network control software for such distributed computing becomes easy to use.
5. *Special Purpose Computers.* The opportunity now exists to develop special purpose highly

parallel computers in which what is implemented in software today to be implemented in hardware at the chip level. These computers would be considerably less general in their applications.

The Graphical and Array Processing System

The GAPS [12] is an asynchronous, multi-tasking, high-performance, scientific parallel processing system consisting of an APTEC 2400 Computer connected to a VAX 11/780 with 12 megabytes of additional fast memory and about 3 gigabytes of online disc storage. The system was constructed to develop practical expertise in multiprocessing systems, to develop an interactive, user-friendly system designed to accommodate interactive graphics, and to provide extra, inexpensive computer time for computationally intensive problems. The GAPS now contains six Numerix MARS 432 array processors which have a maximum performance of 30 megaflops each. These array processors can be programmed in Fortran and have an extensive library of vectorized routines and synchronization software. Several high-resolution color graphics monitors, including a Tectronix 4115B, a Metheus Ω3600, and an IRIS 4D workstation, are connected to the GAPS. These graphics systems have distributed processing capabilities such as zooming and rotating images.

The APTEC is a high-speed computer (24 megabyte/s) that uses programmable data interchange adapters (DIAs) to transfer data among the VAX, the array processors, and other GAPS components. The ability of the VAX to control data flow is superb, but its ability to actually move the data is limited. In practice, 0.25 megabyte/s is the maximum information transfer rate the VAX can sustain, and this severely limits the overall system performance. A very high-speed data switchyard is necessary behind the VAX to display high-resolution fluid dynamics simulations at the rate at which they are calculated. The APTEC, working at its 24 megabyte/s potential, can keep up with the 12 megabyte/s speed required to show 12 frames per second of a 1024 × 1024 pixel array.

The GAPS currently has 180 megaflops of array processors. The NRL Cray X-MP/12 is rated as a 210 megaflop system at full use. When the array processors are processing a two-dimensional FCT benchmark code, simultaneously pixelating the data arrays as it computes them, and sending every third or fourth frame to the Tektronix 4115B for monitoring, only about 25% of the APTEC data bus capacity is being used. When the six 30 megaflop array processors are used together on the benchmark, they give about 80% of the performance of the CRAY. The APTEC could support up to a dozen array processors.

An important part of our research has been to develop optimally efficient algorithms that can take advantage of an asynchronous multi-tasking parallel architecture such as the GAPS. We have implemented algorithms that can take advantage of more parallelism than available in the pipeline and vector register architectures used in current supercomputers and GAPS. The Reactive Flow Model (RFM), currently implemented on the GAPS, can run for very long periods of time on relatively high-resolution fluid simulations. This model can also take full advantage of the parallelism inherent in computers such as the Connection Machine and large hypercubes. Because the GAPS runs in the background without affecting the operation of the VAX, days or weeks of computer time can be devoted to a problem. The GAPS effectively gives us an extra 100 hours of CRAY time every week.

Results from GAPS simulations using RFM can be displayed as they are calculated using VOYEUR, a high-bandwidth graphics package that can select data from the VAX, from the GAPS array processors, the CRAY, or the Connection Machine. For example, if during the course of the calculation we need an enlargement of a region in the flow, or want to change the scale or even the quantities plotted, we can create this graph and put it on the screen where it will also be updated in time as the calculation evolves.

Several problems have been simulated using the RFM implemented on GAPS, including transition to turbulence, flame holding at supersonic speeds, detonations, and stability analysis of flow in ramjets. Simulations with very high resolution (200 × 900 finite-difference grid) and with over a million timesteps have been completed. A three-dimensional reactive flow model and a three-dimensional molecular-dynamics model are currently under development using the same parallel-processing techniques.

The Connection Machine

The Connection Machine [13] consists of thousands of individual scalar processors connected by hypercube communications into a single-instruction, multiple-data (SIMD) architecture. This means that all of the processors perform the same operations on different data as the stream of instructions is broadcast to them by a conventional scalar computer. There are two Connection Machines at the Naval Research Laboratory, one with 16,384 (16K) processors and one with 8,192 (8K) processors. Communication to and control of these processors is through a front-end computer that may be either a VAX, a Symbolics, or a Sun. Each individual processor can be subdivided in powers of two into a number of virtual processors. This subdivision is performed by software instructions and the actual number of virtual processors used for a problem is determined by the size of the grid and limited by the physical variable and scratch storage required at each grid point. Floating-point arithmethic is executed by Weitek chips, each of which does pipelined processing of the floating-point operations for 32 of the scalar processors.

The user programs in one of several languages specially adapted to parallel processing: *Lisp (Star Lisp), C* (C Star), or a parallel extension of FORTRAN 8X. Compiling programs written in these extended languages produce a series of calls to the resident library of PARIS (PARallel Instruction Set) microcoded routines that control the CM processors and the front-end computer's interaction with the CM. There is also a lower-level microcode capability called CMIS (Connection Machine Instruction Set). CMIS is more complicated to program than C* or PARIS but does produce some increase in speed over PARIS.

On-line color graphics are available to display the results of computations and to make video tapes (though the hardcopy capability is lagging). The Connection Machine has an integrated framebuffer allowing colored pixels to be written simultaneously from each processor directly to program-specified locations on the screen. There is now a good set of two-dimensional graphics routines that can display most computations fast enough to make movies as the computations themselves are evolving. Auxiliary capabilities also exist to produce three-dimensional images that include zooming, panning, and rotation. As yet, publication-quality hardcopy can only be generated by transferring the output files to other computers and using graphics available on those computers. The graphics shown in this paper were produced using the NCAR post-processing routines on a VAX. Color graphics have been generated by photographing directly from the color monitor connected to the framebuffer or through the GAPS routine, VOYEUR, that allows automatic conversion to the Dicomed.

We have found that the Connection Machine is an excellent computer on which to carry out large, multidimensional, complex FCT computations [14]. The timings are clearly competitive and in fact can be a factor of four or five better than the Cray YMP for the computations tested. The Connection Machine is relatively user friendly and easy to program and the staff at the Naval Research Laboratory has been particularly helpful. One of the problems with SIMD computers, however, is the inability to do different computations in the boundary elements from those necessary in the interior. With ingenuity, it is possible to work around this limitation with efficient tricks.

We have tested some of the options for realistic boundary conditions in our models, but by no means tried all possibilities [14]. While sometimes the codes are correspondingly more complex, this complexity is not prohibitive and the resulting models are still very efficient. Using one or another of the available options for boundary conditions, it is possible to do just about everything we want with better speeds than can be obtained on other computers available to us. This flexibility would be extended further by an increase in the amount of memory available per processor.

There are several other types of computations which we have also converted to the Connection Machine and these will be reported in other papers. We have added chemical-reaction models to the fluid codes to compute the properties of propagating detonations and other reacting supersonic flows. This addition is straightforward because these are local processes and no interprocessor communication is needed. Along totally different lines, we have converted several MLG particle codes to the Connection Machine and have found similar good performance.

5. Computations of Detonations

Weak and Strong Ignition behind Reflected Shocks

Shock-induced ignition of detonations occurs in one of two modes, depending on the stoichiometry and thermodynamic conditions of the shocked gas. At lower temperatures, the ignition is weak, or mild. Many locations where chemical reactions begin appear behind the shock, and they gradually develop into an explosion. At higher temperatures, ignition is strong, or sharp. A single reaction center causes the abrupt appearance of a secondary shock induced by the explosive reaction. These modes have been studied extensively by Strehlow, Oppenheim, Edwards, Soloukhin, and their collaborators. Their experiments were performed in shock tubes in which they studied ignition behind incident and reflected shocks. Typical diagnostics were pressure measurements and Schlieren photography. Here we summarize selected results of numerical simulations which provide some insight into the differences of the behavior of these modes.

The behavior of shocks and detonations in reactive mixtures of hydrogen and oxygen was computed using a numerical model that combined FCT with CHEMEQ to integrate the set of reactive-flow conservation equtions with a complete model of the chemical kinetics [15]. These results were then compared to shock-tube experiments in regimes where there is strong ignition [16]. The problem simulated, an incident shock moving into an unreacted mixture of hydrogen, oxygen, and argon. The passage of the shock raises the temperature and pressure of the reactive gas. When the shock hits the solid wall, it reflects and raises the temperature and pressure of the gas behind the shock even higher. After the chemical induction time has elapsed, the gas ignites first at the reflecting wall because the mixture there is heated for the longest time. Some time after the gas has ignited, a reaction wave is established that may eventually catch up with the reflected shock. In the idealized picture of this problem, the reflected shock brings the gas behind it to rest. The material behind this shock, at the reflecting wall, which has been heated the longest, should be the first to react after a suitable period equal to the chemical induction time of the gas.

Results from calculations are shown in Figure 1 along with a Schlieren photograph from the experiment. In order to compare the simulations most effectively with the experiment, we show the time-dependent behavior of two quantities: the maximum pressure in the system and the fluid velocity at the reflecting wall. The maximum pressure at each timestep is constant until the incident shock is reflected; then it jumps to a higher value. It increases again when chemical energy is released and the reaction wave starts, and it reaches a maximum at the time when the reaction wave and reflected shock merge. In the ideal problem, the fluid velocity behind the reflected shock would be zero. Just after the incident shock is reflected from the wall in the simulations, the fluid velocity oscillates wildly to maximum values and then settles quickly to a very small value until the reaction wave starts, at which time the velocity jumps up again. Then the fluid velocity near the wall quickly decays again, and rises to a substantial value when the wave generated from the merging of the shock and reaction wave reaches back to the wall.

Comparing the times marked on the Schlieren photograph and the calculation in Figure 1, we see that the time calculated for the formation of the reaction wave is in good agreement with the experiment. There is apparent disagreement between the calculations and experiment in the individual times calculated for merging of the shock and reaction wave and the reflection of this wave from the wall. Part of this could be attributed to nonideal effects at the wall in the experiment. However, the sum of these numbers is very close to that determined experimentally. In fact, there is some ambiguity in determining these quantities from the Schlieren photograph. Figure 2 shows the calculated position of the reflected shock front, reaction wave, transmitted detonation, and contact discontinuity as a function of time. In addition to the information about the macroscopic flow properties, the calculation also produces number densities of all of the individual species as a function of position and time.

Now consider the weak ignition case. Here we would not expect the agreement between the experimental results and a one-dimensional calculation that was shown above. Experimentally, a distinct reaction wave, as we have seen in the strong ignition case, does not appear at reproducible times at the reflecting wall of the shock tube. The time of ignition is blurry and indistinct. Compared to the strong ignition experiment, the time scales are much larger as a result of the considerably longer chemical induction time. However, it is instructive to see just how "wrong" the one-dimensional model is in this case.

Based on the results of the strong ignition simulation, we expect the one-dimensional simulation to show ignition at the wall at about 1550 μs, the chemical induction time of the gas at the conditions behind the reflected shock. At that time, a reaction wave would form and eventually catch up with the shock front. The Schlieren photograph for this case shows that ignition occurred at about 600 μs after shock reflection, about a factor of two too soon. Note that the change of induction time with temperature is much steeper curve than it was for the strong ignition and therefore the induction time is a much more sensitive function of temperature. Figure 3 shows the positions of the various fronts and discontinuities in the problem as a function of time. The reaction wave is defined as that place where the ratio of OH to Ar becomes significant and energy

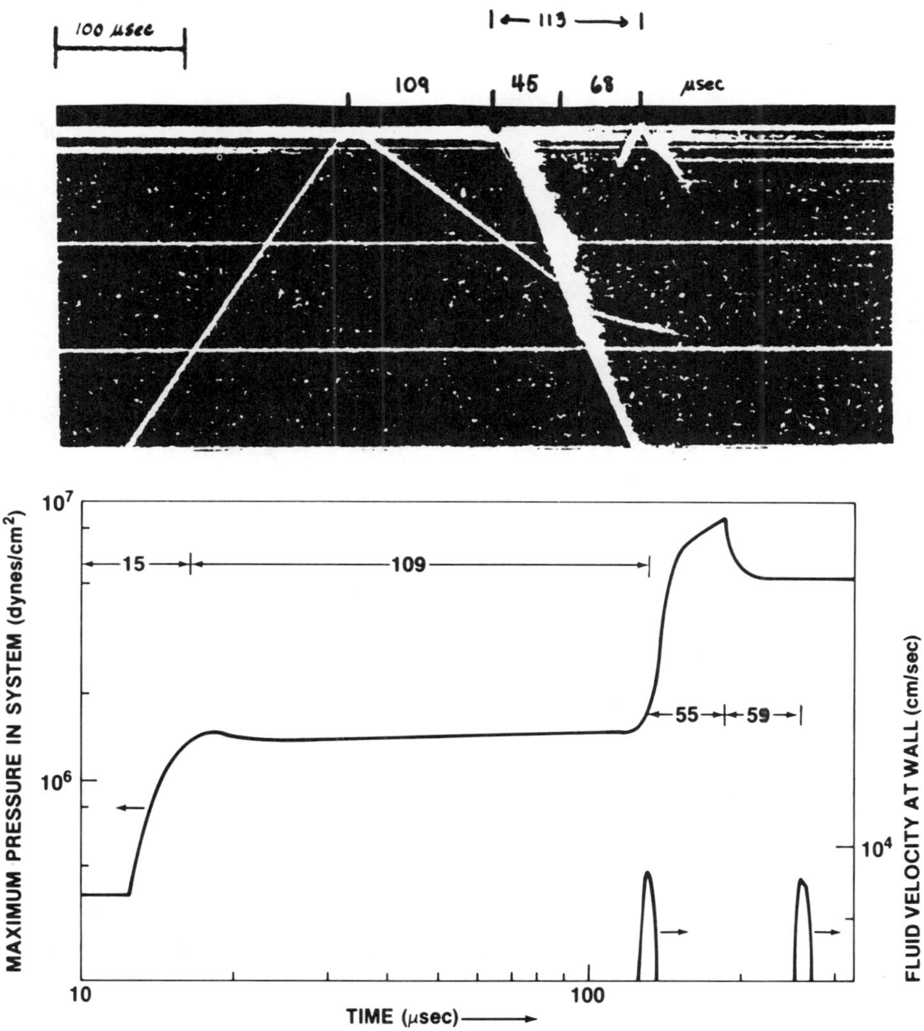

Figure 1. Comparisons of calculation and experiment for the strong ignition case. (a) Schlieren photograph [15,16], (b) calculated time-dependent behavior of two quantities: the maximum pressure in the system modelled and the fluid velocity at the reflecting wall.

has just begun to be released. We first note that the incident shock reflection occurs at about 100 μs into the calculation. The first indication of energy release occurs at 750 μs after shock reflection. By 1120 μs a detonation has formed in the region behind the shock.

The curious feature here is that the time to ignition is reduced from the chemical induction time in the one-dimensional calculation, just as it is in the experiment. Resolution tests of the calculation, in which the computational cell sizes near the reflecting wall are increased, also give

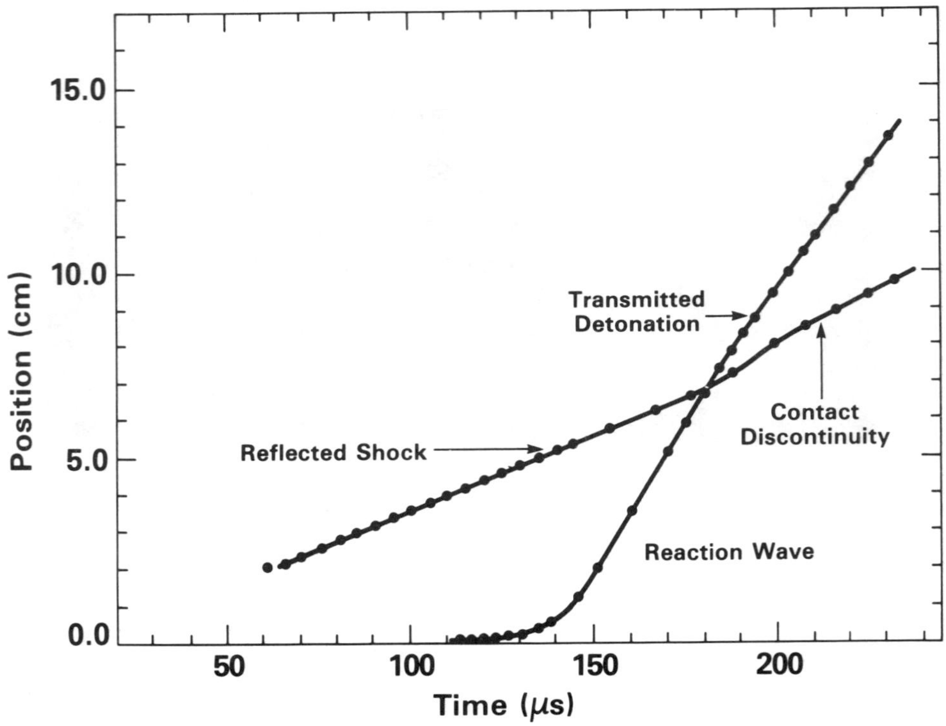

Figure 2. The calculated position of the reflected shock front, reaction wave, transmitted detonation, and contact discontinuity as a function of time for the strong ignition case.

results which show much shorter ignition times than would be expected. A close inspection of the temperature and velocity profiles from the computations [15] show that ignition has not started at the wall in the calculation, but that a hot spot forms at some distance from the wall.

The explanation for these results in the calculations lies in both the physical and numerical deviations of the calculations from idealized conditions. First, the calculations are fully compressible solutions of the conservation equations and so include the presence of sound waves. Although in the idealized shock-tube problem the gas behind the reflected shock comes fully to rest, there are always acoustic waves present behind the reflected shock in a compressible medium. There are also numerical effects that generate noise. The reactive mixture is very sensitive to acoustic disturbances [14] so that small fluctuations in temperature or pressure greatly affect the induction time. This strong sensitivity, the increased ease by which a system ignites due to fluctuations, is at the heart of what causes weak ignition. Extensios of these calculations that focus on weak ignition behind incident shocks [17,18] have shown that a number of hot spots may form before ignition occurs.

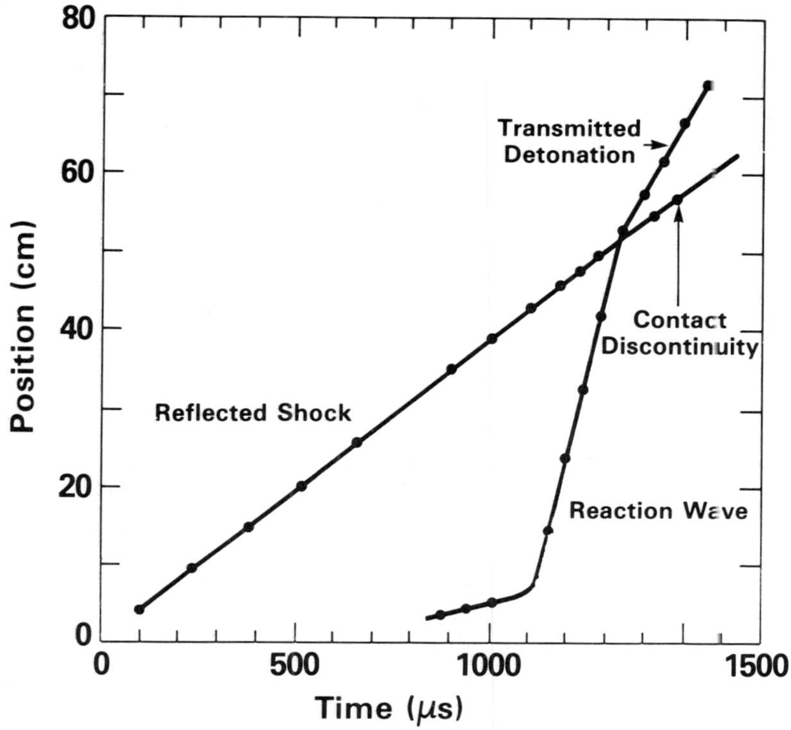

Figure 3. The calculated position of the reflected shock front, reaction wave, transmitted detonation, and contact discontinuity as a function of time for the weak ignition case.

Multidimensional Detonation Structure

It has become well known in the last twenty years that the front of a self-sustained propagating detonation in gases is not uniform and planar. Its structure is complex and multidimensional, involving interactions between incident shocks, Mach stems, transverse waves, and boundaries of the regions through which the detonation is moving [19]. The triple points formed at the intersection of the transverse wave with the Mach stem and the incident shock trace out patterns that are called detonation cells. These structures may be obscured for certain material conditions or if the detonation is heavily overdriven, but they seem to occur consistently in self-sustained gas phase detonations. Extensive experimental data show that the size and regularity of this cell structure is characteristic of the particular combination of initial material conditions, such as composition, density and pressure. In particular, it has been shown [20] that in the hydrogen-oxygen system, fuel-lean mixtures and mixtures diluted with greater then 40% argon show a very regular structure. As the mixture becomes less dilute, the structure becomes less and less regular. When the diluent is nitrogen instead of argon, the structures are also irregular.

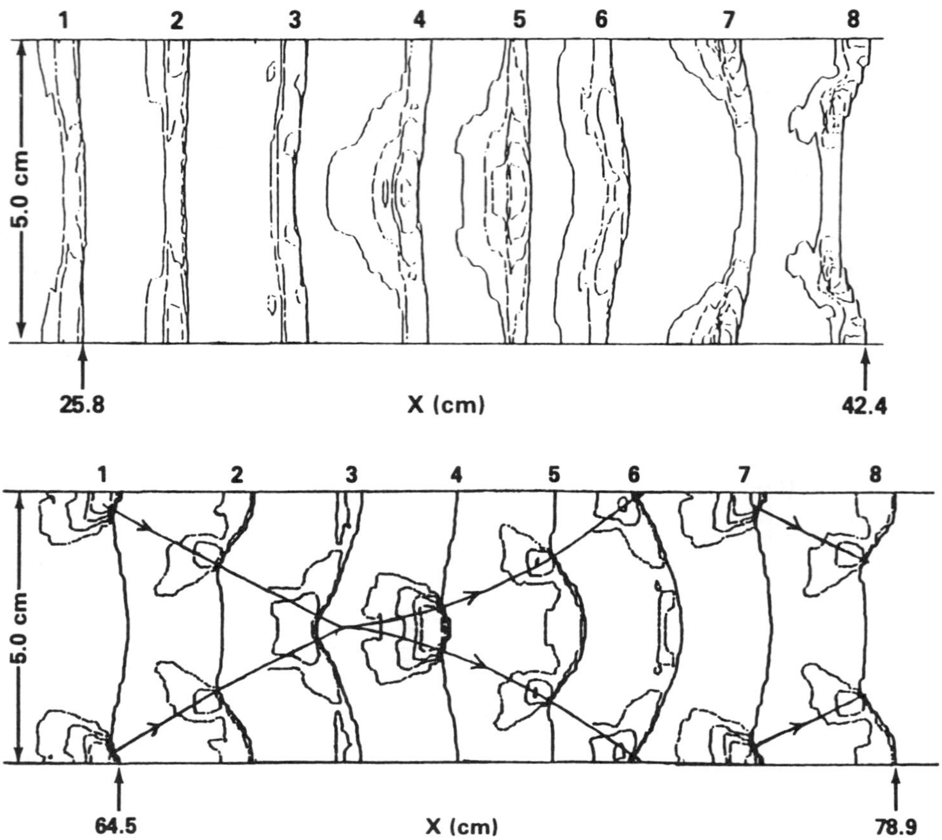

Figure 4. Upper figure: pressure contours at 10 μs intervals showing the effects of perturbing a two-dimensional planar detonation. Lower figure: later-time pressure contours.

Numerical simulations have been used to help explain detonation structure, to look at those features of the structure which determine the cell size, to study the interaction of the structure with the geometry, and to study those factors that can cause irregularity in cell structures (see, for example, [21–23]). Most of this work has been for $H_2 - O_2$ mixtures highly diluted with argon at initial pressures substantially below one atmosphere. More recent work has included studies of the structure of detonations in liquid nitromethane.

Planar detonation fronts may be unstable to perturbations in the transverse directions. Given an initially planar front, it tends to depart from a one-dimensional configuration. The upper diagram in Figure 4 contains a series of pressure contours from a calculation of the early effects of perturbing a planar detonation front [22]. The calculation was initialized by placing an elliptical pocket of hot, unburned gas behind a planar detonation. The pocket, which was merely used as a device for initiating the perturbation, burns slowly and sends out pressure disturbances in all directions. These waves interact with the incident shock front causing it to curve outwards. The pressure waves also reflect from the side walls of the channel and move transverse to the incident shock front, as seen in frame 2 of the figure. These pressure waves are strengthened due to collisions with each other and further increase the curvature of the incident shock front, as can be seen in frame 7. After a short time, a portion of the incident shock reflects from the side walls of the channel. Frame 8 shows that a Mach reflection takes place since a pair of triple points are seen. The reflected shock waves, which are the transverse waves, are initially weak but are later strengthened when the incident shock accelerates due to the increased rate of chemical reactions caused the transverse waves.

The further evolution of the pair of triple points is shown in the lower diagram in Figure 4. The path and the direction of movement of the triple points are indicated by the lines with arrows. In the first frame, the transverse waves are moving towards each other and away from the wall. By the fourth frame, they have collided with each other and are moving away from the center of the channel. In frame 7 they are again moving towards the center after colliding with the walls. Frames 7 and 8 are similar to frames 1 and 2 and show that a configuration with a repeating pattern has been established.

Detonation Cells and Unreacted Pockets

The first numerical simulations of the evolution to the triple point structure in detonations was done for low-pressure, stoichiometric H_2–O_2 mixtures heavily diluted with Ar [24]. These calculations demonstrated that computer simulations could, in fact, reproduce the types of shock and reaction front structures characterizing detonations. Figure 5 shows patterns of triple points at the front of detonations in dilute H_2–O_2 mixtures for three different tube heights [22]. These figures show that as the height of the tube increases, a structure typical of a detonation cell is seen.

One initially unexpected and potentially important phenomenon found in the experiments and simulations is the formation of unreacted pockets of material behind the detonation front [25, 22]. Experiments first showed such pockets, but they were not explained until they were observed in numerical simulations. Figure 6 is a series of contours from a calculation similar to that shown above, but in a very narrow tube. The figures on the left show the extent of reaction. The detonation propagates to the right into the unreacted gas mixture and fully reacted gas is on the lefthand side. The various shades in between represent different degrees of reactedness and together comprise the detonation front. In the third frame, we see a detached pocket of unburnt gas behind the detonation front. The figures on the right show the temperature of the gases in the

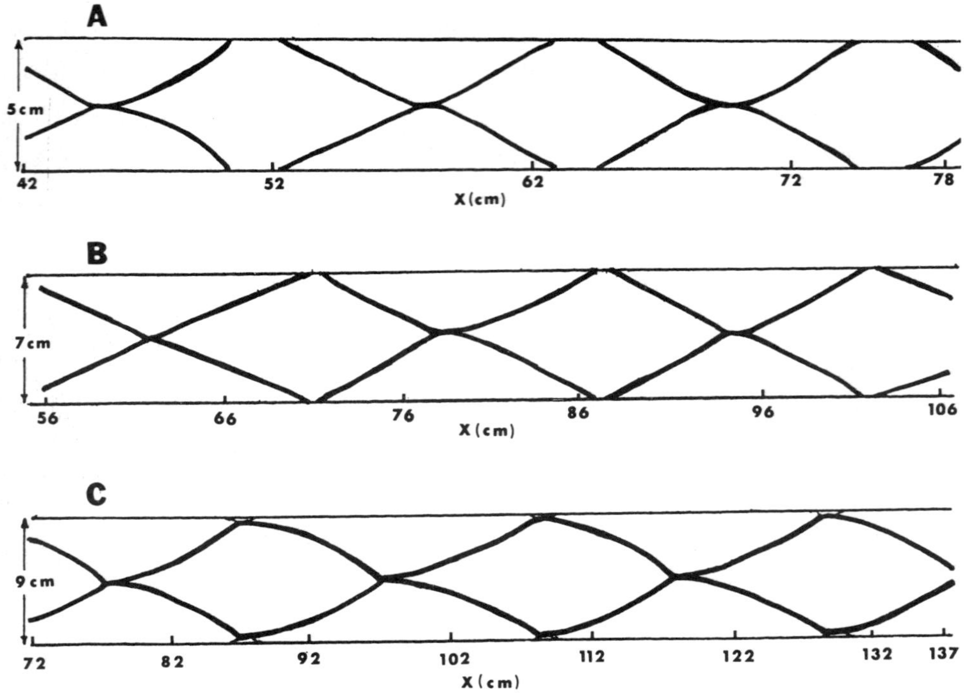

Figure 5. Comparison of triple point paths for the same mixture in tubes of increasing height.

various regions. The figures shows a cold, unreacted pocket cut off by interacting shock waves.

The formation of the unburned pockets can be traced directly to the curvature of the transverse shock waves. When two transverse waves collide or one hits a wall, the interaction can cut off a portion of unreacted, cold material. If the material in the pockets burns slowly enough, the process effectively draws energy out of the detonation and can provide a mechanism for detonation extinction. If the pockets burn rapidly, they can generate new pressure pulses that perturb the system and cause new structures to form at the detonation front. This is one possible mechanism for generating the irregular cellular pattern observed in most gases and liquids.

These findings are curious because they show how an initially homogeneous material can develop an extremely inhomogeneous structure as shocks moves through it. These findings are important because the existence of unreacted pockets and their properties can determine whether a detonations lives or dies.

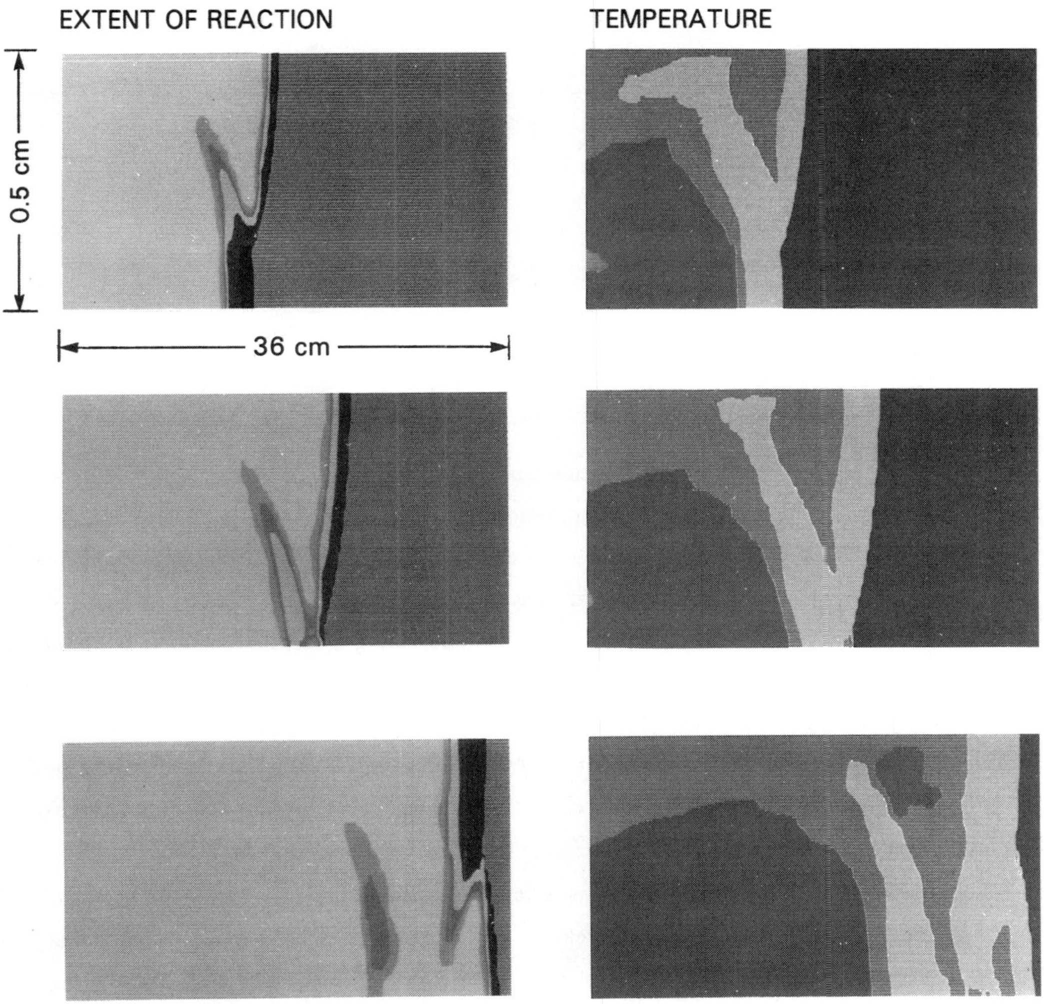

Figure 6. Extent of reaction and temperature contours for a detonation propagating in a small tube. The detonation is propagating from left to right. Cold, unreacted pockets form and are left behind the detonation front.

6. Computations of Premixed Laminar Flames

Importance of Contributing Processes

With so many processes contributing to the description of a flame, it is important to understand their relative importance under different circumstances. Here we summerize the work in [26], a systematic study of the relative importance of various diffusive transport processes to the burning velocity in rich, lean, and stoichiometric hydrogen-air mixtures. These computations were performed with the one-dimensional flame model, FLAME1D [27], that uses the ADINC

Table 2. Burning Velocities of H_2-Air Mixtures

Case	Processes On	Lean	Stoichiometric	Rich
(a)	All	36 cm/s	185 cm/s	44 cm/s
(b)	Thermal Conduction Ordinary Diffusion	40	195	48
(c)	Thermal Conduction	25	85	0
(d)	Ordinary Diffusion Thermal Diffusion	0	115	38

algorithm with a specially designed Lagrangian adaptive rezoning method, a detailed 25-step hydrogen-oxygen chemical reaction scheme, describing interactions of H_2, O_2, H_2O, H, O, OH, HO_2 and H_2O_2 [28], the full iterative solution of the diffusion equations for the individual species, and algorithms for thermal conduction and thermal diffusion.

In the calculations described now, the model was configured with an open boundary at one end to simulate an unconfined system. All of the calculations were performed in a planar (Cartesian) geometry. The initial temperature and pressure of the mixture were 298 K and 1 atm, respectively. The flame was initiated by superimposing a Gaussian temperature profile. The central temperature and width were chosen so that the added initial energy was above the minimum ignition energy required for the flame to evolve into a steady profile.

For either thin or planar flames, the instantaneous normal burning velocity can be calculated from the flame velocity if we know the velocity of the unburned gases ahead of the flame. As shown in Figure 7, for planar flames, the velocity of the unburned gases ahead of the flame is constant. Hence the burning velocity can be unambiguously determined as the difference between the flame velocity and the flow velocity ahead of the flame,

$$V_{\text{burn}} = V_{\text{flame}} - V_{\text{fluid}} \ . \qquad (14)$$

The three diffusive transport processes considered were thermal conduction, molecular (ordinary) diffusion, and thermal diffusion. Calculations were performed with various combinations of these turned on or off for three mixtures of H_2, O_2, and N_2 which are fuel-lean 0.65:1:3.76, stoichiometric 2:1:3.76, and fuel-rich 15:1:3.76.

Table 2, which summarizes these results for four cases, shows that the relative importance of thermal diffusion does not change from lean to rich mixtures. Cases (a) and (b) shows that the effect of thermal diffusion is to lower burning velocities 6–11%. Hydrogen moves fast and its behavior dominates the diffusion properties of the system. It diffuses from the flame into the cold material by molecular diffusion. Thermal diffusion diffuses hydrogen from the lower-temperature region to the higher-temperature region. Thus, the processes have an opposite effect. Thermal diffusion inhibits hydrogen movement into unreacted material and therefore when there is no thermal diffusion, the burning velocity increases. In stoichiometric mixtures, thermal conduction and molecular diffusion have about equal importance in determining the flame velocity. This is not true in lean or rich mixtures. In the lean mixture, thermal conduction is much more important; without it no steady burning velocity is reached. In the rich mixture, molecular diffusion is more important and without it, the flame does not propagate.

Evolution to Cellular Flame Structure

Flames are subject to a number of different instabilities that originate in the competing effects of the controlling physical processes. Thermo-diffusive, Landau, and Rayleigh-Taylor instabilties lead, for example, to phenomena such as cellular structures at the flame front, differences in the shape and behavior of upward-propagating and downward-propagating flames, and the transition of laminar to turbulent flames. These instabilities themselves are usually analyzed by a linear stability analysis of simplified sets of equations describing the balance between selected processes. However, the nonlinear effects are those which govern the behavior after the initial onset of the instability, and sometimes the combined effects of a number of the controlling processes cannot be separated from each other in any simple way. In addition, sound waves in the system can act as both a means of propagating information and as a trigger of instabilities in the reacting flow.

Recent detailed two-dimensional time-dependent simulations of hydrogen-oxygen flames have started the process of isolating and studying the importance of various physical and chemical processes in determining the structure and stability of a multidimensional flame. These calculations use the FLIC2D model [29], which uses BIC-FCT, the full 25-step chemcial reaction mechanism, and includes molecular diffusion, thermal conduction, and gravity. The computations to date were performed in a reference frame moving at the one-dimensional laminar burning velocity. This burning velocity, and the initial unperturbed conditions of the flamefront, are determined from analogous one-dimensional calculations.

Figure 8, taken from the work by Patnaik et al. [30], shows the evolution of a lean zero-gravity hydrogen flame with the perturbation in the center, as shown in the first frame. In time, the structure at the front grows into a cellular shape very similar to that seen experimentally. By the end of the computation, the cellular structure appears to be splitting again, but the computation had to be terminated because the flame was moving out of the finely gridded region and coming too close to the upper and lower walls. Simular computations in a rich hydrogen flame

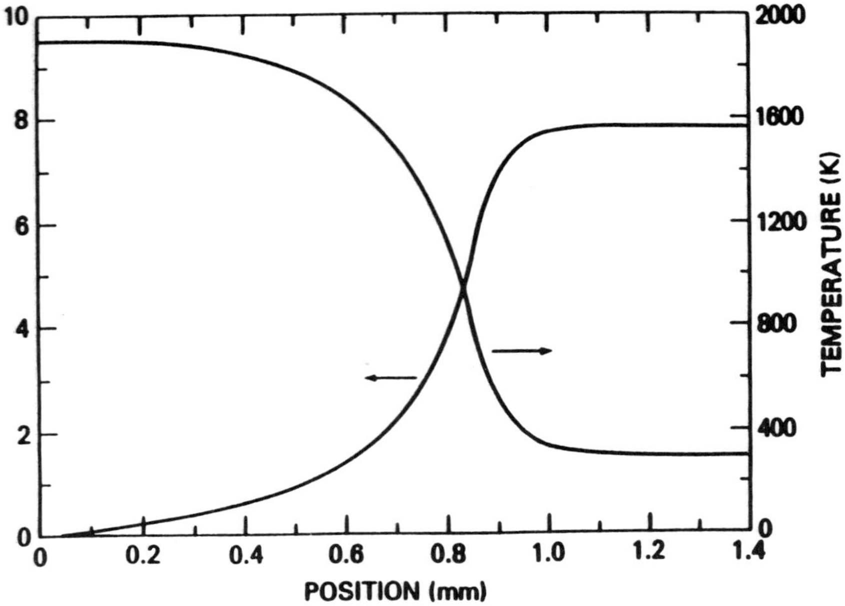

Figure 7. Velocity and temperature in a planar flame propagating in an H_2:O_2:N_2/2:1:4 mixture.

Figure 8. Temperature and OH number density contours for a flame propagating in a mixture of H_2:O_2:N_2 in the ratio 1.5:1:10, at the burning velocity of 12 cm/s, in the reference fram of the flame. The hot burned material (high temperature, high OH densities) are on the left, and the cold, unburned material (low temperature, low Oh density) are on the right. The initial perturbed flamefront develops progressively more structure.

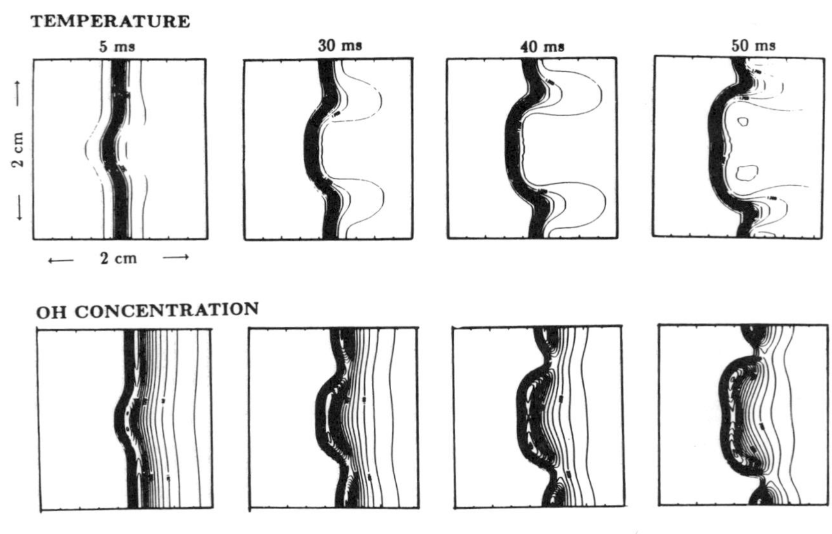

show that the instability is damped. Further simulations in the lean mixture with the hydrogen

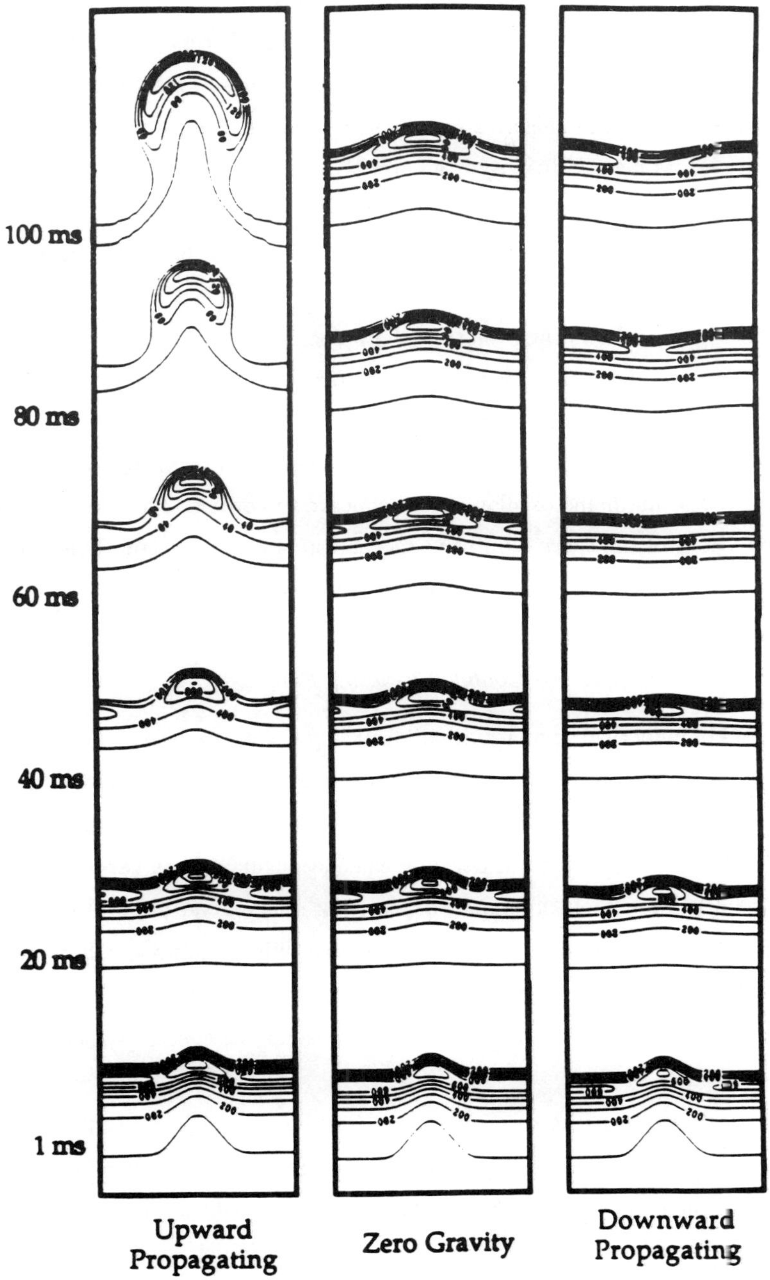

Figure 9. Comparison of OH number density contours in flames ain a hydrogen-oxygen-nitrogen mixture in the ratio 1:1:10.

diffusion coefficient set equal to the oxygen diffusion coefficient show that the instability dies; but when the oxygen coefficient is set equal to the hydrogen coefficient, the instability persists. The results of the calculations support the idea that the diffusional-thermal instability is an important controlling mechamism in producing cellular structure in flames. This mechanism assumes that the process is controlled by the deficient component. For lean hydrogen mixtures, hydrogen is the limiting reactant and its mass diffusivity is greater than the thermal diffusivity of the mixture.

It is known from flammability studies that a flame propagating upward in a tube propagates in a wider range of mixture stoichiometries and dilutions than a flame propagating downward. This suggests a dependence on gravity that can be investigated with the detailed flame model described above. Figure 9 shows the evolution of a flame propagating upward, downward, and in zero gravity for the same lean mixture of hydrogen, oxygen, and nitrogen. The differences here are dramatic: the zero-gravity case shows a cellular structure forming; the upward-propagating flame becomes more and more curved until a buble is formed and seems to continue to grow; and the downward-propagating flame oscillates from concave to convex. These results are explained based on combinations of the thermo-diffusive mechanism and a Rayleigh-Taylor instability [31].

7. Complex Fluid Flows

An Unstable, Subsonic, High-Speed Confined Reacting Flow

An extensive series of numerical simulations performed by Kailasanath (e.g., [32]) have shown that acoustic interactions in confined flows are crucial for determining the generation and merging of large-scale structures in the flow. These simulations use two-dimensional planar and axisymmetric versions of the explicit FCT algorithm. Selected calculations have been carried over hundreds of thousands of timesteps to obtain fluctuation spectra of selected variables. In calculations over such a large physical domain with complex and changing fluid structures, it becomes prohibitive to resolve the structure of the flame front itself. In this case, a model of energy release based on the hydrogen-oxygen reaction mechanism is used to consider the effects of energy release on the flow.

Figure 10 shows a complicated axisymmetric system in which a gas flows out of a long cylindrical inlet into a chamber of larger diameter. The exit condition from this larger chamber is choked flow so that the flow becomes sonic at the throat of an exit nozzle. The top portion of the figure shows a schematic of the chamber and the bottom portion shows a series of instantaneous streamlines of the flow in the cross-hatched region. Lines originating at the centers of vortices have been drawn between the streamlines to show the evolution of the vortex structures.

This calculation describes a subsonic high-speed slightly preheated mixture of hydrogen and air entering into a combustor. The mixture evolves into a quasi-periodic pattern as vortices are

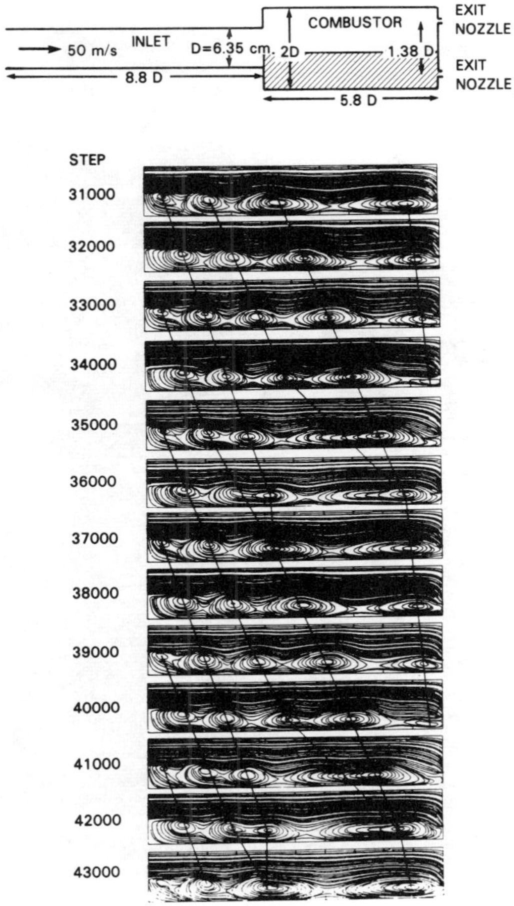

Figure 10. Top: Schematic of an axisymmetric flow with the line of symmetry through the center. Bottom: Calculated instantaneous velocity streamlines in the hatched region of the schematic. Lines through the streamlines connect the centers of vortex structures [33].

shed, merge, and exited through the nozzle. Temperature contours before ignition are essentially uniform, but show the effects of compressibility. There is a strong coupling between the vortex structures formed in the shear layer and the acoustics in a ramjet combustion chamber. For example, the flowfield has an overall repetition cycle of 6000 steps, corresponding to a low-frequency mode of $\approx$ 150 Hz. In addition, the first vortex rollup appears with a frequency of $\approx$ 450 Hz. The low-frequency mode, corresponding to the quarter wave mode of the inlet, controls the overall merging pattern in the chamber. The higher frequency mode corresponds to the first longitudinal acoustic mode of the chamber.

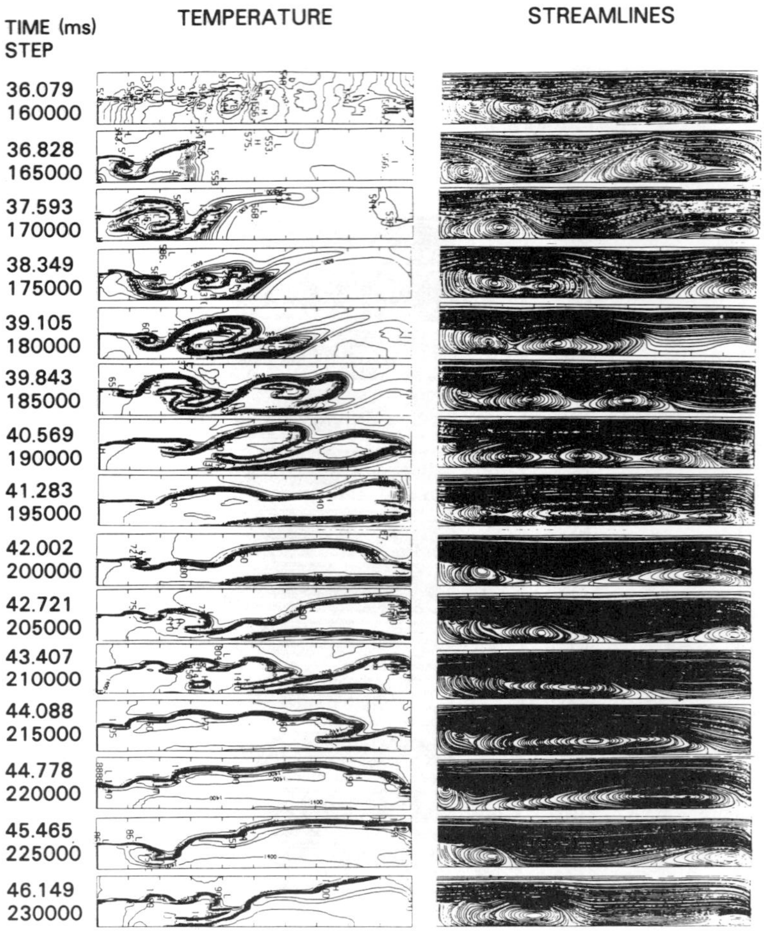

Figure 11. Instantaneous streamlines and temperature contours from a calculation of a reacting flow initiated in a cold flow similar to that shown in Figure 10 [33].

Figure 11 is a continuation of a calculation similar to that in Figure 10, but after a time the mixture was "ignited" in a small region around the inlet [33]. Chemical reactions and heat release are allowed to occur according to an induction parameter model. The instantaneous streamlines and temperature contours show the flame front moving down the chamber and a quasi-steady pattern is eventually set up. The flame front is indicated in the temperature contours both by the the high temperatures and the dark lines caused by closely spaced contours. In time, the roll-up in the shear layer causes the reaction front to curve downward and engulf the cold mixture which subsequently burns. The entrainment velocities are fast enough in these calculations that the laminar flame velocity can be neglected. As the reaction moves downstream, a new vortex forms near the step between timesteps 175,000 and 180,000. This mixes the burned gases with the incoming mixture and acts as an ignition source. The flow field undergoes a cycle of roughly 25,000 timesteps, or 3.463 ms. The low-frequency mode persists in the reacting flow and its amplitude is larger.

Rayleigh-Taylor Instability

The Rayleigh-Taylor instability describes how a dense fluid, initially supported over a lower-density fluid, becomes fluid-dynamically unstable, fragments, and falls through the lighter fluid. In a laboratory, the system eventually approaches a lower-energy state with the lighter fluid on top and the heavier fluid on the bottom. The rate of interpenetration, that is, the rate at which the final state is approached, and the ensuing turbulent mixing rate between the two fluids is crucial to many fluid systems with gravity, to combustion systems, to inertial confinement fusion, and in many geophysical situations.

The rate at which the two fluids interpenetrate and mix has been a subject of intense scientific speculation for decades. Our research has focussed extensively on the high-energy laser and pulsed power aspects of the Rayleigh-Taylor instability (e.g., [33]) but the general fluid dynamic question of how fast two fluids of different density interpenetrate has recently been tackled from a more general and fundamental point of view to understand the complex mechanisms which control compressible fluid turbulence [34].

To limit the computational scope of the problem, we consider a two-dimensional doubly periodic geometry. To effect this, the problem must be formulated in a way that removes the secular pressure gradient that characterizes the equilibrium configuration of a heavy fluid unstably supported by a light fluid. This average pressure gradient can be removed by considering a problem which has no net force. The two fluids are accelerated in opposite directions with two different gravitational forces that are inversely proportional to the mass density. This could be be viewed as two oppositely charged fluids in an electrostatic field or as the usual Rayleigh-Taylor problem in an accelerated frame of reference. By ensuring that the total force on the system is zero, the net momentum of the system is constant for all time and can be taken as zero. The two fluids are being pulled through each other, however, and will have a nonzero net interpenetration drift.

Four scalar ideal gasdynamic equations are solved for conservation of mass, two components of momentum, and energy. Two additional fluid equations,

$$\frac{\partial f_1(x,y,t)}{\partial t} + \mathbf{v} \cdot \nabla f_1 = 0 \tag{15}$$

$$\frac{\partial f_2(x,y,t)}{\partial t} + \mathbf{v}\cdot\nabla f_2 = 0, \tag{16}$$

are solved for the advective transport of the two distinct species number-density fractions, $f_1(x,y,t) = n_1/(n_2+n_1)$ and $f_2(x,y,t) = n_2/(n_2+n_1)$. The f_1 and f_2 sum to unity and are advected without the compression term. The overall mass density, $\rho(x,y,t)$, is defined by

$$\rho(x,y,t) = \rho_1 n_1 + \rho_2 n_2 \tag{17}$$

where

$$n(x,y,t) = n_1(x,y,t) + n_2(x,y,t). \tag{18}$$

A series of simulations of this problem have been performed using FCT algorithms on GAPS in a doubly periodic square domain. Several cases have been run varying the density ratio between the two fluids, the strength of the opposed accelerations acting on the two fluids, and the system size. Figure 12 shows the spatially averaged upward velocity of the light fluid, ${V_y}^{avg}$ (the solid line), and the root mean square vertical velocity, ${V_y}^{rms}$ (the dashed line), as a function of time. In this case, the lower density averages 1 kg/m^3 and the higher density averaves 10 kg/m^3. The lighter fluid feels an upward acceleration of $G_1 = 10^5 m/s^2$; the heavier fluid is accelerated downward with $G_2 = 10^4$ m/s^2. The system size for this calculation was 100 $\times$ 100 cells of size 1 cm by 1 cm.

The horizontal solid line marks the light fluid vertical drift of 20.0 m/s averaged over the 30,000 timesteps of this simulation. Significant fluctuations in the system-averaged vertical drift are observed with short periods when the lighter fluid actually moves downward on average throughout the system. The accelerations are ten thousand and one thousand times the normal acceleration of gravity. This is enough to accelerate the heavier fluid to a velocity of three or four times the speed of sound and the light fluid much faster. Nevertheless, nonlinear Kelvin-Helmholtz and mixing interactions between the oppositely moving fluids keeps the interpenetration drift velocity orders of magnitude smaller.

The light fluid velocity shows large excursions about the average drift, typically with an rms speed of 100 to 150 m/s. Thus transient local flows with Mach number over 0.3–0.5 are not uncommon. This means that the average interpenetration drift of the two fluids is much smaller than the speed at which most of the fluid is moving. Strong vortices and shear flows occur on all scales in this problem but the mixing of the two fluids is correspondingly fine grained so almost as much light fluid is moving downward as upward at any time. Shear of the two fluids at an interface tends to mix the fluids in a coarse-grained way, as shown in Figure 13 for three different system sizes. This tendency to mix on finer and finer scales is counterbalanced by a tendency

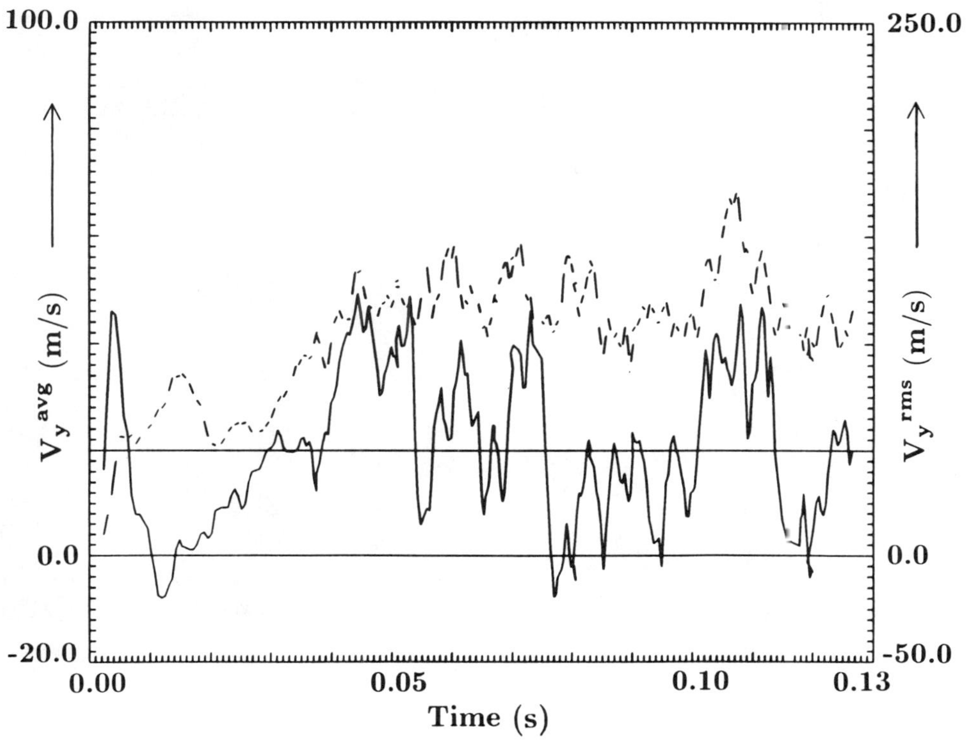

Figure 12. Average interpenetration drift velocity versus time for differentially accelerated fluids in a doubly periodic geometry. The spatially average light fluid drift and the rms fluctuation speed about this drift are show for a 2:1 density ratio and an acceleration 10^4 times normal gravity.

for like fluids to coagulate under the influence of the opposed accelerations. One forcing factor in this coagulation is the radial pressure and density gradients in a strong vortex which tend to spin the heavier fluid out. Thus coherent vortex structures tend to drift in the direction of the lighter fluid.

When the gravity and the density ratio are varied, the resulting situation is very similar. The mass ratio was varied in three calculations using values 2:1, 10:1 and 50:1 with the lower density held fixed at 1.0 kg/m^3. The system size was taken as 100×100 cells and $G_1 = 10^4\ m/s^2$. The corresponding interpenetration drifts, averaged over 30,000 timesteps, are 5.2 m/s, 7.8 m/s, and 10.7 m/s. A set of three 100×100 calculations were performed using a species mass ratio of 10:1

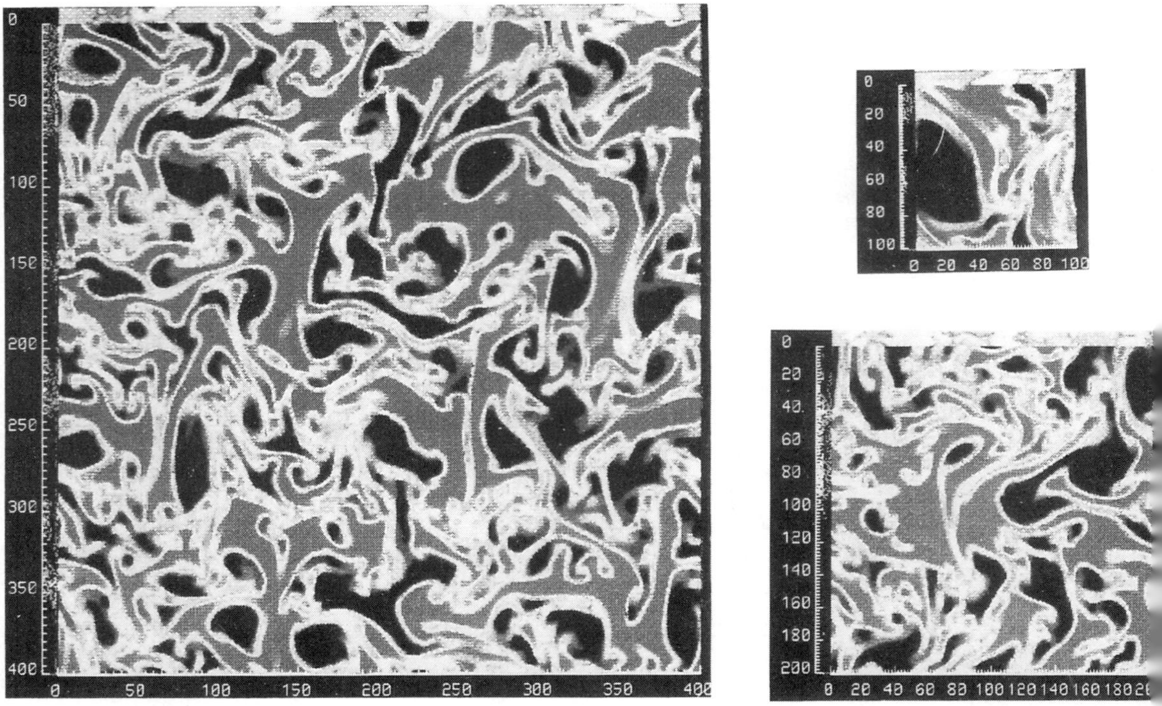

Figure 13. Heavy species number density contours for a doubly periodic simulation of the asymptotic Rayleigh-Taylor interpenetration of two fluids.

with the acceleration of the light fluid taking the values 10^4 m/s^2, 3×10^4 m/s, and 10^5 m/s^2. The corresponding drift velocities of the lower-density fluid, 11 m/s, 14 m/s, and 20 m/s, roughly show a square root scaling of the interpenetration drift with acceleration. Another series of three calculations, with the species mass ratio of 1:2 and with corresponding accelerations fixed at 10^4 m/s^2 and -5×10^3 m/s^2, were performed with system sizes of 100×100 cells, 200×200 cells, and 400×400 cells. The corresponding drift velocities were 5.2 m/s, 8 m/s and 12 m/s, also supporting a square root scaling of interpenetration drift speed with system size.

This approximate square root scaling is reasonable because the only characteristic velocity which can be constructed out of the one spatial scale in the problem, the system size, and the acceleration is

$$V_{char} = (G_1 L_{sys})^{1/2} . \tag{19}$$

What is surprizing is that the velocity actually calculated from this expression for the 400×400 cell case mentioned above, $V_{char} = (10^4 \times 4)^{1/2} = 200$ m/s, is a factor of twenty larger than actually measured from the simulations.

This disparity can be explained by the near uniform coarse-grained mixing indicated above and is well illustrated in the heavy species number density plots of Figure 13. The overall macroscopic interaction seems to involve slightly denser mixtures moving in one direction with slightly less dense mixtures drifting the other way. The density imbalance results in a much smaller effective differential acceleration which reduces the interpenetration drift appreciably.

8. Discussion

The costs in computer time and memory are formidable for numerical simulations of reacting flows. Major costs are in integrating the ordinary differential equations describing the chemical reactions, evaluating diffusive transport processes, and in the solution to the convective transport when there are irregular boundaries or obstacles. Dealing with these problems requires improving numerical algorithms and other software, and taking advantage of larger, faster, more accurate, and friendlier computers. Over the last few decades, improved algorithms have contributed to simulation capability at least as much as hardware development. Together, hardware and software advances have increased our capability by orders of magnitude. Early one-dimensional gas-dynamics calculations were resolved with fifty to one hundred omputational cells. Such problems can now be performed in three dimensions, about four to five orders of magnitude more computation, in the same time as the previous one-dimensional calculations.

This paper has covered a broad range of topics in reacting flows. We first described the general general gas-phase reactive flow problem and then describes some recent developments in scientific computing that will make a significant impact on simulating these complex flows. The types of advances discussed are in models of the physical systems themselves, computer architectures, numerical methods, and the complexity of the problems that can be solved. New supercomputer architectures, that allow vectorization and multiprocessing, make it necessary to develop algorithms that are optimal on the new machines and to rethink how the software is written for them. Such algorithms are now becoming available for computational fluid dynamics and reactive flows. These advances in computers and numerical methods allow us to solve complicated multidimensional problems with more realistic boundary conditions, more complicated chemical models, for longer times, and at higher resolutions.

Acknowledgements

The authors wish to thank the Instituto per le Applicazioni del Calcolo "Mauro Picone" for the invitation to present these lectures and the stimulating environment for discussion at the Institute. This work was sponsored by the Office of Naval Research through the Naval Research Laboratory. The authors would also like to acknowledge the extensive research contributions of our colleagues, Drs. David Book, K. Kailasanath, Gopal Patnaik, Rainald Löhner, Theodore R. Young, David Fyfe, Raafat Guirguis, Sam Lambrakos and Stephen Zalesak in the work summarized above.

The authors, Elaine S. Oran and Jay P. Boris, are in the Laboratory for Computational Physics and Fluid Dynamics at the Naval Research Laboratory, Washington, D.C. 20375, USA.

References

1. Oran, E.S. and J.P. Boris, 1987, *Numerical Simulation of Reactive Flows*, Elsevier, New York.
2. Young, T.R., and J.P. Boris, 1977, A Numerical Technique for Solving Stiff Ordinary Differential Equations Associated with the Chemical Kinetics of Reactive-Flow Problems, J. Phys. Chem. 81, 2424-2427; Young, T.R., 1979, CHEMEQ: Subroutine for Solving Stiff Ordinary Differential Equations, Naval Research Laboratory Memorandum Report 4091, Naval Research Laboratory, Wash., D.C., 20375, 1979.
3. Boris, J.P., 1971, A Fluid Transport Algorithm that Works, in *Computing as a Language of Physics*, pp. 171–189, International Atomic Energy Agency, Vienna; Boris, J.P., and D.L. Book, 1976, Solution of the Continuity Equation by the Method of Flux-Corrected Transport, *Methods in Computational Physics*, vol. 16, Academic Press, New York, p. 85–129.
4. E.S. Oran, J.P. Boris, and E.F. Brown, *Fluid-Dynamic Computations on a Connection Machine – Preliminary Timings and Complex Boundary Conditions*, AIAA Paper 90-0335, American Institute of Aeronautics and Astronautics, Washington, DC, 1990.
5. B. van Leer, 1979, Towards the Ultimate Conservative Difference Scheme. V. A Second-Order Sequal to Godunov's Method, *J. Comp. Phys.* 32: 101–136.
6. P. Colella, and P.R. Woodward, 1984, The Piecewise Parabolic Method (PPM) for Gas-Dynamical Simulations, *J. Comp. Phys.* 54: 174–201.
7. S.R. Chakravarthy, A. Harten, and S. Osher, 1986, Essentially Non-Oscillatory Shock-Capturing Schemes of Arbitrarily-High Accuracy, AIAA Paper No. 86-0339, AIAA 24th Aerospace Sciences Meeting, AIAA, New York; A. Harten, 1983, High Resolution Schemes for Hyperbolic Conservation Laws, *J. Comp. Phys.* 49: 357–93.
8. G. Patnaik, R.H. Guirguis, J.P. Boris, and E.S. Oran, 1987, A Barely Implicit Correction for Flux-Corrected Transport, *J. Comp. Phys.* 71, 1–20.
9. R. Löhner, K. Morgan, and O.C. Zienkiewicz, 1985, An Adaptive Finite Element Procedure for Compressible High Speed Flows, *Comp. Meth. Appl. Mech. Eng.*, 51: 441–465; R. Löhner, K. Morgan, M. Vahdati, J.P. Boris, and D.L. Book, 1986, FEM-FCT: Combining Unstructured Grids with High Resolution, *J. Comp. Phys.*, submitted.
10. J.P. Boris, 1986, A Vectorized "Near Neighbors" Algorithm of Order N Using a Monotonic Logical Grid, *J. Comp. Phys.* 66: 1–20; S. Lambrakos, and J.P. Boris, 1987, Geometric Properties of the Monotonic Lagrangian Grid Algorithm for Near Neighbors Calculations, *J. Comp. Phys.*, to appear; S.G. Lambrakos, J.P. Boris, E.S. Oran, I. Chandrasekhar, and M. Nagumo, A Modified Shake Algorithm for Maintaining Rigid Bonds in Molecular Dynamics Simulations of Large Molecules, *J. Comp. Phys.*, 85, 473–486, 1989.
11. M. Peyrard, S. Odiot, E. Oran, J. Schnur, and J. Boris, *Microscopic Model for Propagation of Shock-Induced Detonations in Energetic Solids*, *Phys. Rev. B.*, 33, 2350–2362, 1986; S.G. Lambrakos, M.

Peyrard, E.S. Oran, and J.P. Boris, *Molecular-Dynamics Simulations of Shock-Induced Detonations in Solids, Phys. Rev.* B, 39, 993–1005, 1989.

12. J.P. Boris, 1987, Supercomputing at the U.S. Naval Research Laboratory, *Proceedings of the 1986 SPIE Meeting on Optical Computing.*
13. *Connection Machine, Model CM-2 Technical Summary*, Thinking Machines Corporation, Cambridge, MA, 1989.
14. E.S. Oran, J.P. Boris, and E.F. Brown, *Fluid-Dynamic Computations on a Connection Machine – Preliminary Timings and Complex Boundary Conditions*, AIAA Paper No. AIAA-90-0335, American Institute of Aeronautics and Astronautics, Washington, DC, 1990.
15. Oran, E.S., T.R. Young, J.P. Boris, and A. Cohen, 1982a, Weak and Strong Ignition. I. Numerical Simulations of Shock Tube Experiments, *Comb. Flame* 48, 135–148; Oran, E.S. and J.P. Boris, 1982, Weak and Strong Ignition. II. Sensitivity of the Hydrogen-Oxygen System, *Comb. Flame* 48: 149–161.
16. Cohen, A., and J. Larsen, 1967, B.R.L. Report No. 1386, Ballistics Research Laboratories, Aberdeen, Maryland.
17. Kailasanath, K., and E.S. Oran, 1983, Ignition of Flamelets behind Incident Shock Waves and the Transition to Detonation, *Comb. Sci. and Tech.* 34: 345.
18. Edwards, D.H., G.O. Thomas, T.L. Williams, 1981, Initiation of Detonation by Unsteady Planar Incident Shock Waves, *Combust. Flame* 43, 187.
19. Fickett, W., and W.C. Davis, 1979, *Detonation*, University of California Press, Berkeley, CA; Strehlow, R.A., 1984, *Combustion Fundamentals*, McGraw-Hill, New York;
20. Strehlow, R.A., 1969, The Nature of Transverse Waves in Detonations, *Astro. Acta* 14: 539–548.
21. Taki, S. and Fujiwara, T., 1981, Numerical Simulation of Triple Shock Behavior of Gaseous Detonation, *Eighteenth Symposium (International) on Combustion*, pp. 1671–1681, The Combustion Institute, Pittsburgh, PA; Guirguis, R., E.S. Oran, and K. Kailasanath, 1986, Numerical Simulations of the Cellular Structure of Detonations in Liquid Nitromethane — Regularity of the Structure, *Comb. Flame* 61: 199–209.
22. Kailasanath, K., E.S. Oran, J.P. Boris, and T.R. Young, 1985, Determination of Detonation Cell Size and the Role of Transverse Waves in Two-Dimensional Detonations, *Comb. Flame* 61, 199–209;
23. Guirguis, R., E.S. Oran and K. Kailasanath, 1986, Numerical Simulations of the Cellular Structure of Detonations in Liquid Nitromethane – Regularity of the Cell Structure, *Comb. Flame* 65: 339–365; R. Guirguis, E.S. Oran, and K. Kailasanath, *The Effect of Energy Release on the Regularity of Detonation Cells in Liquid Nitromethane, Proceedings of the 21th Symposium (International) on Combustion)*, 1987, pp. 1659–1668, The Combustion Institute, Pittsburgh, PA. R. Guirguis, E.S. Oran and K. Kailasanath, *Numerical Simulations of the Cellular Structure of Detonations in Liquid Nitromethane - Regularity of the Cell Structure,Combustion and Flame*, 65: 339-366, 1986.
24. Taki, S. and T. Fujiwara, 1978, Numerical Analysis of Two-Dimensional Nonsteady Detonations, *AIAA J.* 16, 73-77.
25. Oran, E.S., T.R. Young, J.P. Boris, J.M. Picone, and D.H. Edwards, 1982b, A Study of Detonation Structure: The Formation of Unreacted Gas Pockets, *Nineteenth Symposium (International) on Combustion* pp. 573–582, The Combustion Institute, Pittsburgh, PA.

26. K. Kailasanath and E.S. Oran, in *Complex Chemical Reaction Systems*, eds. J. Warnatz and W. Jager, 243–252, Springer-Verlag, New York (1987).

27. K. Kailasanath, E.S. Oran, and J.P. Boris, *A One-Dimensional Time-Dependent Model for Flame Initiation, Propagation, and Quenching*, NRL Memorandum Report 4910, Naval Research laboratory, Washington, DC, 20375 (1982).

28. T.L. Burks and E.S. Oran, *A Computational Study of the Chemical Kinetics of Hydrogen Combustion*, Naval Research Laboratory Memorandum Report 4446, Naval Research Laboratory, Wash., D.C., 20375 (1981).

29. G. Patnaik, K.J. Laskey, K. Kailasanath, E.S. Oran, and T.A. Brun, *FLIC – A Detailed, Two-Dimensional Flame Model*, to appear as NRL Memorandum Report, Naval Research Laboratory, Wash., D.C., 20375

30. G. Patnaik, K. Kailasanath, E.S. Oran, and K.J. Laskey, *Twenty-Second Symposium (International) on Combustion*, 1517–1526, The Combustion Institute, Pittsburgh (1988).

31. K. Kailasanath, G. Patnaik, and E.S. Oran, *Effect of Gravity on Muldidimensional Laminar Premixed Flame Structure*, IAF-88-354, International Astronautical Federation, Paris (1988).

32. K. Kailasanath, J.H. Gardner, J.P. Boris, and E.S. Oran, *J. Prop. Power*, 3, 525–533 (1987); K. Kailasanath, J.H. Gardner, J.P. Boris, and E.S. Oran, *J. Prop. Power*, 5, 165–171 (1989).

33. M.H. Emery, J.H. Gardner, J.P. Boris, A.L. Cooper, Vortex Shedding Due to Laser Ablation, *Physics of Fluids*, Vol 27 (5), pp 1338–1340, May 1984.

34. J.P. Boris, Compressibility in Turbulence Generation and Mixng, Proceedings, International Workshop on the Physics of Compressible Turbulent Mixing, Princeton University, Princeton NJ, 24–27 October 1988, to be published by Springer-Verlag.

A F GHONIEM

Numerical simulation of combustion in reacting shear flow

I. INTRODUCTION

Numerical simulation is defined as the integration of the unsteady, unaveraged partial differential equations governing chemically reacting flow without resorting to turbulence closure modeling [31]. Due to the obvious complexities of these equations, including the fact that they describe multidimensional, elliptic, multiscale, multispecies, variable temperature and pressure phenomena, the task is enormous. Thus, in each case, one must simplify the equations in accordance with the physical system of interest before attempting the simulations. The simplification may involve limiting the domain to two dimensions, reducing the chemical model to few species and few kinetic steps, restricting the variation of the temperature or density to a small range, confining the simulation to a small domain and applying simple boundary conditions, or using phemonenological models to describe certain aspects of the process of interest. Despite these limitations, we show in this short review that numerical simulations have been most revealing when applied to study complex combustion processes.

Numerical simulation has been developed for two purposes. (1) To investigate systems which can not be described be simple equations that yield to analytical tools of analysis. Due the convective nonlinearity of Navier-Stokes equations, especially at high Reynolds number, and the exponential nonlinearity of the kinetics equations, combustion equations are in most cases, not amenable to analytical solutions and numerical integration becomes necessary. (2) To provide predictive tools for practical systems, minimize the experimentation involved in developing new concepts, and assist in constructing simple models for complex phenomena. At this stage, computer power has not reached the level where it is possible to predict, in detail, a reacting flow in a practical combustor. However, computer technology has not matured yet as evident from recent developments, e.g., parallel architecture and RISC, and it is early to predict what the limit would be. It is likely that, as numerical algorithms and computer hardware become more powerful, most current limitations will be relaxed allowing simulations of practical systems to be obtained.

Several numerical methods have been constructed and applied to the numerical simulation of reacting flows including finite difference methods [31], spectral methods [33] and vortex methods [7]. The first two algorithms are based on the discretization of the governing equations on a fixed Eulerian grid and advancing the solution in time by updating the values of the gas dynamics variables according to the discretized equations. These methods have been successful in investigations involving several reacting systems. More developments are required to allow grid-based systems to simulate: (1) flows with sharp fronts that move in space; (2) flows in complex geometry; and (3) processes at high Reynolds and Damkohler numbers. In vortex methods, which are inherently Lagrangian, derivatives of the primary variables, the vorticity and scalar gradients, are represented by discrete elements that move along particle trajectories [11]. Thus, these methods are naturally adaptive and do not require the discretization of the convective derivatives. In this paper, we review few applications of vortex methods to reacting flow and show how their results have been used to compliment analytical and experimental studies of reacting systems.

Most applications of numerical simulation to reacting flows have aimed at investigating some of the fundamental mechanisms of turbulence and turbulence-combustion interactions. These include solutions describing the evolution of turbulence on turbulent-like structures in shear flow [18-21], solutions of vortex-flame interactions [13], shear layer-combustion interactions [22,25], and separating flow-flame interactions [7,15]. These mechanisms include mixing enhancement by strong entrainment into large eddies, reaction extinction due to vorticity-generated shear, turbulence suppression by heat release, generation of combustion instability by pressure-vorticity-flame coupling, and the generation of flame instabilities. Most combustion models have been limited to thin flame models or one-step kinetics. In this paper, we will review some of the solutions obtained for reacting shear flow and separating flow cases with different levels of complexity.

The paper is organized as follows. The formulation describing a two-component reacting system at low Mach number is reviewed in Section II. The basic concepts of the vortex method and the transport element method in free space are summarized in Sections III. In Section IV, V and VI, we review results obtained for three different reacting flows: a three-dimensional reacting shear layer with no heat release, a two-dimensional jet with heat release and a premixed flame stabilized in a separating flow. The paper is closed by some general remarks in Section VII.

II. FORMULATION

In combustion systems, heat release at low Mach number leads to the generation of an irrotational velocity field, $\nabla\phi$, which when superimposed on the vorticity-induced soleniodal velocity, u_ω, represents the total velocity [4,7]. The low Mach number approximation is employed to filter out the pressure waves and to render the pressure independent of the density in unconfined flows [6,28]. Employing the velocity decomposition, the governing equations of combustion can be written as follows:

$$u = u_\omega + \nabla\phi \qquad (1)$$

$$\nabla^2 \phi = \frac{1}{T}\frac{dT}{dt}, \tag{2}$$

$$\frac{d\omega}{dt} = \frac{1}{\rho^2} \nabla \rho \, x \, \nabla p - \omega (\nabla . u) + (\omega . \nabla) \, u + \frac{1}{R_e} \nabla^2 \omega, \tag{3}$$

$$\frac{dT}{dt} = \frac{1}{P_e} \nabla^2 T + \quad Q \, W, \tag{4}$$

$$\frac{dc_k}{dt} = \frac{1}{P_e L_e} \nabla^2 c_k + W_k \ , \tag{5}$$

$$\rho T = constants \tag{6}$$

where $u = (u,v,w)$ is the velocity, $x = (x,y,z)$ is the position vector, t is time, $\omega = \nabla x u$ is the vorticity, T is temperature, ρ is density, c is species concentration, W is the rate of creation/destruction of species, Q is the enthalpy of reaction, R_e, P_e and L_ϵ are the Reynolds, Peclet and Lewis numbers, respectively. Here, we assume that: (1) the transport properties are constant except for mass diffusivity which is replaced by a density-averaged value; and (2) molecular weights of reactants and products are the same. The pressure gradient is obtained from the momentum equation: $\nabla p/\rho = - a + 1/(\rho R_e) \nabla^2 u$, where $a = du/dt$. Both velocity components, u_ω and $\nabla \phi$ must satisfy the velocity boundary condition in the direction normal to the boundary of the domain. Note that Eq. (2) is an expression of the continuity equation while Eq. (3) is an expression of the Navier-Stokes equation in terms of the vorticity.

III. NUMERICAL SCHEMES

In vortex methods, the vorticity field is discretized into a number of vortex elements of finite and overlapping cores [9,18]:

$$\omega(x,t) = \sum_{i=1}^{N} \omega_i \, h^n f_\delta(x-\chi_i(X_i,t)), \tag{7}$$

where ω_i is the vorticity of an element, N is the total number of vortex elements, h is the average distance between the centers of neighboring elements, δ is the core radius of a vortex element, $f_\delta = 1/\delta^n f(r/\delta)$ is the core function describing the distribution of vorticity associated with an element, and n is the number of space dimensions of the system, $n = 2$ and 3 for two and three dimensional flows, respectively.

The velocity field of a distribution of vortex elements is obtained by substituting Eq. (2) into the Biot-Savart law and performing the integration. The result is:

$$u_\omega(x,t) = \sum_{i=1}^{N} \omega_i h^n K_\delta(x-\chi_i(X_i,t)), \tag{8}$$

$$\frac{d\chi_i}{dt} = u(\chi_i(X_i,t),t), \tag{9}$$

$$\frac{d\omega_i}{dt} = \omega_i \cdot (\nabla u)_i \tag{10}$$

where $K_\delta(x) = K(x)\kappa(r/\delta)$,

$K(x) = (y,-x)/2\pi|x|^2$ for $n=2$, $K(x) = -x/4\pi|x|^3$ for $n = 3$,

while $\kappa(r) = 2(n-1)\pi \int_0^r f(r')\, r'^{(n-1)} dr'$.

For an incompressible flow, the velocity field is supplemented with a component, $\nabla\phi_p$, to satisfy the no-flow boundary conditions on the solid boundaries of the domain. In a variable density flow, baroclinic generation of vorticity changes the strength of the elements according to the first term in the right hand side of Eq. (2) [9-11]. To evaluate this term we need $\nabla\rho$. This is computed using the transport element method. Before summarizing this method, we mention that three schemes have been suggested to simulate the diffusion operator in Eq. (2): (1) random walk in which a random element drawn from a distribution which satisfies the diffusion equation is added to the location of all the vortex elements each time step [1,8]; (2) core expansion, also in accordance with the solution of the diffusion equation in a strained flow [11,26]; and (3) vorticity exchange between elements according to the discrete form of the diffusion operator obtained by using Eq. (7) [32]. It has been found that different algorithms are suitable for different applications. Bounded flows are best simulated using the random walk algorithm. The core expansion is most convenient in two dimensional simulations in free space, while the vorticity exchange method is most suitable in three dimensional computations.

In reacting flow at low Mach number, combustion heat release produces an additional velocity field due to volumetric expansion [22]. This field is represented by Eq. (2). In vortex methods, the solution of this equation can be written, similar to the summation to Eq. (8), in the following form:

$$\nabla \phi(x,t) = \sum_{i=1}^{N} \frac{1}{T_i} (\frac{dT}{dt})_i \, h_i^n(t) \, \nabla G_\delta(x-\chi_i(X_i,t)). \tag{11}$$

where $G_\delta = G * f_\delta$ is the convolution and G is the Green function solution of Eq. (2).

Vortex simulation is motivated in part by the fact that at high Reynolds number, vorticity is concentrated within relatively small zones such as boundary and shear layers and recirculation zones utilizing vortex elements which move with the material elements one can concentrate the computational effort within these zones. Similarly, in most cases of combustion, reaction occurs within thin zones across which the gradients of the scalars, e.g., temperature and species concentration, are finite. Since outside these zone the gradients vanish and the scalars attain uniform values, it should be convenient to use scalar gradients, instead of the scalars, in numerical simulation. This is precisely the basic concept of the transport element method which has been developed to integrate the system of equations (1)-(6) in a combustion process [23,25]. The fundamental concept of the method is briefly summarized next.

In the transport element method, the gradient, $g = \nabla s$ where s is either T or and of c_k, is discretized in a form similar to that expressed by Eq. (7). The value of g_i is then updated along a particle trajectory. The change of the gradient along a particle trajectory is obtained by taking the gradient of Eq. (5). The new equation is integrated each time step to find the value of the gradient at the center of the transport elements. The values of the scalar are obtained using a Green's function formulation which expresses the relationship between the gradient and the scalar concentration. The method takes advantage of the relationship between the changes of the scalar gradients along particle trajectories and the distortion of the flow map when exposed to a strain field [6].

The application of the vortex and transport element methods to three combustion systems, a reacting shear layer, a reacting jet and a premixed, step stabilized combustor, is summarized next. Some observations on these simulations, and their importance to the study of turbulent combustion are described.

IV. COMBUSTION IN A THREE-DIMENSIONAL SHEAR LAYER

In this exapmle, we investigate the effect of the flow field on the combustion process while neglecting the effect of combustion of the dynamics, i.e. we consider a chemical reaction in which heat release is vanishingly small [20]. The main concern is thus the dynamics of mixing in a shear flow and how it affacts the rate of product formation. The flow is initially formed of two streams of unequal velocities, U_1 and U_2, and each stream carrying one of the reactants, c_1 and c_2. When the two streams are brought in contact downstream a splitter plate, shear layer dynamics, mixing, and combustion start. Some of the essential dynamics of this flow can be captured, while greatly reducing the computational labor, if one lets the computational window move with the mean velocity of the two streams while imposing periodic boundaries across the domain [19]. This is shown in figure 1. The flow is initially perturbed in the streamwise and the spanwise directions.

Figure 2 shows the evolution of the shear layer as revealed by the deformation of a material plane initially aligned along the midsection of the shear layer. This plane is represented by a mesh, which is also used to discretize the initial vorticity lying on this plane. The figure shows three dimensional perspectives of this plane. The growth of the streamwise perturbation, without apparent changes in the spanwise direction is clearly seen at $t = 4$. This is due to the Kelvin-Helmholtz instability mechanism which, as the layer extends in the cross-stream direction, generates a flow in the streamwise direction. The roll-up of the plane in the streamwise direction and around a spanwise axis, as the vorticity is concentrated around the midsection of the domain, and the formation of a "large coherent structure" in which most of the material initially within the shear layer is entrained is clearly seen at $t = 8$. The structure, or vortex core, extends in the spanwise direction with small deformation generated by the growth of the initial spanwise perturbation. The deformation of the core in the spanwise direction is due to the translative instability mechanism. The Kelvin-Helmholtz and the translative instability are linear mechanisms which are excited by the smallest perturbation present in the flow. However, the two-dimensional mechanism, the roll-up instability proceeds faster until the cores form. Following their maturation, the cores then become subject to spanwise deformation by the translative instability.

It is clear from the computational results that until $t \sim 10$, the flow remains almost two dimensional and most of the entrainment is in streamwise direction (this conclusion is exploited in the next two examples to limit the computations to two dimensions). The strong strain field associated with roll-up leads to the formation of thin layers of vorticity which connect neighboring large cores, the braids. Before $t = 12$, most of the entrainment is into the cores, while the braids remain flat. For $t > 12$, the growth of the translative instability along the cores leads to the deformation of the braids in the spanwise direction. This deformation is the beginning of roll-up of the braids, in alternate directions, around the boundary points and midpoints of the spanwise length of the domain. This secondary motion is associated with the formation of "vortex rods" of alternating signs of streamwise vorticity which extend in the streamwise direction. Entrainment in the spanwise direction is accomplished by the formation of strong, streamwise vortex structures of length scale equal to one-half the initial perturbation. It is noted that spanwise entrainment into the rods starts after the streamwise entrainment into the cores ends.

Detailed account of the evolution of the vorticity field, the entrainment of passive scalars, and the effect of nonuniform density across the layer have been obtained [19-21]. The important conclusion is that the formation of large-scale structures due to the intrinsic instability mechanisms lead to substantial enhancement of entrainment of fluids from the two sides of the layer into large eddies where mixing proceeds by molecular diffusion. As shown above, mixing enhancement occurs first in the large cores, then in the smaller rods. The apparent cascading of scales into smaller structures as successive instabilities mature lead to mixing enhancement throughout the entire space between the two streams, as shown at the late stages in figure 2. In actual computations of shear layer flow, it is very expensive to follow the cascading process beyond the stages shown in figure 2 since the number of elements exceeds affordable limits. However, one can investigate the essential mechanisms in this cascade process by focusing on one relatively simple structure and carrying out the computations to later stages.

An example of such studies was presented in [18] where the late stages of the evolution of a vortex ring, which can be considered as a vortex core produced by the roll-up of an unstable jet, were computed using the same numerical scheme. The major conclusion of this study is that at the late stages, following the maturation of the large scale azimuthal instability around the ring axis, vortex lines experience successive folding in the form of hairpins of ever decreasing size. The folding of hairpin vortex elements eventually fills out the space with smaller eddies until a molecular scale is reached, a process conventionally referred to as inertial cascading of turbulence. Results of this simulation provide a mechanistic description of the cascade process and can be used in the future to construct a model for the effect of small-scale dynamics on the large scales of the flow.

Before leaving this section, it is important to emphasize that much work remains to be done in the numerical simulation of turbulence. As indicated, for economic reasons, one is forced to limit the range of scales that can be captured by the simulation. While it is clear from the above discussion that large scales are responsible for most of the entrainment which proceeds mixing, small scales, which determine the rate of mixing by molecular diffusion, are, in most cases, beyond what can be afforded in the simulation of a complex flow. This artificial dichotomy of scales is not founded on clear physical principles and work is needed to determine what are the appropriate small scale, if any, at which one can stop the simulation without scarificing important processes. Another question is whether one can represent the effect of scales, beyond a certain small threshhold, on the large scales using some "substructure models" that are derived from a comparison between a simulation in which "all" scales are accounted for and another simulation in which only a limited range of scales are represented. These subscale models are different than classical turbulence models since they are restricted to a certain range of scales.

The effect of mixing enhancement within the large scales on the rate of product formation is obtained by repeating the above computation for a flow in which the rate of reaction is $W = D_a c_1 c_2$. Figure 3 shows the product concentration and reaction rate across a plane parallel to the streamwise direction and located in the middle of the domain along the spanwise direction. These results were obtained for $P_e = 1000$ and $D_a = 10$, and are shown using six shades of gray covering that entire range of each variable, c_p or W. Both quantities are shown to emphasize the connection between the dynamics of the flow, as represented by vorticity concentration, and the progress of the combustion process, as measured by either quantities. The most apparent results is that product concentration follows the evolution of the vorticity field. One can actually use product concentration plots to determine the relative magnitude of vorticity in the field. Caution should be excersized when applying this similarity since product concentration is a scalar quantity while vorticity is a vector. The figure shows the presence of the cores in the middle of the domain by the spiraling of the product around its center and the appearance of the rod, which wrap around the outer edges of the core, at the late stages.

An outstanding question in turbulent combustion is what is the appropriate relationship between the reaction rate and the flow dynamics. The results shown above, and other results obtained over a wide range of Damkohler number, indicate that this relationship is not unique and is strongly dependent on the Damkohler number. For instance, it was found that at low Damkohler number, the reaction zone is confined within the cores of the eddies where intense mixing occurs. At high Damkohler number, the reaction zone moves outwards from the centers of the eddies to their outer edges as time progresses. It was also found that, in all cases, a strong similarity exists between product concentration and vorticity, independent of the Damkohler number. It is thus suggested that vorticity should be used as the appropriate dynamical property in formulating turbulent combustion models. One may be able, utilizing the observed similarity between vorticity and product concentration and implementing other conservation relations, to obtain a model of combustion in shear flow.

The complexity of mixing in shear flow and the need for high numerical resolution in the simulation reacting shear flow are evident in the results shown in figure 4 where the product concentration and reaction rate are shown across a plane parallel to the spanwise direction and taken halfway in the computational domain along the streamwise direction. Here also the reaction zone migrates out from the center of the eddy to the outer edges as the high rate of chemical reaction at $D_a = 10$ causes a fast transformation of mixed reactants, while high product concentration always exit within the eddies. Due to the translative instability described above, the eddy core is twisted along the spanwise direction. The formation of strong streamwise vortices within the cores due to the translative instability is clearly seen in these plots. These plots emphasize the similarity between product concentration and vorticity as the flow evolve through different stages.

Results of these simulations show that product distribution in high Reynolds reacting flows is dominated by the convective currents generated by the large-scale structures which form due to the evolution of flow instabilities. Strong strain field accompanying this process induce severe distortion of the flow field. The latter must be accurately captured by the numerical scheme to avoid the accumulation of artificial diffusion. In particular, at high Damkohler number, the scales of the reaction zone are different from those corresponding to the flow structure.

V. INITIAL STAGES IN A REACTING JET

The purpose of this study are to: (1) investigate the effect of exothermic energy deposition due to chemical reaction on the dynamics of shear flow [23,24]; and (2) describe the dependence of the reaction zone structure on the Damkohler number [12,17]. We choose a two-dimensional, planar, symmetric "fuel" jet, with velocity U_j and concentration c_j, issuing into a stationary atmosphere of "oxidizer," with concentration c_O. The jet is characterized by two symmetric shear layers on both sides of its centerline. We use a compressible flow model and assume that the physical properties of both streams are the same. Similar to the first case, only the temporal problem, in which the computational window is allowed to move with the average velocity of the two streams, $U_j/2$, is considered. We also limit the analysis to the initial stages where the flow is to two dimensional. The temperature ratio of the product to the reactants is $T_P/T_R = 4$, the normalized activation energy is $T_a = 10$, the pre-exponential factor is $A_f = 750$ for the following source term $W = A_f\, c_O\, c_j\, exp\, (-T_a/T)$. The Peclet number is 1000 and the Lewis number is one. The jet is initially perturbed according to a varicose mode with perturbation $\varepsilon = 0.05\ \lambda$ where λ is the perturbation wavelength in the streamwise direction [14].

Figure 5 shows the distribution of product concentration, vorticity and jet fluid concentration at three nondimensional times $t = 5, 7.5$ and 10. As the flow evolves, product concentration within the shear layer changes from a uniform distribution in the streamwise direction to a highly concentrated distribution within the large eddies. Product distribution within the braids is small at the early stages, dropping down to almost zero at the later stages. The shear layers rollup generates strong entrainment currents which draw fluid from the braids into the cores, thus reducing the braid thickness. The reduction of the thickness of the reaction zone within the braids is confirmed by figure 6 which shows the reaction rate contours for this case and for a case with zero activation energy, or fast chemistry in which $W = D_a c_j c_O$ and $D_a = 0.5$. The reaction rate contours are shown for both cases to help illustrate the mechanism of reaction extinction within these zones of high strain rates. As mentioned earlier, the thinning of the braids due to the strain field strongly reduces the reaction zone thickness and decreases the time available for mixing within these areas. In the finite-activation energy case, the reaction rate is even slower than in the zero-activation energy case. This is because when product concentration within the braids is reduced, the temperature is also reduced leading to a cooler zone and a reduction of the temperature-dependent reaction rate.

Figure 5 shows a strong similarity between product concentration and vorticity. Clearly, the flow and combustion are driven by convection. The same observation was made over a wide range of Damkohler numbers and for zero and finite activation energy reactions [14,17,23,25]. In all cases, product concentration is high within the cores and very low within the braids. This is not true for the reaction rate. As shown in Figure 7, for $A_f = 150$ and $D_a = 0.063$, the reaction rates in this case of slow chemistry are highest, for all times, within the eddy core. In this slow chemistry case, reactants are first drawn into the eddy cores, mixed by diffusion and then forced to react forming products. In the fast chemistry case, figure 6 shows that the reaction zone moves from the eddy cores to their outer edges as time progresses, i.e, the zone of product formation moves outwards where surfaces of contact between reactants exist. However, due the entrainment field, product are drawn into the centers of the eddies. As mentioned in the discussion of the shear layer, results strongly suggest that the proper dynamic variable to be used in modeling turbulent combustion is the vorticity, and that product concentration can be expressed as simple function of vorticity.

Results in figures 6 and 7 suggest that for slow chemistry, combustion occurs in zone of almost uniform, distributed reaction located within areas of intense vorticity. In this case, the rate of combustion is controlled by the rate of advection of reactants into the eddy core. For fast chemistry, combustion occurs in thin reaction zones, which resemble ideal flamelets, that exist around areas of intense vorticity. In this case, chemistry is fast enough to cause immediate burning of co-existing reactants. The distinction between the two regimes is made here using the Damkohler number since we hold the flow dynamics the same. It is important to mention, however, that the distinction is not sharp. In the high Damkohler number case, we encounter distributed combustion zones within the eddy cores at the early stages of rollup while this reaction zone develops near the end of the simulation.

The effect of heat release on the growth of the large eddies has been investigated for the infinite-rate (temperature-independent) and finite-rate (temperature-dependent) kinetics. We found that, in most case, heat release tends to suppress the rate of growth of the large eddies. At infinite-rate kinetics, the effect is stronger, especially when the initial amplitude of perturbation is small, $\varepsilon = 0.01\ \lambda$. In this case, rollup starts later than heat release since while the former waits for the perturbation to grow into the nonlinear stages, the latter starts immediately because there is no "chemical delay". In the case of infinite-rate kinetics with small initial perturbation, we found that heat release establishes a layer of low-density products within the shear layer before substantial rollup has occurred. A low-density layer of products reduces the vorticity of a shear layer (since the circulation is conserved while the area expands) and reduces the growth rate of its instability. This effect can be reduced if the initial perturbation is high, $\varepsilon = 0.05\ \lambda$. In this case, as shown in figure 7, the layer rolls up to form eddies before substantial amount of heat has been deposited. The moderate effect of heat release on the growth of the eddy is still noticeable; its size is smaller than that corresponding to the nonreacting case [25].

A curious observation, also shown in figure 7, is that the effect of heat release on the finite-rate kinetics case is smaller than its effect on the infinite-rate kinetics solution. This, however, is consistent with the notion that the effect of heat release is strongly connected with the time scale on which it occurs. If rollup starts before a substantial amount of heat has been deposited, then energy release does not affect the growth of the eddies. This is the case at finite-rate kinetics since, in this case, there is a "chemical delay" associated with the rise of the temperature to a level that allows the reaction rate to proceed faster than the flow. In the finite-rate kinetics case, chemical delay caused by the slow rise in temperature allows the shear layers to roll up before any substantial heat release has caused the formation of the low-density layer that causes the delay in growth.

The study of reacting shear flow at high Reynolds and Damkohler numbers requires numerical schemes which can accommodate multiscale phenomena, both in space and in time. As the Damkohler number increases, the thickness of the reaction zone becomes smaller than that of the layer of strong shear. Moreover, at high Reynolds number, the stretch and convolution of the zones of high shear is strong. The schemes used in this study, being Lagrangian and adaptive in principle satisfy these requirement. It is important to mention, however, that as Reynolds number and Damkohler number increase, more computational elements and smaller time steps are needed to resolve these scales.

VI. INSTABILITY OF A DUMP COMBUSTOR

The previous two examples were concerned with different forms of reacting shear flow in unconfined space between a stream of fuel and a stream of oxidizer. Another problem in turbulent combustion is that of premixed combustion between a stream of homogeneous reactants a stream of combustion products; i.e. premixed combustion stabilized on a shear layer at constant pressure. This problem has been analyzed computationally using the vortex method in [10,11,22]. In this section, however, we review results of premixed combustion stabilized on a sudden expansion in the channel. In this case, the variation of the pressure across the part of the channel where combustion occurs is important and may be responsible for the growth of different instability modes within the combustion chamber. The purpose of the following analysis is to reveal the essential mechanisms associated with these instability modes and to determine the most probable frequencies of the instabilities. The components of the model are schematically shown in figure 8. The analysis is motivated by an attempt to explain the different modes of large amplitude, low frequency oscillations encountered in premixed dump combustors when operated at equivalence ratios near stoichiometry.

The analysis starts with the computation of the nonreactiong flow in a channel with a cavity [30]. The cavity models the sudden expansion and sudden contraction encountered in the combustion chamber of a jet engine or a ramjet engine. In nonreacting flow, it has been established that as the Reynolds number increases beyond a certain transition value, the flow becomes unsteady [8,34]. The amplitude and frequency of the unsteady flow depends on the geometry of the channel, the Reynolds number and the inlet velocity distribution. In figure 9, we show a sample of these results, obtained for what we call a long cavity. The plots show the vortex elements and their instantaneous velocity vectors. The Reynolds number is 2500 and the length/depth ratio is four. Results reveal that the dominant mode of oscillation is associated with the shedding of large vortices from the recirculation zone downstream the sudden expansion. The Strouhal number, based on the shedding frequency, the incoming flow velocity and the cavity depth, is O(0.1). This value varies between 0.08 - 0.2 depending on the cavity geometry and inlet flow conditions, and agrees with a large number of experimental studies of vortex shedding in wake flow. Since this value was determined without any external disturbance, we call this frequency the natural frequency of the combustor.

It is important to mention here that results like figure 9 do not show actual turbulent flow. The latter is a three-dimensional phenomenon in which vortex stretching and cascading processes play important roles. Two-dimensional results, as in figures 5-9, illustrate the behavior of post-transitional flows in which an initially laminar flow has been forced to change its state due to the growth of perturbations via some natural instabilities. The new state, dominated by large-scale vortex structures which enhance mixing in a similar way as that in real turbulence, is a precuser to some more complex transitional state, as seen in Section IV. The importance of analyzing these two dimensional post-transitional states, which are cheaper to compute than three dimensional flows, is that many phenomena, similar to the one discussed here, are nearly two dimensional and are controlled by the large-scale structures which develop as a result of two-dimensional transition. Large-scale structures, especially in confined separated flows, are system dependent, i.e., are a function of the configuration and dynamic conditions. Thus, they must be computed on a case-by-case basis.

The reacting flow model formulated to study this combustion instability is described by a schematic diagram in figure 8. Within the combustor section, the low Mach number model is used, and premixed combustion is described using the thin flame approximation [13]. Thus, the combustor itself is assumed to be acoustically compact, and the flame is assumed to be an active interface between reactants and products. The flame moves by advection and by self propagation at a laminar burning velocity which depends on the conditions in the reactants and the local curvature of the front. The random vortex method is applied to compute the flow field [1-3],SLIC is used to track the flame interface and volumetric sources are utilized to model the effect of volumetric expansion across the flame front [7]. The inlet boundary conditions of the combustor section, in terms of the inlet flow rate, $Q_{in}(t)$, are determined by matching the conditions at the combustor inlet section, defined by the pressure $p_{in}(t)$, with the conditions in an infinite reservoir in which the pressure is held constant. The exit boundary conditions of the combustor section, in terms of the pressure p_{ex}, are determined by assuming that the flow downstream the combustor and until the atmosphere is reached, is one dimensional and compressible.

To investigate the response of the system to acoustic forcing, the downstream section was replaced by an exit boundary condition where the pressure is either held constant or forced at a given small amplitude and frequency. Two forcing cases were simulated. In the first case, high frequency forcing, corresponding to a Strouhal number of 0.4 which is four times the natural frequency of oscillation of the combustor, was applied at the exit boundary of the combustor. In the second case, the pressure was forced at the natural frequency, Strouhal number of 0.1. In the constant exit pressure case, it was observed that as the rate of heat release was increased by changing the density ratio across the flame, the oscillation of the flow and the flame front were reduced, i.e., combustion stabilized the flow and decreased the rate of growth of the instability. It was concluded, similar to the reacting shear layer case, that volumetric expansion induces a favorable pressure gradient which stabilizes the flow field if the exit pressure was maintained constant.

When the flow was forced at frequencies higher that the natural frequencies, similar observations were made. Although the flame exhibited oscillations at the forcing frequency, as shown in figure 10, increasing the heat release rate stabilized the low frequency modes. The high frequency mode never posed any serious challenge to the stability of the combustor. In the same figure, the variation of the inlet flow rate and pressure, and the volumetric release rate within the combustor are shown. Forcing at low frequency, close to the natural frequency, led very different behavior. As the rate of heat release is increased, flame oscillation at the forcing frequency increased and at some critical values and the flame, as shown in figure 11, exhibits strong tendency to propagate over the step into the incoming stream where a flow reversal is also detected. For values of heat release rate higher than this critical value, the flame moved into the incoming channel and the simulation was terminated. Figure 10 shows the inlet pressure and flow rate, and the volume release rate for the same case.

The simulations show that the coupling between the vorticity field, the energy deposition across the flame and the pressure across the combustor is necessary for the growth of instability. In most cases, the instability frequency is close to the natural wake instability mode of the nonreacting flow in the combustor. However, this frequency may vary due to the nonlinear coupling with the system acoustic. If forced at frequencies higher than the natural frequency, combustion in the system is stabilized without a noticeable decrease in the heat release rate.

VII. CLOSURE

Results of numerical simulations of three different reacting shear flows have been reviewed in this paper for the purpose of introducing this tool as a means of investigating the interaction between shear flow and combustion in flows of fundamental and practical interest. In each case, a physical model which limits the number of processes included in the simulation is used. Each model is based on what is perceived as the most important effects in the particular flow under invesigation. This is necessary since attempting to perform simulations using the equations governing the general three dimensional, unsteady, reacting flow is prohibitively expensive. The major limitations at this stage are the smallest length scale accommodated by the numerical discretization scheme and the level of complexity of the chemical kinetics model.

Due to these limitations, numerial simulations should be regarded as controlled experiments whose results are used to inquire into the physics of turbulent reacting flow. While the flow field in some relatively simple systems can be fully predicted using the type of numerical simulation described in this paper, it is important to remember that many systems can be simulated if carefully formulated assumptions are applied to the physical model. On the other hand, results of numerical simulation should be subjected to various tests for validation purpose. This includes comparisons with quantitative experimental data, results of linearized and approximate analytical theories, and results of flow visualizations. These validation studies help identifying the weak points in the models and in the numerical algorithms. With the rapid growth of computer technology, in hardware and software, less restrictive models will be used and more complex systems will be simulated.

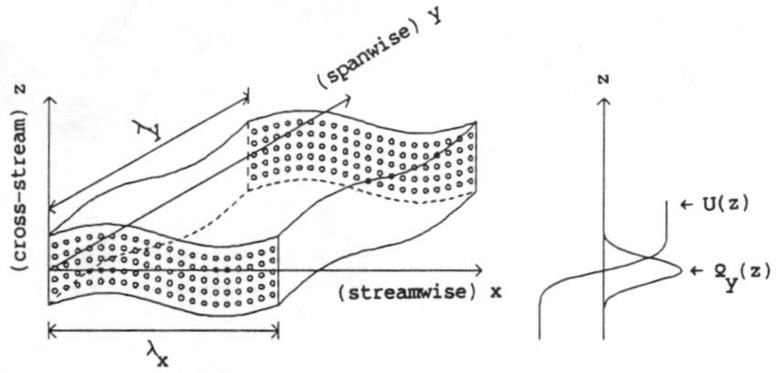

Figure 1. A schmetic diagram showing the model used to study a three dimensional reacting shear layer. In this case, x is the streamwise direction, y is the spanwise direction and z is the cross-stream direction.

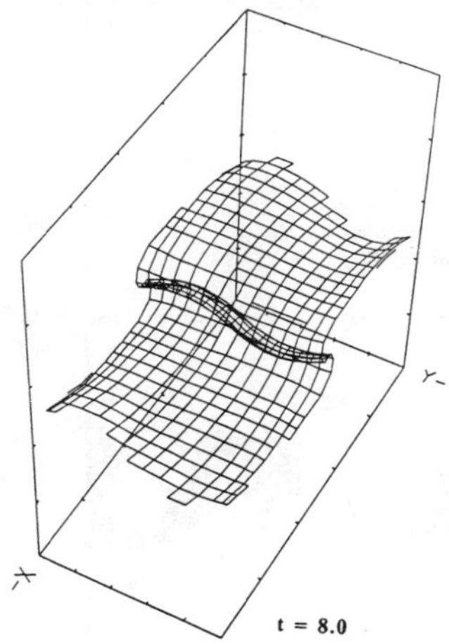

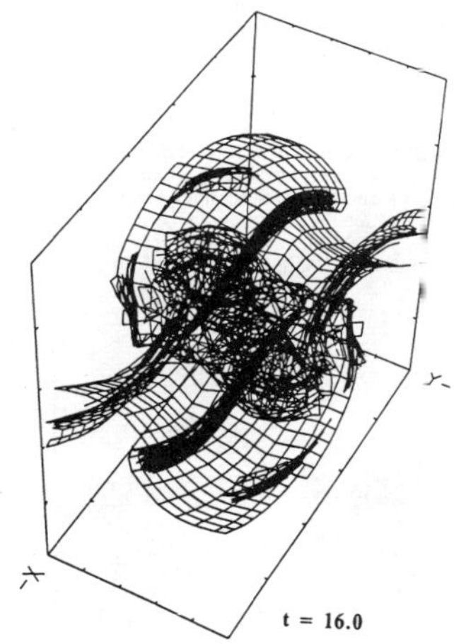

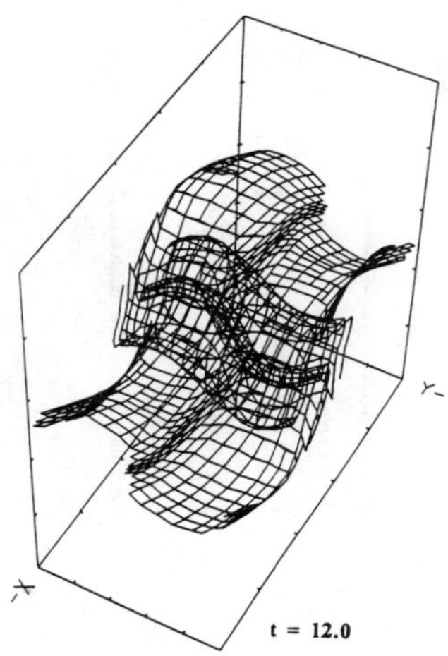

Figure 2. Three-dimensional perspective views of a surface initially coinciding with an x-y plane located in the middle of the domain. At time $t = 0$, the layer is perturbed in the streamwise direction, at the most unstable wavelength, and in the spanwise direction, at one-half of this wavelength. The amplitudes of the perturbations are equal to 2% of the streamwise wavelength.

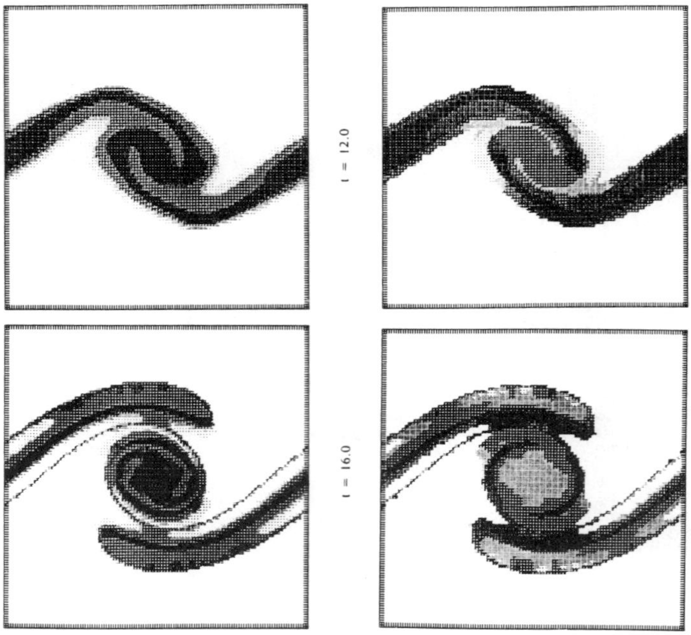

Figure 3. Results of a three-dimensional simulation of doubly periodic reacting shear layer between two reacting stream. The figure shows the product concentration (left) and reaction rate (right) on an x-z spanwise plane located at the middle of the domain. Results are shown in gray scale in which the maximum value is indicated by black and the minimum value is shown in white. The gray scale for the product concentration is fixed, $0 < c_P < 1$, while the gray scale for the reaction rate is floating, $W_{min} < W < W_{max}$.

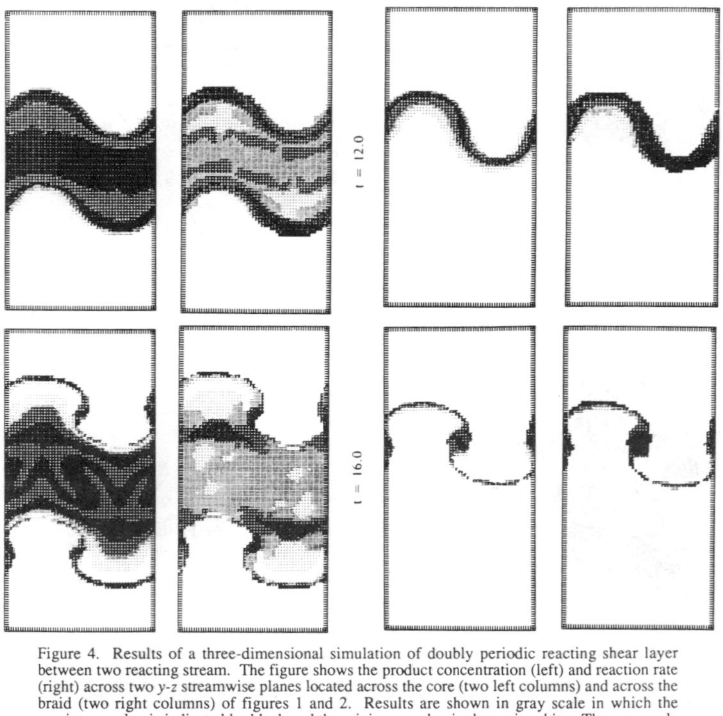

Figure 4. Results of a three-dimensional simulation of doubly periodic reacting shear layer between two reacting stream. The figure shows the product concentration (left) and reaction rate (right) across two *y-z* streamwise planes located across the core (two left columns) and across the braid (two right columns) of figures 1 and 2. Results are shown in gray scale in which the maximum value is indicated by black and the minimum value is shown in white. The gray scale for the product concentration is fixed, $0 < c_P < 1$, while the gray scale for the reaction rate is floating, $W_{min} < W < W_{max}$.

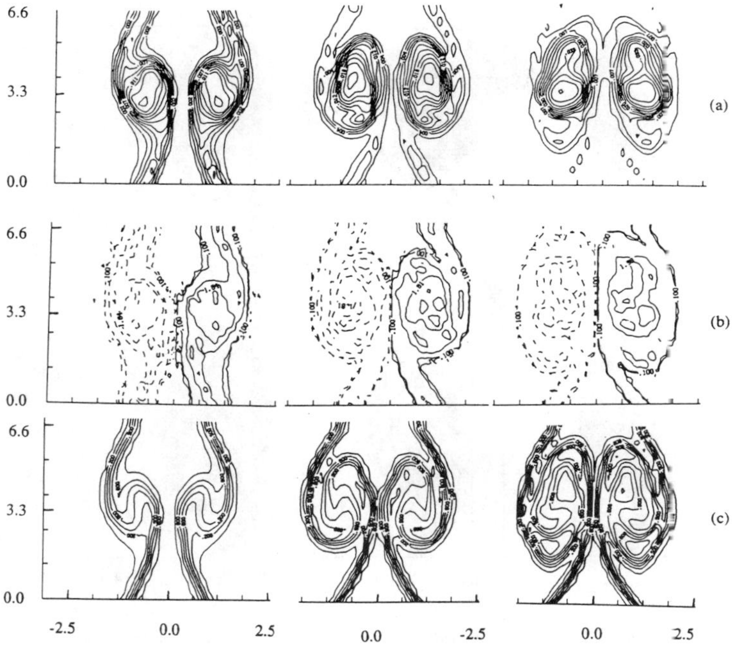

Figure 5. (a) Product; (b) vorticity; and (c) jet fluid concentrations for the initial stages of a reacting jet issuing in an open oxidizing atmoshpere at high Damkohler number with finite-rate kinetics, $T_a = 10$ and $A_f = 750$.

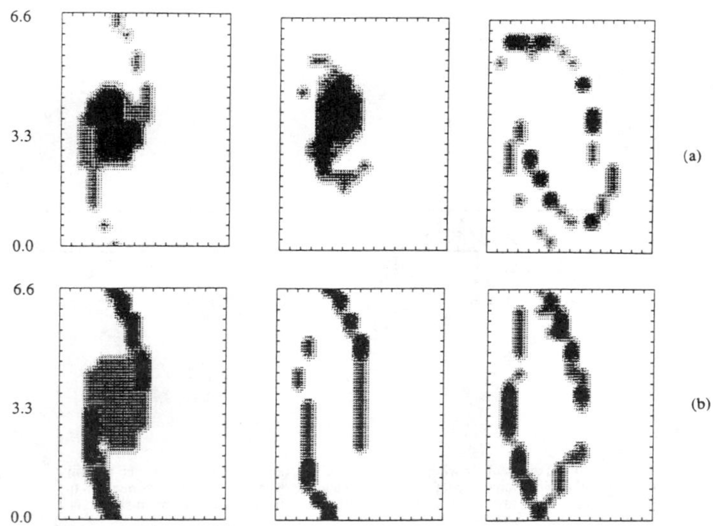

Figure 6. Reaction rate contours, depicted in different levels of gray, for high Damkohler number at: (a) finite-rate kinetics with $A_f = 750$; and (b) infinite-rate kinetics with $D_a = 0.5$, shown only for the right hand side of the jet. The gray scale divides the range of of W into essentially three zones.

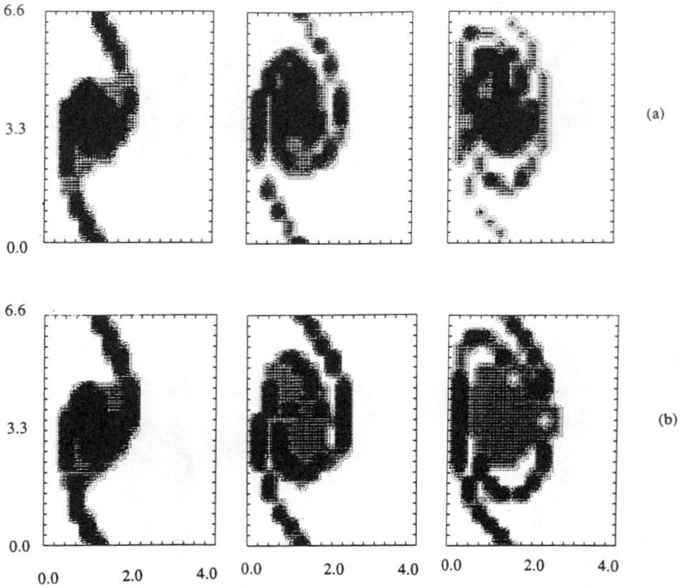

Figure 7. Reaction rate contours, depicted in different levels of gray, for low Damkohler number at: (a) finite-rate kinetics with $A_f = 150$; and (b) infinite-rate kinetics with $D_a = 0.063$, shown only for the right hand of the jet. The gray scale divides the range of of W into essentially three zones. Note that due to the selection of A_f and D_a, product formation and volumetric expansion are higher for case (b) than for case (a).

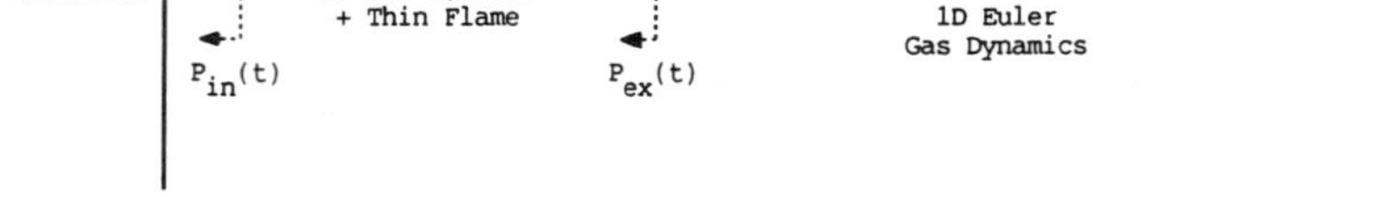

Figure 8. A schematic diagram showing the model used to study the coupling between pressure oscillations up and downstream the combustor section and the dynamics of combustion within the combustor section. The upstream section is an infinite reservoir and the downstream section is a long pipe of constant cross section.

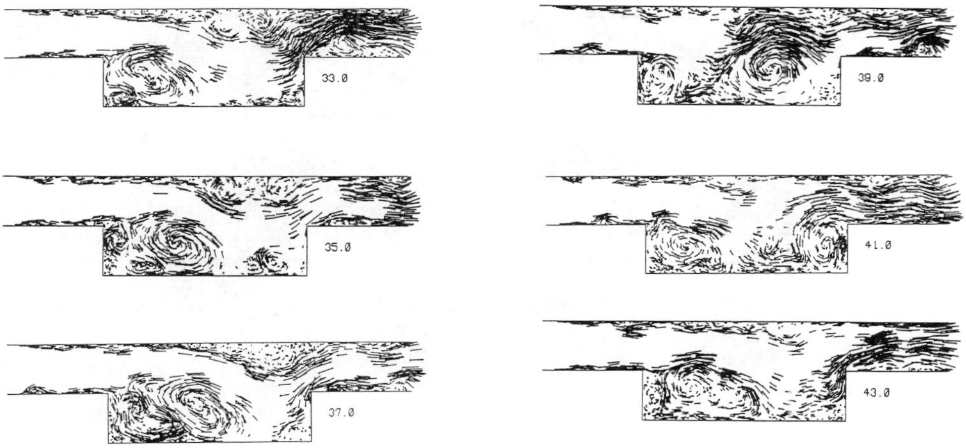

Figure 9. Large scale eddy shedding in a nonreacting flow inside a channel with a cavity. The flow is represented by the vortex elements and their instantaneous velocity vectors.

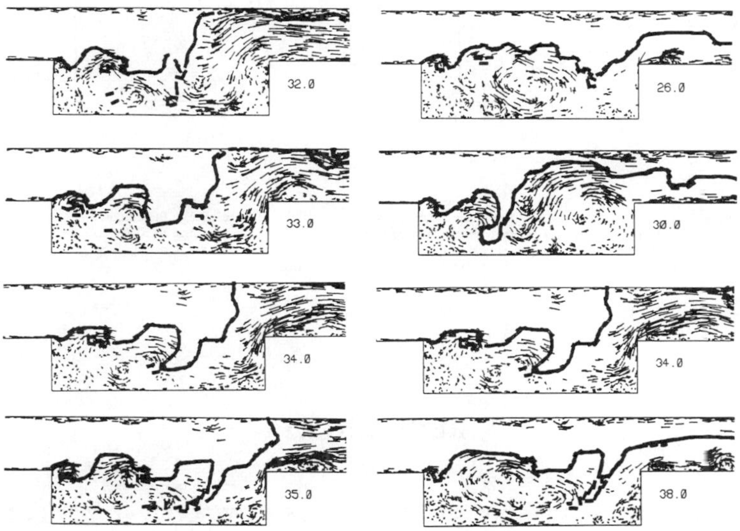

(a)

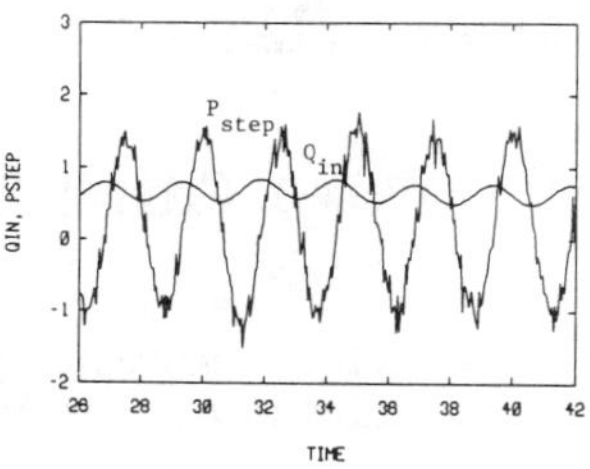

(b)

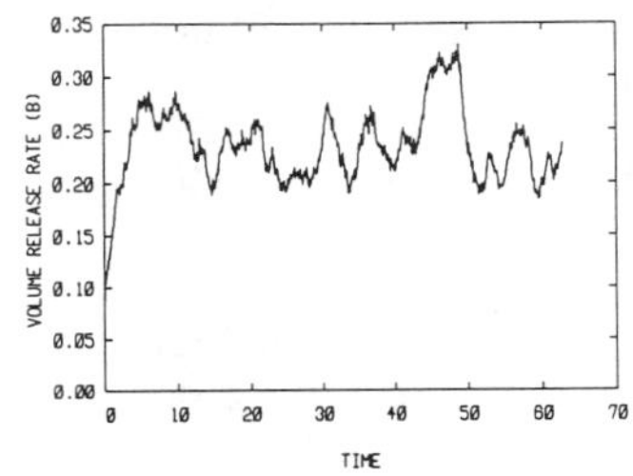

(c)

Figure 10. (a) The flow field and the flame front of combustion at low heat release in the channel shown in figrue 9. The exit pressure is modulated at high frequency corresponding to Strouhal number $fD/U = 0.4$. (b) The pressure at the step and the inlet flow rate for the same case. (c) The rate of heat release within the compustor.

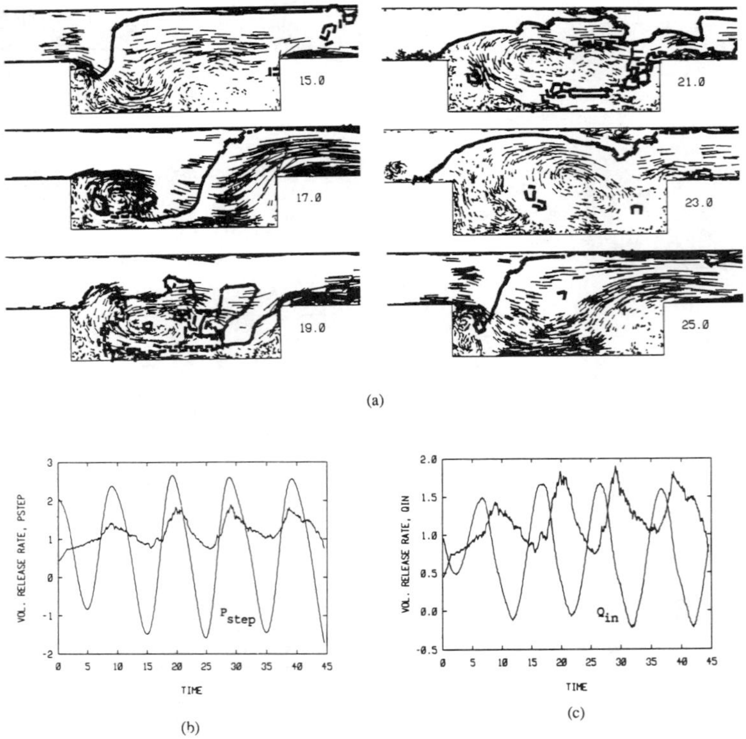

Figure 11. (a) The flow field and the flame front of combustion at high heat release in the channel shown in figrue 9. The exit pressure is modulated at low frequency corresponding to Strouhal number $fD/U = 0.1$. (b) The pressure at the step and the heat release rate within the combustor. (c) The rate of heat release within the compustor and the inlet flow rate.

REFERENCES

1. Chorin, A.J., Numerical study of slightly viscous flow, J. Fluid Mech., **57**, 1973, 785-7794.

2. Chorin, A.J., Vortex models and boundary layer instability, SIAM J. Sci Stat. Comput., **1**, 1980, pp. 1-21.

3. Chorin, A.J., Vortex sheet approximation of boundary layers, J. Comput. Phys., **27**, 1987, pp. 423-442.

4. Ghoniem, A.F., Computational Methods in Turblent Reacting Flow, in Reacting Flows: Combustion and Chemical Reactors, ed. by Ludford, Lectures in Applied Math., AMS Pub 1986, pp-199-265.

5. Ghoniem, A.F., Vortex methods in two and three dimensions, a review and some extensions, Proceedings of the First AIAA/ASME/SIAM/APS National Fluid Dynamics Congress, Cincinnati, OH, July 1988, p. 658.

6. Ghoneim. A.F. Vortex Simulation of Reacting Shear Flow, Chapter X in Numerical Methods in Combustion, ed. by E. Oran and J. Boris, AIAA Pub., to appear.

7. Ghoniem, A.F., Chorin, A.J. and Oppenheim, A.K., Numerical modelling of turbulent flow in a combustion tunnel, Phil. Trans. R. Soc. Lond., **A304**, (1982) pp. 303-325.

8. Ghoniem, A.F. and Gagnon, Y., Vortex simulation of laminar recirculating flow, J. Comput. Phys., **68**, 1987, pp. 346-377.

9. Ghoniem, A.F., Heidarinejad, G. and Krishnan, A. Numerical simulation of a thermally-stratified shear layer using the vortex element method, J. Comput. Phys. **79**, 1988, pp. 135-166.

10. Ghoniem, A.F., Heidarinejad, G. and Krishnan, A., On mixing, baroclinicity and the effect of strain in a chemically reacting shear layer, the AIAA 26th Aerospace Sciences Meeting, January 11-14, Reno, Nevada, AIAA-88-0729.

11. Ghoniem, A.F., Heidarinejad, G. and Krishnan, A., Turbulence-combustion interactions in a reacting shear layer, in Turbulent Reactive Flows, ed by Borghi and Murthy, Lecture Note in Engineering, 40, Springer-Verlag, New York, pp. 638-671 (1989).

12. Ghoniem, A.F., and Heidarinejad, G., Effect of Damkohler number on the reactive zone structure in a shear layer, Combust. Flame, **83**, (1991) pp. 1-16.

13. Ghoniem, A.F. and Knio, O.M., Numerical simulation of flame propagation in constant volume chambers, Proceedings of the 21st Syposium (International) on Combustion, the Combustion Institute, Pittsburgh, PA. (1986) pp. 1313-1320.

14. Ghoniem, A.F., Knio, O.M. and Krishnan, A., Lagrangian Simulation of the Initial Stages in A Reacting Jet, The 23rd Symposium (International) on Combustion, July 1990, France, proceeedings in press.

15. Ghoniem, A.F. and Najm, H., Numerical simulation of the coupling between the vorticity and pressure oscillation in combustion instability, AIAA/ASME/SAE/ASEE 25the Joint propulsion Conference, Monterey, CA/July 1989, AIAA-89-2665.

16. Ghoniem A.F., and Ng K.K., Numerical study of a forced shear layer, Phy. Fluids, **30**, 1987, pp. 706-721.

17. Hiedarinejad, G., Numerical Simulation of the Reacting Shear Layer Using the Transport Element Method, Ph.D. Thesis, Department of Mechanical Engineering, M.I.T., 1989.

18. Knio, O.M. and Ghoniem, A.F.. Numerical study of a three-dimensional vortex method, J. Comput. Phys., **86**, 1 (1990) pp. 75-106.

19. Knio, O.M., and Ghoniem, A.F., Three dimensional vortex simulation of the rollup and entrainment in a shear layer, AIAA-89-0574, and J. Comput. Phys., in press.

20. Knio, O.M. and Ghoniem, A.F., Three-dimensional Lagrangian simulation of a reacting shear layer, AIAA-90-0150, AIAA Journal, in press..

21. Knio, O.M., Three Dimensional Lagrangian Simulation of Reacting Shear Flow, Ph.D. Thesis, Department of Mechanical Engineering, M.I.T., 1990.

22. Krishnan, A. and Ghoniem, A.F., Origin and manifestation of flow- combustion interaction in a premixed shear layer, Proceedings of the 22nd Symposium (International) on Combustion, The Combustion Institute, Pittsburgh, PA, pp. 665-675 (1988).

23. Krishnan, A. and Ghoniem, A.F., Vorticity-combustion interactions in a reacting jet, in Dynamics of Deflagration and Reactive Sysems: Flames, Vol. 131, in Progress in Astronautics and Aeronautics, pp. 237-256.

24. Krishnan, A., and Ghoniem, A.F., Numerical simulation of the structure of a heated jet in a cold environment, AIAA-89-0485, 27th Aerospace Sciences Meeting, Jan. 9-12, 1989/Reno, NV.

25. Krishnan, A., Vorticity-Combustion Interaction in Reacting Shear Flow, Sc.D. thesis, Department of Mechanical Engineering, M.I.T., 1989.

26. Leonard, A., Vortex methods for flow simulation, J. Comput. Phys., **37**, 1980, pp. 289-335.

27. Leonard, A., Computing three dimensional incompressible flows with vortex elements, Ann. Rev. Fluid Mech., **42,** 1985, pp. 532-559.

28. Majda, A. and Sethian, J.A., The derivation and numerical solution of zero Mach number combustion, Combust. Sci. Tech., **42**, 1985, p. 185.

29. Martins, L.F., Vortex Computations of Axisymmetric High Reynolds Number Flows in Complex Domains, Ph.D. thesis, Department of Mechanical Engineering, M.I.T. 1990.

30. Najm, H. and Ghoniem, A.F., Numerical simulation of the convective instability in a dump, AIAA/SAE/ASME/ASEE 23rd Joint Propulsion Conference, San Diago, CA, 1987, AIAA-87-1874.

31. Oran, E.S. and Boris, J.P., Numerical Simulation of Reacting Flow, Elsevier, 1987, xxi + 601 p.

32. Raviart, P.A., Particle numerical models in Fluid Dynamics, in Numerical Methods for Fluid Dynamics II, ed by K.W. Morton and M.J. Baines, Claredon Press, Oxford, 1986, pp. 231-254.

33. Riley, J.J. and McMurtry, P.A., the use of direct numerical simulation in the study of turbulent, chemically reacting flows, in Turbulent Reactive Flows, ed by Borghi and Murthy, Lecture Note in Engineering, 40, Springer-Verlag, New York, pp. 486-514 (1989).

34. Sethian, J.A. and Ghoniem, A.F., Validation study of vortex methods, J. Comput. Phys., **74**, 1988, pp. 283-317.

35. Williams, F.A., Combustion on Theory, second edition, Benjamin/Cummings Publ Co., 1985, xxiii + 680 p.

ACKNOWLEDGMENT. This work is supported by the Air Force Office of Scientific Research Grant AFOSR 84-0356, The National Science Foundation Grant CBT-8709465, and the Department of Energy Grant DE-FG04-87AL44875. Computer support is provided a grant from the John von Neumann National.Supercomputer Center.

Ahmed F. Ghoniem
Massachusettts Institute of Technology
Cambridge, MA 02139

6 Heterogeneous combustion

C BRUNO

Gas-surface interaction and catalytic effects

1. INTRODUCTION

Wall boundary conditions (= BCs) need to be specified to close the system of differential equations related to practical applications of combustion theory. These include flows in gas turbines and ramjets combustors, internal combustion engines, and power generators. Wall BCs are also important wherever gas–phase chemical reactions in general are important; this is the case of nozzle flows, hypersonic gasdynamics, and plasma torches, to give a few examples.

In all these situations, the first question is usually whether the surface is reactive (catalytic), or inert. If the former is true, the second is, which are the reacting species, their kinetics, the corresponding rates and the products formed. Finally, if reasonable answers to these two questions can be found, the question is to build a realistic model for surface reactivity that can be implemented in the form of standard BCs (i.e., Dirichlet, Neuman, or mixed).

Present knowledge of surface reactions fundamentals is scarce, even for simple reactions (e.g., recombination of atomic species to form diatomic molecules). Progress in this area is hampered by the difficulty and cost of adequate diagnostics; besides, even 'simple' surface recombinations, such as the $O + O \rightarrow O_2$, or $N + N \rightarrow N_2$ termination steps, are beginning to show an unsuspected degree of complexity and a much tighter coupling between gas and surface than previously assumed [1]. The challenge posed to surface physicists and modelers is to reduce this complexity to manageable mathematics. Looking at more complex reactions, such as combustion of fuel–air mixtures over solid catalysts (catalytic combustion), it is clear that there is no hope at the moment of resolving in detail their surface kinetics; rate expressions are by necessity empirical fits derived from what measurements are available, and this situation is not going to improve anytime soon.

A new promising tool in this respect is ab–initio calculations, that have been used successfully to predict CO–air oxidation rates over Pt [2–4,18]. They could , plausibly, predict recombination rates, provided the energy treatment be kept semi–classical (full QM treatment is still too lengthy and expensive even with supercomputers). Time and cost still prevent ab–initio calculations of catalytic oxidation of even simple hydrocarbons.

What follows is meant to be a primer in the area of surface reactions of interest to combustion. Basic assumptions are: continuum regime (Knudsen number $\ll 10^{-2}$), and Navier–Stokes equations apply; surfaces are assumed macroscopically impervious (i.e., nonporous, with 'pores' meaning cavities of size $\gg$ crystal lattice size). This last assumption, may be unrealistic at the lower end of the range of temperatures of interest to catalytic combustion, where, due to low sintering rates, surface area may be several orders of magnitude larger than geometric (or apparent) area.

No assumption needs to be made as to the flow regime (i.e., laminar or turbulent) when building a surface reactivity model; this because the characteristic 'distance from the wall' associated with reactivity is of the same order of molecular size. The relative importance of surface reactions, hower, depends on the macroscopic Reynolds number of the flow, since it affects transport of reactants toward the surface.

In Section 2 reactive surface BCs will be derived. Section 3 contains rate theories for simple reactions, needed to implement BCs, and the impact of accomodation factors on surface-gas coupling. Section 4 describes briefly what is known about more complex oxidation reactions typical of catalytic combustion. Selected applications and examples are reported in Section 5.

2. CATALYTIC BOUNDARY CONDITIONS

Current understanding of the interaction between a gas and a surface is based on a physical mechanism consisting of three steps [14-17]: upon striking a surface, gas molecules or atoms may attach themselves ("stick") to locations on the surface where bonds surface–species may be formed ("sites"). This step is called adsorption. Weak bonds (Van der Waals forces) lead to physical adsorption, or physisorption; strong bonds (surface–orbital interaction) lead to chemical adsorption, or chemisorption.

Once on the surface, species may move ("surface diffusion") and collide among themselves; they may also be struck by other, or same, gas species. In both cases there may be reactions forming surface–bound products. This second step is called reaction.

Finally, the products formed may use part of the reaction heat to leave the surface; this is the desorption step, through which a surface reaction eventually releases new species into the gas.

In principle, surface reactivity depends on the details of these three steps. One's expectation is that, at least in a certain range of conditions, one of these steps becomes rate–controlling and the problem of finding surface reaction rates be simplified. As it will be seen later, this is indeed the case. In any event, BCs at the interface will use a surface rate w_i whose functional dependence needs not to be specified for the BCs to hold.

When surface reactions are present, BCs at a solid interface may be derived from their general form [5]. Taking a control volume $Vol = \delta A \cdot \delta h$ straddling the interface

(Fig. 1), assuming the surface does not move in the laboratory frame of reference, and neglecting inertial effects, the BC for the momentum equation becomes

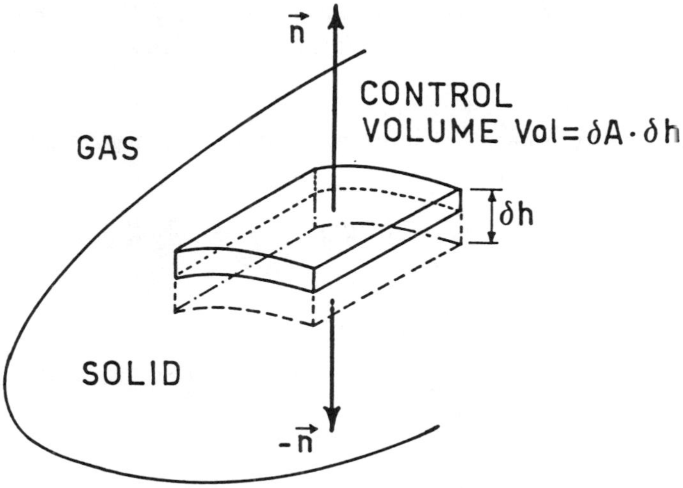

Fig. 1 Control Volume for Energy Boundary Condition

$$(\vec{V} \cdot \vec{n})_g \equiv (\vec{V} \cdot \vec{n}) = V_{blowing} = V_b \qquad (2.1)$$

If the surface is burning and regressing, due to surface reactions, and r_b is the regression rate, V_b is given by

$$V_b \rho_b = (\rho r_b)_{solid}$$

The BC for the i–th species equation is

$$\int_s \rho Y_i(\vec{V}_i + \vec{V}) \cdot \vec{n}\, dS + \frac{d}{dt}\int_{Vol} \rho Y_i d\, Vol = \int_{Vol} w_i d\, Vol \qquad (2.2a)$$

where S is the surface bounding Vol, and w_i the rate of change of species i (mass per unit volume and time). $\vec{V}_i$ and Y_i are the diffusion velocity and mass fraction of species i, respectively.

In the limit of $Vol \rightarrow 0$ by $\delta h \rightarrow 0$, the w_i volume integral becomes the surface integral of w_i', the surface rate of change of species i. If the accumulation of species i at the interface is negligible, the first volume integral on the LHS tends to zero. This may not be true when a clean surface is exposed to reactants, since it will take time for one or more layers of reactants to cover ("fill") the surface. However, when reactants have filled all available surface locations ("sites"), this term is actually negligible. In this case the BC, dropping the superscript for w_i', becomes

$$\rho(\vec{V_i} + \vec{V}) \cdot \vec{n} = w_i \qquad (2.2b)$$

expressing the balance between convective + diffusive transport and the formation or consumption due to surface reactions. If there is no blowing at the surface, $\vec{V} = 0$ and the BC tells that all i species transported by diffusion must originate from (or be destroyed by) surface reactions.

The BC for the energy equation is similarly derived, and, neglecting energy accumulation within *Vol*, is [5,7]

$$\int_S \rho(e + V^2/2)\, \vec{V} \cdot \vec{n}\, dS = -\int_S \vec{q} \cdot \vec{n}\, dS - \int_S \vec{V} \cdot \bar{\bar{T}} \cdot \vec{n}\, dS \qquad (2.3a)$$

where

$$\vec{q} = -k \nabla T + \rho \sum_{i=1}^{N} h_i Y_i \vec{V_i} + \vec{q}_R, \qquad \textit{heat flux vector}$$

$$h = \sum_{i=1}^{N} Y_i h_i = \sum_{i=1}^{N} Y_i \left(\int_{T_0}^{T} C_{p_i} dT + \Delta h^\circ_{fi} \right) = e + p/\rho, \qquad \textit{enthalpy}$$

$$\bar{\bar{T}} = \left[p + \left(\frac{2}{3}\mu - \kappa \right) (\nabla \cdot \vec{V}) \right] \bar{\bar{U}} - \mu[\nabla \vec{V} + \nabla \vec{V}^T], \qquad \textit{Newtonian stress tensor}$$

The diffusion thermoeffect has been neglected [6]; $\bar{\bar{U}}$ is the unit tensor, e the internal energy, k the thermal conductivity, κ the bulk viscosity coefficient, and $(\,)^T$ indicates the transposed of the tensor $(\,)$. The contributions to $\vec{q}$ are due, respectively, to molecular heat conduction, to energy transported by diffusion (including heat relase due to surface reactions), and to the gas radiating toward the surface, $\vec{q}_R$ (at sufficiently high T).

In the limit, and assuming the Mach number $M_b \equiv \left|\vec{V}\right| / a_w \equiv V_b / a_w \ll 1$, both the KE contribution to total energy and the stress tensor work may be neglected, and the BC imposes to the energy flux from the gas, $\vec{q}_g$, to equal that into the surface, $\vec{q}_s$:

$$\vec{q}_g \cdot \vec{n} = \vec{q}_s \cdot \vec{n} \qquad (2.3b)$$

The $\vec{q}_s \cdot \vec{n}$ term is called also "thermal load" q_w; explicitly, the energy BC reads

$$q_w = k\partial T/\partial n - \rho \sum_{1}^{N} h_i Y_i \vec{V_i} \cdot \vec{n} + q_R - \varepsilon_s \sigma T_s^4 \qquad (2.3c)$$

To use this BC some assumption must be made for q_w, unknown unless the solid temperature field is also solved for. Common assumptions are:

– adiabatic wall, i.e. $q_w = 0$; this implies wall radiation $\varepsilon_s \sigma T_s^4$ balances conduction, enthalpy transport by diffusing and recombining species, and radiative flux from the gas to the wall (radiative equilibrium);

– fixed $T_w \equiv T_s$;

– fixed q_w, if known in some way (e.g.: for combustion chamber walls of internal combustion engines, where coolant temperature is known).

Since formation enthalpies Δh°_{fi} are included in the definition of h_i, no energy surface source term needs to appear, contrary to what done for the species BC. The enthalpy of formation to use inside the definition of h_i depends on gas–surface interaction. In fact, when exothermic surface reactions occur, the heat may not necessarily be transfered entirely to the surface. Fast surface reactions (see Section 2) lead to desorption of out of equilibrium (excited) products, whose vibrational or electronic temperature may be higher than T_w. Examples are recombination of H or N on metals to form H_2 or N_2 [8–10], or O to form long–lived $O_2(^1\Delta_g)$ (singlet delta) [11]. Such processes may have great importance (e.g., for re–entry vehicles thermal protection systems [7]), and are accounted for by defining an "energy accomodation factor β" as the percentage of reaction energy actually deposited on the surface.

In physical terms, nonequilibrium species (marked by a *) will carry away from the surface energy h_i^* with a diffusion velocity $\vec{V}_i^*$ that will depend in general on species excitation [19]. Thus the net heat flux deposited on the surface by diffusion will be

$$q_w = -\rho \sum_1^N h_i Y_i \vec{V}_i \cdot \vec{n} - \left(-\rho \sum_1^N h_i^* Y_i \vec{V}_i^* \cdot \vec{n} \right) \qquad (2.3d)$$

where $(h_i - h_i^*)/h_i \equiv \beta_i$ is the accomodation factor.

If $\vec{V}_i^* \simeq \vec{V}_i$ (usually the case at not too high temperature) the net flux will be

$$q_w = -\rho \sum_1^N (h_i - h_i^*) Y_i \vec{V}_i \cdot \vec{n} = -\rho \sum_1^N h_i \beta_i Y_i \vec{V}_i \cdot \vec{n} \qquad (2.3e)$$

No general, first-principles derivations of β exist; there are however phenomenological models as far as O, N recombination on metals [11,23,32]. Qualitative agreement with experimental data for N on W is reasonable [23,32], and indicates a gradual shift from the E.-R. to the L.-H. mechanism with increasing T_w (decreasing coverage). Ab-initio calculations could be used to predict β, provided surface bond potentials were known [33].

When $\beta \neq 1$ excited desorbing species, carrying extra (nonequilibrium) energy migrate in the gas; their presence requires writing energy conservation equations for each of the excited species. Their relaxation (de–excitation) processes must be modeled; current models [11,12,13] are not always satisfactory, especially at temperatures typical of strong shock or detonation waves.

The simplest model of relaxation, assuming nonequilibrium internal energy e^* (and h^*) to be vibrational (V) and equal for all species, goes back to Landau and Teller work [20]. The model assumes a relaxation e_V^* source linearly proportional to nonequilibrium, with a characteristic relaxation time t_V that depends on translational temperature T; the conservation equation for e_V^* is an Euler equation:

$$\frac{De_V^*}{Dt} = (e_V - e_V^*)/t_V$$

$$t_V = \frac{K_1 T^{5/6} \exp(K_2/T)^{1/3}}{p[1 - \exp(T_V/T)]}$$

where K_1, K_2 and T_V depend on the molecule [20]. The equilibrium e_V is given by (harmonic oscillator theory)

$$e_V = R \int_{T_o}^{T} \frac{(T_V/2\tau)^2}{\sinh^2(T_V/2\tau)} d\tau$$

If $V - V$ exchanges between species pairs are allowed, this basic model has been modified; see for instance [21]. Common to all formulations is the empirical description of the source term for e_V^* or $e_{V_i}^*$: fundamental knowledge in this area is still missing [13].

To solve the problem defined by equations (2.1)-(2.3) surface reaction rates and accomodation factors are needed. Their modeling will be examined in the following sections.

3. SURFACE RATES

These depend on the reaction mechanism. Currently accepted for simple ractions (recombinations) are the Langmuir-Hinshelwood (L.-H.) and the Eley-Rideal (E.-R.) mechanisms. Applied, for istance, to O recombination, both follow a 3-step sequence

	L.-H.	E.-R.
adsorption:	$O + s \rightarrow O_s$	$O + s \rightarrow O_s$
reaction :	$O_s + O_s \rightarrow O_{2s}$	$O_s + O \rightarrow O_{2s}$
desorption:	$O_{2s} \rightarrow O_2 + 2s$	$O_{2s} \rightarrow O_2 + s$

where s stands for a free site, i.e., a preferential location for the bond formation between species and surface. Both mechanisms may take place at the same time; hopefully one will dominate. Likewise, for a given mechanism, one hopes that one of the steps is controlling.

For many atom recombinations over metals it can be shown [22] that the adsorption step is the slowest. Using kinetic gas theory, the adsorbing atom i flux, $J_{i,ads}$, will be proportional to the atom i flux reaching the surface, J_i, through a sticking coefficient, or probability, s_o:

$$J_{i,ads} = s_o n_i \bar{c}/4 \qquad (3.1a)$$

$$s_o = P\exp(-E_{ads}/RT) \le 1 \qquad (3.1b)$$

where n_i is the atom number density and $\bar{c}$ the space average speed. For atoms, P varies between 0(1) and 0(10^4) going from noble (VIII group) to transition metals.

The adsorbed mass flux will be

$$w_{i,ads} = j_{i,ads} = s_o \rho Y_i \frac{\bar{c}}{4} = P\exp(-E_{ads}/RT)\rho Y_i \frac{1}{4}\left(8\pi \frac{R}{W_i} T\right)^{1/2} \qquad (3.2)$$

where W_i is the molecular weight. The temperature T is that common to gas and surface at their interface.

The same simple theory may be used to model recombination reactions. The flux of impinging i atoms, J_i, is

$$J_i = n_i \frac{\bar{c}}{4} = \frac{n_i}{4}\left(\frac{8}{\pi}\frac{k_B}{m}T\right)^{1/2} = \frac{n_i}{4}\left(\frac{8}{\pi}\frac{R}{W_i}T\right)^{1/2} \qquad (3.3a)$$

where m is the atom mass and k_B the Boltzmann constant. The corresponding mass flux j_i is given by

$$j_i = \rho Y_i \frac{\bar{c}}{4} = \rho Y_i \left(\frac{1}{2\pi}\frac{R}{W_i}T\right)^{1/2} \qquad (3.3b)$$

If the adsorption step controls recombination, the recombining i atoms flux $j_{i,rec}$ (equal to $-w_i$, the surface rate appearing in the BC (2.2)) must equal the adsorption rate $j_{i,ads}$, which is first order in gas concentration of species i. The simplest (phenomenological) model for $j_{i,rec}$ is therefore [55]

$$j_{i,rec} = -w_i = K_{wi}\rho Y_i \qquad (3.4)$$

where the "catalyticity" of the given surface for species i recombination, K_{wi}, includes our ignorance of the recombination fundamentals and will depend on surface nature and temperature. Other models are possible that include the surface in some detail [23,56]. The recombination probability γ_i is defined as

$$\gamma_i \equiv j_{i,rec}/j_i = K_{wi}/\left(\frac{1}{2\pi}\frac{R}{W_i}T\right)^{1/2} \le 1 \qquad (3.5)$$

Therefore an upper limit is placed on K_{wi} : $K_{wi} \le \bar{c}/4$. If atoms i form a diatomic molecule l, $j_{i,rec} = -w_i = -j_l = w_l$. In terms of number fluxes, $J_i = -2J_l$. Equation (3.4) must be used inside (2.2).

Values of γ for $O \to O_2$ and $N \to N_2$ recombinations over many materials may be found [23–30], mostly at room temperature. High temperature data show γ to depend on T_w via an Arrhenius rate, with an activation energy in the 2–20 Kcal/mole range. The higher values are surprising, in view of the nature of the adsorption process that should control the overall rate. Lower values may reflect either the effect of diffusion to the wall (the dependence on $\sim T^{1.7}$ of the diffusion coefficient is essentially indistinguishable from $\exp(-E/RT)$, when E = 4–8 Kcal/mole), or perhaps a weak physisorption barrier due to Van der Waals forces. This is more likely to be the case with dissociative adsorption (e.g., H_2 on Cu, [16]), rather than with recombination. For porous walls, pore diffusion leads to an apparent Ea approximately 1/2 of the Ea of the surface reaction [22, p.194; 31].

Modeling of β, to be used inside (2.3d/e), is far more involved than that of γ, since the atom-surface bond(s) must be specified in detail. In [23] β is modeled (for N recombining on metals) as the energy released by adsorbing atoms and molecules, minus that required to desorb via E.-R. and L.-H. mechanisms, divided by the total dissociation energy. This definition requires knowledge of the surface coverage determining γ, so that γ and β are, in fact, linked by the coverage. Results indicate moderate agreement with experimental data.

4. COMPLEX SURFACE REACTIONS

Modeling reactions other than simple atom recombinations remains a formidable task.

Two classes of such reactions are of interest in combustion, namely radical or intermediates recombination on combustor or vessel walls (usually chain killing steps leading to formation of final products) and heterogeneous oxidations of fuels (e.g., H_2, C_xH_y) in the presence of O_2 or air ("catalytic combustion"). Chemistry fundamentals are still missing in both classes.

Some information is available for the rate of destruction of certain species (e.g., OH, HO_2, H_2O_2, H, O) on glass walls. Varying these rates over two orders of magnitude, to account for their uncertainty, has an effect on explosion limits prediction for H_2/O_2 mixtures [34]. The fact that many of the rates are identical, irrelevant of the species destroyed, and the overall formulation of the reaction itself, are proof of our ignorance of these processes. Nevertheless, these rates may still be used to get a sense of surface effects magnitude when there is a question of their impact on gas-phase kinetics [35]. For light species recombinations on glass vessels, extant issues are the presence of 'dangling bonds' over the SiO_2 surface and, at high enough T_w, interstitial diffusion of species such as H, H_2, O, O_2 and OH down to some microns below the geometrical surface. Diffusion transport is $\sim \exp(-E/RT_w)$; therefore the probability of having recombination reactions in the bulk of the glass increases exponentially with T_w [36].

A surface kinetics algorithm, similar to that used in formal kinetics to account for sets of reactions involving many species, but limited to like species recombinations (or destructions), can be found in [34].

As for catalytic combustion, apart from early applications [37–38], its systematic study began in the mid '70s (see [39–44] for reviews). Its interest is mainly due to the fact that heterogeneous oxidation reactions reach equilibrium faster than the gas-phase analogs, due to the lower apparent activation energy (for HC fuels, typically a difference of 15–20 Kcal/mole). Properly controlling combustion temperature by means of the equivalence ratio, it is possible to stably burn lean mixtures with negligible formation of NO_x, UHC and CO.

With regard to surface-gas interaction, detailed knowledge is pratically absent, mostly due to the difficulty of measuring what happens on a surface that may be at $T_w \geq 1500K$ and at $P \sim 1 - 20$ atmosphere. Most data reported are therefore overall fuel F consumption rates, i.e. experimental fits $d[F]/dt = [F]^a[O_2]^b \exp(-Ea_w/RT_w)$ [42, 44–45].

While these can be used to impose BCs for the fuel, there remains the problem of BCs for intermediates (e.g., C_2H_2, or radicals such as CH_3, C_2H_5, etc.) produced by possible simultaneous homogeneous (gas-phase) reactions. BC for O_2, instead, can be imposed by stoichiometry, as the fuel is nearly always present in less than stoichiometric proportions to keep maximum temperature within acceptable limits. Therefore, for a C_xH_y fuel, the molar rate of O_2 consumption due to direct catalytic fuel conversion to CO_2 and H_2O (usually the case) is $(x + y/4)$ larger than that of the fuel. However, if gas-produced intermediates are also oxidized at the surface, total O_2 consumption will have to take this into account [39].

Lack of good information on this aspect of catalytic combustion may be bypassed entirely when T_w is so high that kinetic control yields to diffusion-controlled regime. Species BCs at the wall, using Fick's law for simplicity, are

$$\rho D_i \frac{\partial Y_i}{\partial y} = w_i \quad [= K_{wi} \rho Y_i \quad for\ first\ order\ reactions] \qquad (4.1)$$

equating the diffusion rate of species i to its rate of consumption or formation. The characteristic time for diffusion is $O(\delta^2/D_i)$, while the surface kinetics time is $O(\delta/K_{w_i})$, the time needed by species i to cross the layer of thickness δ and disappear due to the first order kinetics. A surface analog of the Damköhler number of second kind for gas phase reactions [46] may be defined as

$$D_{a_w i} = K_{w_i}\delta/D_i = diffusion\ time/surface\ reaction\ time \qquad (4.2)$$

(see also [47]). When $D_{a_w i} \to \infty$ the reaction is 'fast', the rate is controlled by mass diffusion of species i, $\sim T^{1.7-1} \sim T^{0.7}$, and decreases with pressure as $1/p$. For $D_{a_w i} \to O$, the rate is controlled by surface kinetics, has an Arrhenius dependence on T and is pressure-independent. In the limit $D_{a_w i} = O$ the surface reaction is frozen, and the BC (2.2) reduces to

$$\partial Y_i/\partial y = O \qquad (4.3)$$

Notice that

$$\frac{\rho D_i \partial Y_i/\partial y}{w_i} = 1 = (D_i/K_{w_i})\partial \ln(Y_i)/\partial y$$

is always true, but this may occur either when $(D_i/K_{w_i}) \ll 1$ and $\partial \ln(Y_i)/\partial y \gg 1$ (diffusion controls) or when the opposite is true (kinetics controls). Qualitative Y_i profiles in the two cases are therefore as in Fig. 2.

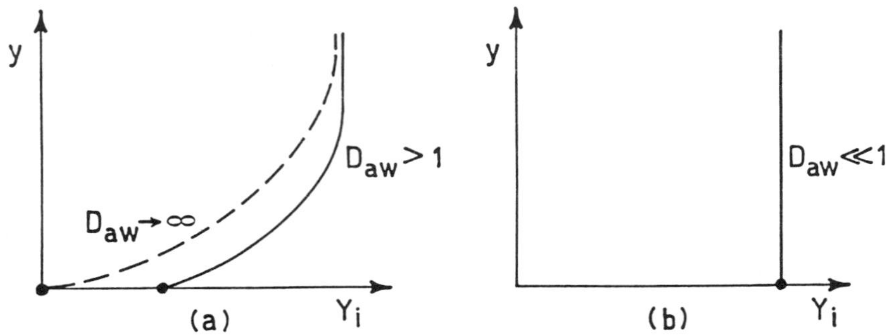

Fig. 2 Species Profiles vs. Distance from the Surface y as a Function of Regime. (a) Diffusion Controlled; (b) Surface Kinetics Controlled.

If i is a fuel species and the mixture is oxidizer-rich, a $D_{a_w i} = \infty$ (surface equilibrium) BC yields.

$$Y_i = O \qquad (4.4)$$

In any event, for $D_{a_w i}$ large enough, the problem of finding accurate surface kinetics data is entirely avoided.

Comments in [40], and recent near-surface measurements [48], indicate that catalytic combustion may take place not only via surface reactions, but also via gas-phase oxidation of fuel species by surface-produced radicals, such as O. Were these results to be definitely confirmed, the resulting picture would be that of a three-layered structure, with oxidations occurring simultaneously on the surface, in a thin gas region penetrated by short-lived O radicals, and in the bulk gas by straight gas-phase kinetics.

Finally, attention has recently focused on soot combustion in airstreams over catalysts, with the purpose of cleaning diesel exhaust. An introduction to this new area is in [49].

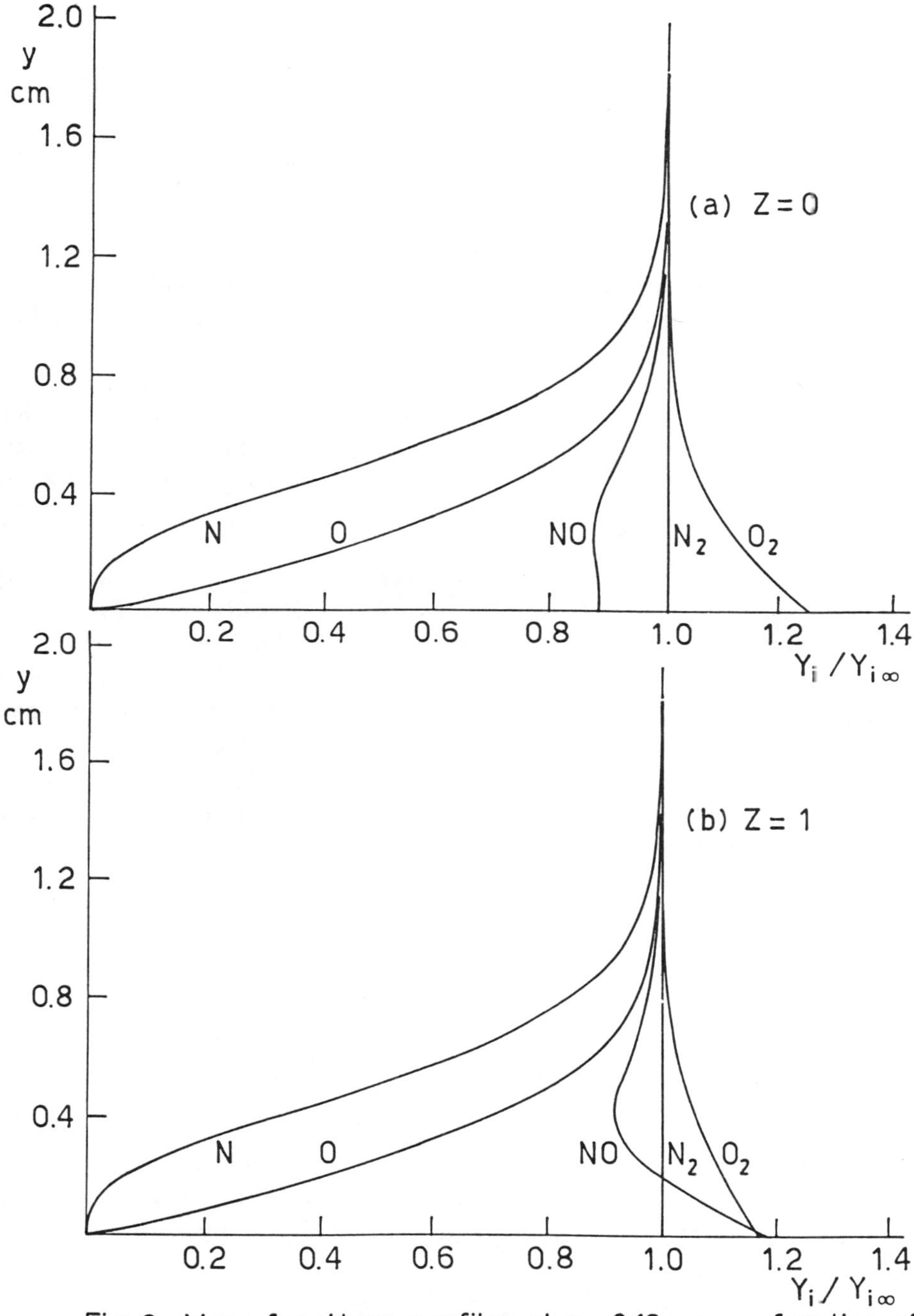

Fig. 3 Mass fractions profiles at x = 0.48 m as a function of Z.

T_∞ = 3000 K

T_w = 1500 K

p_∞ = 0.49 atm

u_∞ = 350 m/s

Equilibrium air assumed at ∞

5. APPLICATIONS

Two examples will be mentioned: surface NO formation and catalytic combustion of CO/air mixtures over Pt.

Surface NO may be formed by "cross recombination" of N and O atoms. The likely mechanism is of the Eley-Rideal type [50–51], with both O, N being possible adsorbates. At low enough T_w the gas concentration of N is negligible and the reaction step reduces to $N_s + O \rightarrow NO_s$.

Reactive, 2-D, Navier-Stokes equations, with imposed T_w, are used to solve a high temperature flow of reacting nonequilibrium air over a flat plate. The BC for NO states (see equation (3.4)) that $w_{NO} = -K_{wNO}\rho Y_O$, with $K_{wNO} = Z \cdot K_{wO}$, or $\gamma_{NO} = Z \cdot \gamma_O$, and $0 \leq Z \leq 1$.

Since experimental data are still missing, Z is used as a parameter. $Z = 1$ means that $O_s + O \rightarrow O_2$ and $N_s + O \rightarrow NO$ are equally probable. Results are in Figures 3–4. Figure 3 shows species mass fractions vs. y (normal to the plate) at $x = 0.48\,m.$ for $Z = 0$ and $Z = 1$. γ_O data were taken from [52]. The overall effect on heat transported by diffusion, $q_d(Z)$ (the integral over the plate length of the second term on the right-hand side of (2.3c)), normalized using $q_d(Z = O)$, is in Figure 4; q_d increases with Z, and percentually more at lower T_w. For T_w high enough, competition between gas-phase NO reduction and surface NO formation lowers q_d (and total thermal load) when $Z > O$. Such predictions await experimental confirmation.

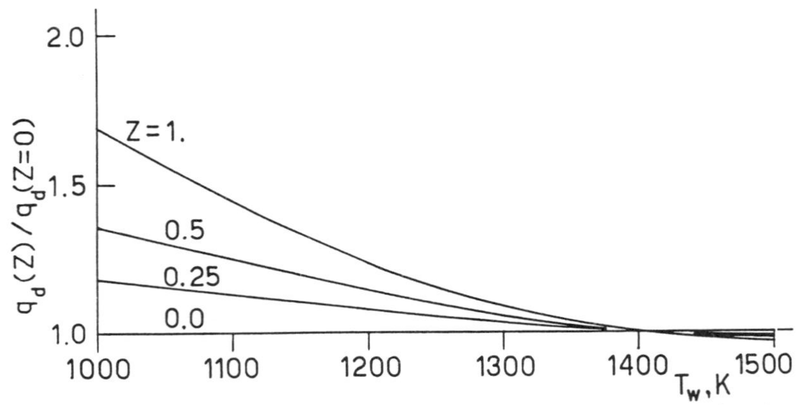

Fig. 4 Ratio between the integrated diffusional heat transfer for $Z \neq 0$ and $Z = 0$ as a function of wall temperature.

Catalytic combustion of CO/air on Pt has been previously investigated in much detail [40–41, 53–54]. Here heterogeneous ignition will be described [41]. CO is gradually added to air at $T = 500 - 800\,K$, $p = 1$ atm and $u_{inlet} = U = 10 - 70\ m/s$ flowing inside a 1.2 mm dia. channel in a ceramic monolithic honeycomb, until an equivalence ratio $\Phi = 0.11$ to 0.32 is reached. Ignition occurs due to surface reactions; the reactive, 2-D, axisymmetrical Navier-Stokes equations include both finite-rate gas- and surface chemical kinetics, and are solved simultaneously with the 2-D Fourier equation for the solid phase (i.e., the ceramic substrate) until steady-state is reached.

BCs (2.1)-(2.3) couple gas and solid; all equations are solved by successive iterations, in which the entire grid is "swept" always in the axial direction. Finite-volume discretization of the conservation equations yields a linear tridiagonal set. The energy *BC* is cast in the same form of the energy equations for gas and solid; the three together constitute a single tridiagonal set. The solution of the energy equation yields therefore the entire radial T distribution, from the centerline (gas) to the inside of the solid, including T_w. In this way, no iterations are required to make solid and gas temperatures match at the interface.

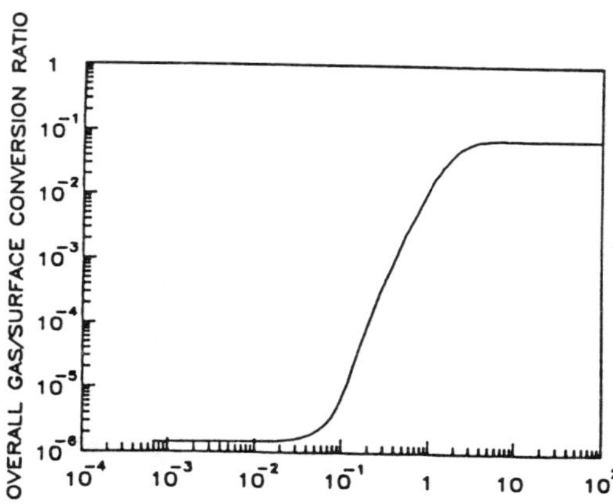

Fig. 5. Overall gas phase to surface conversion ratio during the transient.

Figure 5 shows the ratio between total gas-phase and surface CO consumption as a function of time; steady state is reached after about 3 s. Figure 6 shows steady state predictions of $T_w(x)$ profiles as a function of inlet velocity U and Φ; measured T_w at ten axial locations are also plotted for comparison.

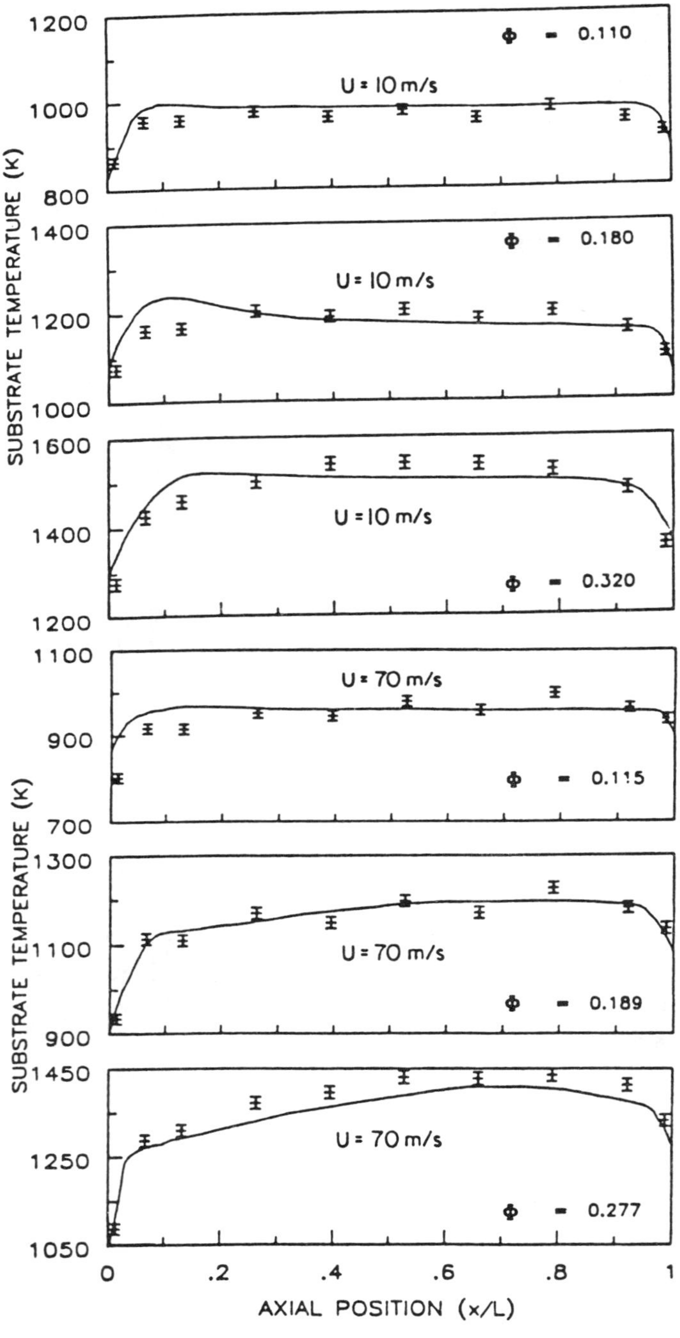

Fig. 6 . Comparison of steady state temperature profiles.

Predicted and measured total CO conversion at the exit ($x = L$), a parameter to characterize combustion completeness, is in Figure 7 for $U = 10\,m/s$ (upper data) and $U = 70\,m/s$ (lower data) as a function of Φ. Finite-rate CO kinetics was used at the surface during the transient until its characteristic time became smaller than diffusion time; when this occurred diffusional control and $Y_{CO} = O$ at the wall were automatically imposed. In view of our ignorance of much of the surface details, agreement is reasonable.

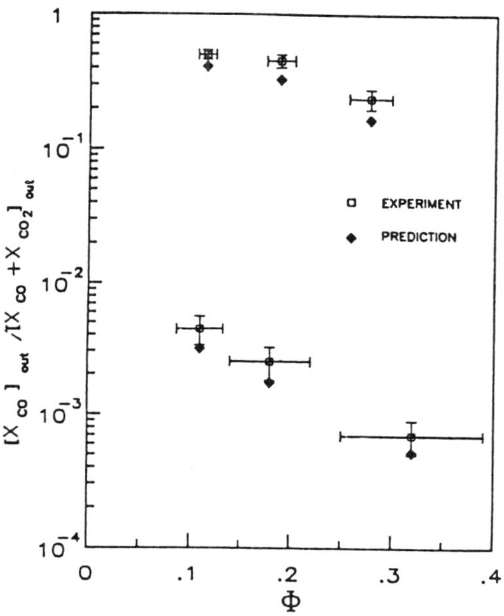

Fig. 7 . Comparison of exhaust emission.

6. CONCLUSIONS

This work is an attempt to bring some measure of order in a field that, due to its physics, is still unsettled. BCs are derived that are fairly general, under the assumptions made, and that do produce either plausible or reasonable results when applied to simple reacting flows. Future progress in this area will have to rely on better and faster surface diagnostic techniques specialized to the particular combustion problem.

ACKNOWLEDGEMENTS

The author wishes to acknowledge the contribution of Dr. L. Guarino to the results shown in Figures 3–4. The author wishes also to acknowledge Mr. A. Angelini for the help provided in preparing the typescript.

REFERENCES

1. Bruno, C., (1989), "Modeling Catalytic Recombination Heating at Hypersonic Speeds", Invited Paper 89–0309, presented at the AIAA 27th Aerospace Sciences Meeting, Reno, Nevada, Jan. 8–12, 1989.

2. Billing, G.D., (1984), "Inelastic Scattering and Chemisorption of CO on a $Pt(111)$ Surface", Chem. Phys., Vol. 86, p. 349.

3. Billing, G.D. and Cacciatore, M., (1985), "Semiclassical Calculation of the Reaction Probability for the Process $C + O \rightarrow CO$ on a $Pt(111)$ Surface", Chem. Phys. Letters, Vol. 113, No. 1, p. 23.

4. Billing, G.D. and Cacciatore, M., (1986), "Semiclassical Calculation of the Probability for Formation of CO_2 on a $Pt(111)$ Surface", Chem. Phys., Vol. 103, p. 137.

5. William, F.A., (1985), "Combustion Theory", Benjamin-Cummings, Menlo Park, p. 13.

6. Bruno, C., (1989), "Real Gas Effects", in: Hypersonics, Vol. I, ed. by J.J. Bertin, R. Glowinski and J. Periaux, Birkhäuser, Boston, p. 303.

7. Bruno, C., (1989), "Nonequilibrium Thermochemistry and Catalysis", in: Proceedings of the 2nd Joint Europe-US Short Course in Hypersonics, US Air Force Academy, Colorado Springs, Colorado, Jan. 16–20, 1989.

8. Comsa, G., "The Dynamical Parameters of Desorbing Molecules", in: Dynamics of Gas-Surface Interaction, ed. by G. Benedek and U. Valbusa, Springer Verlag, Berlin, 1982, pp. 117–127.

9. Thorman, R.P. and Bernasek, S.L., (1981), "Apparatus for the Direct Measurement of the Internal Energy of Heterogeneous Reaction Products", Rev. Sci. Instrum., Vol. 52, p. 553.

10. Brown, L.S. and Bernasek, S.L., (1985), "Vibrational Excitation in the CO_2 Product of the Oxidation of CO on Platinum: Coverage Dependence and Implications on Reaction Dynamics", J. Chem. Phys., Vol. 84, p. 2110.

11. Rosner, D.E. and Feng, H.H., (1974), "Energy Transfer Effects of Excited Molecule Production by Surface-Catalyzed Atom Recombination", J. Chem. Soc., Faraday Trans. Part I, Vol. 70, p. 884.

12. Candler, G., (1989), "Translation-Vibration-Dissociation Coupling in Nonequilibrium Hypersonic Flows", AIAA Paper 89–1739.

13. Park, C., (1985), "Convergence of Computation of Chemical Reacting Flows", in: Thermophysical Aspects of Re-Entry Flows, ed. by J.N. Moss and C.D. Scott, AIAA, New York, p. 478.

14. Zangwill, A., (1988), "Physics at Surfaces", Cambridge University Press, Cambridge, p. 400.

15. Wilkinson, F., (1980), "Chemical Kinetics and Reaction Mechanisms", Van Nostrand Reinhold, New York, p. 219.

16. Wolkenstein, T., (1977), "Physico-Chimie de la Surface des Semi-Conducteurs", Les Presses de l'Université Laval, Quèbec, p. 56.

17. Somorjai, G.A., (1981), "Chemistry in Two-Dimensions: Surfaces", Cornell University Press, Ithaca.

18. Levine, R.D. and Bernstein, R.B., (1987), "Molecular Reaction Dynamics and Chemical Reactivity", Oxford University Press.

19. Brun, R., (1988), "Transport Properties in Reactive Gas Flows", AIAA Paper 88–2655, presented at the Thermophysics, Plasmadynamics and Lasers Conference, San Antonio, Texas, June 27–29, 1988.

20. Vincenti, W.G. and Kruger, C.H., (1965), "Introduction to Physical Gas Dynamics", J. Wiley, New York, p. 204.

21. Park, C. and Yoon, S., (1984), "A Fully-Coupled Implicit Method for Thermo-Chemical Nonequilibrium Air at Sub-Orbital Flight Speeds", AIAA Paper 89–1974

22. Boudart, M. and Djéga-Mariadassou, G., (1984), "Kinetics of Heterogeneous Catalytic Reactions", Princeton University Press, Princeton, p. 38.

23. Halpern, B. and Rosner, D.E., (1978), "Chemical Energy Accomodation at Catalyst Surfaces", J. Chem. Soc., Faraday Trans. I, Vol. 74, p. 1883.

24. Berkowitz, J., (1969), "Catalytic Oxygen Atom Recombination on Solid Surfaces", in: The Structure and Chemistry of Solid Surfaces, ed. by G.A. Somorjai, J. Wiley, New York, p. 80–1.

25. Hacker, D.S., Marshall, S.A. and Steinberg, M., (1961), "Recombination of Atomic Oxygen on Surfaces", J. Chem. Phys., Vol. 35, No. 5, p. 1788.

26. Greaves, J.C. and Linnett, J.W., (1959), "Recombination of Atoms at Surfaces: Part 5. - Oxygen Atoms at Oxide Surfaces", Trans. Faraday Soc., Vol. 55, p. 1346; Part 6. - Recombination of Oxygen Atoms on Silica from $20°C$ to $600°C$", p. 1355.

27. Linnett, J.W. and Marsden, D.G.H., (1956), "The Kinetics of the Recombination of Oxygen Atoms at a Glass Surface", Proc. Royal Soc. London, Vol. 234A, p. 489.

28. Linnett, J.W. and Marsden, D.G.H., (1956), "The Recombination of Oxygen Atoms at Salt and Oxide Surfaces", Proc. Royal Soc. London, Vol. 234A, p. 504.

29. Greaves, J.C. and Linnett, J.W., (1958), "The Recombination of Oxygen Atoms at Surfaces", Trans. Faraday Soc., Vol. 54, p. 1323.

30. Hardy, W.A. and Linnett, J.W., (1967), "Mechanism of Atom Recombination on Surfaces", in: Eleventh Symp. (Internat.) on Combustion, The Combustion Institute, Pittsburgh, p. 167.

31. Satterfield, C.N., (1977), "Mass Transfer in Heterogeneous Catalysis", Dept. Chem. Eng., MIT, Cambridge, p. 138.

32. Halpern, B. and Rosner, D.E., (1982), "Incomplete Energy Accomodation in Surface-Catalyzed Reactions", Het. Atmos. Chemistry, Geophys. Monograph Series, Vol. 26, American Geophysical Union, p. 167.

33. Simonetta, M., (1986), "The Theoretical Approach to Surface Chemistry and Heterogeneous Catalysis", Int. J. of Quantum Chem., Vol. XXIX, p. 323.

34. Maas, U. and Warnatz, J., (1988), "Ignition Processes in Hydrogen Oxygen Mixtures", Comb. and Flame, Vol. 74, p. 53.

35. Panchenkov, G.M. and Lebedev, V.P., (1976), "Chemical Kinetics and Catalysis", MIR Publishers, Moscow, p. 148.

36. Bruno, C., (1989), "Gas/Surface Interaction and Thermal Load Modeling in Hypersonic Flow", paper presented at the NASA-Langley/Princeton University Center for High Speed Gasdynamics, Nov. 21, 1989, Princeton, NJ, to be published.

37. Davy, H., (1840), "Some New Experiments and Observations on the Combustion of Gaseous Mixtures", in: The Collected Works of Sir Humphrey Davy, ed. by J. Davy, Vol. VI, Smith, Elder and Co., London, p. 81.

38. Vollheim, G., (1969), "Catalytic After-Burning. A Method of Purifying Flue Gases", in: Twelfth Symp. (Internat.) on Combustion, The Combustion Institute, Pittsburgh, p. 651.

39. Bruno, C., Walsh, P.M., Yaw, Y., Sinha, N. and Bracco, F.V., (1983), "Catalytic Combustion of Propane/Air Mixtures on Platinum", Comb. Sci. and Technology, Vol. 31, p. 43.

40. Bruno, C., Walsh, P.M., Santavicca, D. and Bracco, F.V., (1983), "High Temperature Catalytic Oxidation for $CO/O_2/N_2$, Ar, He, CO_2/H_2O Mixtures", Int. J. Heat and Mass Transfer, Vol. 26, No. 8, p. 1109.

41. Sinha, N., Bruno, C. and Bracco, F.V., (1985), "Two-Dimensional Transient Catalytic Combustion of CO-Air on Platinum", PCH Physicochemical Hydrodynamics, Vol. 6, No. 4, p. 373.

42. Trimm, D.L., (1983), "Catalytic Combustion (Review)", Appl. Catalysis, Vol. 7, No. 3, p. 249.

43. Kesserling, J.P., (1986), "Catalytic Combustion", in: Advanced Combustion Methods, ed. by F.J. Weinberg, Academic Press, London, p. 237.

44. Ablow, C.M. and Sadamori, H., (1987), "Analytic Model of Combustion in a Catalytic, Fiber-Mat Burner", Combust. Sci. and Techn., Vol. 55, p. 1.

45. Melvin, A., (1988), "Natural Gas", Adam Hilger, Bristol, p. 169.

46. Damköhler, G., (1937), "Einfluss von Diffusion, Strömung und Wärmetransport auf die Ausbeute bei Chemischtechnischen Reaktionen", Der Chemie-Ingenieur, Band III, Teil I, Leipzig, p. 454.

47. Kuo, K.K., (1986), "Principles of Combustion", J. Wiley, New York, p. 654.

48. Griffin, T.A., Pfefferle, L.D., Dyer, M.J. and Crosley, D.R., (1989), "The Ignition of Methane/Ethane Boundary Layer Flows by Heated Catalytic Surfaces", Combust. Sci. and Techn., Vol. 65, p. 19.

49. Various Authors, (1990), "Advances in Diesel Particulate Control", SAE Special Publication SP-816, SAE Int. Europ. Office, Hitchin, UK.

50. Guarino, L., (1989), "Flusso viscoso su parete catalitica nel rientro da orbita bassa", Thesis, Dept. of Aerospace Engineering, Politecnico di Milano, Milano.

51. Guarino, L. and Bruno, C., (1990), "Effect of Surface *NO* Formation on Thermal Load during Re-Entry", paper submitted to the AIAA 29th Aerospace Sciences Meting, Reno, Nevada, Jan. 1991.

52. Scott, C.D., (1981), "Catalytic Recombination of Nitrogen and Oxygen on High Temperature Reusable Surface Insulation", in: Progress in Astronautics and Aeronautics, Vol. 77, ed. by A.L. Crosbie, AIAA, New York, p. 192.

53. Khitrin, L.N. and Solovyeva, L.A., (1959), "Homogeneous-Heterogeneous Combustion of Carbon Monoxide in Narrow Tubes (Channels)", in: VII Symposium (International) on Combustion, Butterworth Scientific, London, p. 532.

54. Voltz, S.E., Morgan, C.R., Liederman, D. and Jacob, S.M., (1973), "Kinetic Study of Carbon Monoxide and Propylene Oxidation on Platinum Catalysts", Ind. Engng. Chem. Prod. Res. Dev., Vol. 12, No. 4, p. 294.

55. Hinshelwood, C.N., (1949), "The Kinetics of Chemical Change", Clarendon Press, Oxford.

56. Jumper, E.W., Wilkins, R.G. and Preppernau, B.L., (1988), "Wall Catalytic Fluorine Recombination in an HF Laser Nozzle", AIAA J., Vol. 26, No. 1, p. 57.

Prof. Claudio Bruno
Dipartimento di Meccanica e Aeronautica
Via Eudossiana, 18
00184 ROMA, Italy
Phone: + 39 6 4687-257

C DI BLASI

Hydrodynamic and combustion processes of laminar pool fires

1. INTRODUCTION

Experimental and theoretical studies on combustion phenomena of condensed fuels are of interest mainly in fire safety science and pollution control. Due to practical implications in building and woodland fires, a rather extensive literature on solid fuel flammability is available (see, for example, reviews [17,18,38,45]). However some research on combustion processes of liquid fuel pool fires has also been conducted. For both types of fuel, it is a matter of heterogeneous combustion which involves phase transition, chemical reactions and transport phenomena. Two main differences exist between the two cases: the first being that solid fuels vaporize according to some kinetic law, while liquid fuels vaporize at the surface according to their equilibrium vapor pressure; the second that liquid fuels present a further mechanism of heat transfer due to convective liquid motions. This paper considers hydrodynamic and combustion processes of laminar liquid pool fires. The main purpose is not to review publications on this subject, as this has already been done in excellent papers [21,22,38,45], but, by means of the analysis of experiments and theories, to point out those aspects which must be taken into account if one wishes to confront the problem through numerical modeling. Even though numerical modeling is indeed becoming a more and more popular tool for understanding physical processes in complex combustion systems, it has been scarcely used in the field of pool fires. Most of the analyses available to date have been made by semiempirical formulae and similarity techiques and, even if the solution is searched numerically, the results are very partial due to the use of many simplifying assumptions. In this paper, after a short presentation of the state of art, a comprehensive mathematical model of hydrodynamic and combustion processes of laminar liquid pool fires will be given. Some numerical results, obtained by one of the most advanced models published to date, will also be discussed

2. EXPERIMENTAL INVESTIGATION

Combustion processes of liquid fuel pools include [22] ignition, flame spread and steady burning. Ignition characteristics and flame spread over liquid fuels strongly depend on the initial liquid temperature, more precisely on the liquid fuel being below or above the flash point [3,21,22,23,24,32,33]. For low initial liquid temperatures (below the flash point), after ignition, a pre-heating of the liquid is still needed in order to get a sufficient amount of vapor fuel for the flame to spread. As the initial liquid temperature is decreased below the flash point, a weak decrease of the spread rate is observed. Changes are also observed in the mode of flame spread, that is a region of uniform flame spread is followed by pulsating flame spread and then, for very low initial liquid temperatures, by uniform flame spread again [3,4]. Convective currents in the liquid have been observed [3,4,6,21,22,23,32,36] as a result of surface tension gradient and buoyancy effects. From experiments [32] it appears that the first effect prevails at normal gravity conditions. Gradients in the surface tension are caused by nonuniform surface temperature distribution. The flame indeed heats the liquid underneath, while upstream of the flame leading edge the liquid temperature is still low. These surface temperature gradients cause surface tension gradients which drive flow from the warmer temperature zone to the cooler temperature zone. The role played by liquid motions at subflash initial liquid temperatures was investigated by artificially varying the viscosity of the liquid [32] and it was found that the spread rate increases almost linearly with the inverse of the viscosity. When the liquid is made very viscous, the spread rate tends to a constant value, indicating that the controlling mechanism of heat transfer is different from that of liquid convection. As for the gas phase, at liquid temperatures much lower than the flash point, no appreciable gas motion was observed at the flame leading edge [25].

As the initial liquid temperature is increased above the flash point, the convective motions in the liquid become weak and the flame spreads faster than the liquid moves, while the velocity at the flame front reaches rather high values [24,25] and affects flame spread characteristics. When the liquid temperature is well above the flash point, after ignition, flame spreads in a pre-mixed mixture since no further vaporization is required. Thus, after a stronger increase and the attainment of a maximum, constant values of the spread rate are observed [3,22] as the initial liquid temperature is further increased. This behaviour, as well as the high values of the spread rates in these conditions (four or five times larger than the corresponding stoichiometric laminar flame speeds), are typical of flame spread through layered fuel/air mixtures [16,29,30,35] and are the consequence of interactions between fluid-dynamic and combustion.

Noticeable changes have also been observed in the flame structure as the initial liquid temperature is varied. At subflash temperatures, a small premixed flame precedes the main diffusion flame [21,30]. For high initial liquid temperatures [16,28] or in flame propagation along the interface between a methane layer and air [27,33], the flame presents a characteristic shape which can be divided into three different regions: a pre-mixed, U-shaped flame (1) is closely followed by a diffusion flame (2) and, at a certain distance, by a convective flame (3).

Initial liquid temperature also affects the ignition of laminar liquid pools [33]. At low liquid temperatures, the ignition delay times are very long, due to the convective liquid motions carrying away the heat transferred by the igniter from the ignition area, while for the initial liquid temperature close to the closed cup flash point, the ignition delay times decrease by one order of magnitude. Ignition delay times also depend on liquid viscosity showing a sharp increase as the liquid viscosity is increased [33].

Flammability characteristics of liquid fuels also depend on other parameters such as: the purity of the fuel, the mode of ignition, the width of the pool, the depth of the fuel layer, the flow conditions. Particularly interesting is the dependence of flame spread over liquid fuels on forced flow conditions [40]. For opposed flow conditions the spread rate gradually decreases as the flow is increased until extinction occurs. For concurrent flows, the spread rate remains almost constant for a while and then becomes nearly equal to the forced flow velocity. Furthermore, the interest for fire-safety issues in a space related environment has motivated the undertaking of studies on the effects of microgravity on flame spread [36].

Analyses of experimental measurements have tried to define the controlling mechanisms of ignition and flame spread across liquid fuels. The use of flash points has proved to be very useful for classifying the different types of spread, however it is insufficient to explain all observed behaviours and sometimes has led to controversial interpretation of experimental results. Two main interpretations of the dependence of flame spread on the initial liquid temperature have been formulated. Akita and coworkers [3,4] have classified flame spread as pre-heat type (initial liquid temperature below the theoretical flash point) and pre-mixed type (initial liquid temperature above the theoretical flash point). The pre-heat type of spread is divided according to three further regions characterized by uniform spread, pulsating spread and, for very low initial liquid temperatures, by uniform spread again. The pulsating spread was attributed to interactions between the flame and the surface flows. The second interpretation is due to Glassman and coworkers [21]. They assume that two main regions exist, one controlled by gas phase processes (high initial liquid temperature) and the other by liquid phase processes (low initial liquid temperature). According to this theory, the pulsating spread separates one region from the other and is the result of flame spread alternating between gas phase controlled type and liquid phase controlled type. Thus the role played by the different physical mechanisms is not clearly understood yet. Particularly important is liquid phase convection. Some recent detailed measurements [28] of liquid phase temperature confirmed indeed the existence of some liquid motion even for uniform spread rates.

Most of the papers on the burning of liquid fuels in open tanks are related to turbulent conditions. The fire indeed becomes turbulent even in relatively small scale phenomena [19]. Large scale (turbulent) steadily burning pool fires are strongly dominated by buoyancy-induced motions. In these conditions fire is a cyclic process controlled by fuel and oxidizer mixing followed by combustion and then by mixing again [31], with radiative heat transfer being of great importance. For laminar burning, natural convection also occurs at the base of the flame, while, in the upper part, diffusive processes are dominant [34]. Radiative heat transfer becomes negligible in comparison with conductive and convective contributions. Experimental measurements [5,27,34] have been made of the burning rates and other properties, such as the flame length and the flame shape, as a function of fuel type, size of the pool, environment conditions, degree of vitiation of air supply. The relation between convective motions in the liquid and the gas phase burning process has been analyzed in [44]. Convective motions in the liquid dominate the burning rate in the early stage of combustion. For longer times, the consumption of the liquid starts to influence the burning rate which decreases as the free board length increases.

3. THEORETICAL ANALYSES

Mathematical and numerical simulation can be very useful to understand the controlling mechanisms of combustion processes of liquid pool fires. On the other hand, the complexity of the mathematical models has prevented computing solutions accounting for all processes involved. Models [39,41-43] have uncoupled the liquid phase processes from gas phase processes which are described only through an imposed heat flux capable of initiating convective motions in the liquid. A very simplified solution of hydrodynamic equations was given in [39]. More detailed analyses of the liquid phase processes have been conducted in [41-43] where both momentum and energy equations are numerically solved through finite difference formulation. Both buoyancy and surface tension effects are accounted for. A thermal source with an assigned spread rate represents gas phase processes. Subsurface flows, shown by a circulating eddy, preceding flame spread were predicted. The dependence of the rate of fluid circulation in the eddy on the Reynolds (fuel layer and spread rate), Prandtl and Grashof numbers and on the surface tension parameter were investigated. A viscous regime at low Reynolds numbers and a boundary-layer flow at high Reynolds numbers were predicted. The varying of the Prandtl number was achieved through thermal diffusivity and, given that convective effects are dominant, only very weak dependence on this parameter was found. Weak dependence on the Grashof number was also predicted. Increase in the surface tension effects makes the eddy grow almost linearly. This causes an increase of the surface boundary layer and the velocity profile gradually changes from a boundary-layer regime shape to a viscous flow one.

More recent analyses [1,2] treated the problem of prevaporization and preignition of enclosed gas and liquid with a nonuniform heating from above with the assumptions that no fuel vaporization and no chemical reactions take place. Momentum and energy balance equations formulated according to the Boussinesq approximation have been solved numerically. In [1], the effects of varying gravity (Grashof number) with and without surface tension variation effects have been investigated. Multicellular circulatory flow in the gas phase was found at very high values of the Grashof number, with convective transport being the dominant mechanism. On the contrary, conductive transport dominates at low gravity levels. The increase of surface tension does not significantly affect the gas phase flow field. In [2] the effects of thermal radiation and thermal inertia of the walls bounding the enclosure were analyzed.

Analyses [1,2,39,41-43] have been able to predict interesting characteristics of flow induced by surface tension variation. However their usefulness to flame spread problems is limited by the absence of combustion. This limitation is removed in [20] where gas phase combustion is modeled through the flame sheet approximation. Two-dimensional, steady equations for momentum, energy and chemical species for the gas phase and for the liquid phase are solved numerically. Gas phase properties, except for density, are constant, while liquid phase buoyancy effects are modeled through the Boussinesq approximation. Steady formulation of gas phase equations is achieved by locating the origin of the coordinate system at the flame leading edge and the liquid is seen to move with the assigned spread rate. A simplified description of the boundary conditions at the liquid/gas interface assumes that the surface temperature is at the flash point under the flame and no vaporization occurs elsewhere. This study has allowed surface tension driven flow recirculation to be predicted both in the liquid phase and in the gas phase ahead of the flame leading edge. However the assumptions made in the model (infinite gas phase kinetics, steady flow and simplified description of vaporization) do not allow spread rates, detailed flame structure, possible oscillatory phenomena, as well as ignition and extinction processes to be computed.

The most complete model formulated to date is that of Ref. [7]. Even though constant liquid phase and gas phase properties, except for density variations described according to the Boussinesq approximation, are assumed, all physical processes of laminar pool fires are considered. Finite gas phase kinetics are assumed to describe combustion which occurs according to a global second order reaction. Unsteady, two-dimensional, fully elliptic equations for the gas phase (momentum, energy and chemical species) and for the liquid phase (momentum and energy) are coupled through the boundary conditions at the interface. Vaporization occurs along the whole liquid/gas interface according to the experimentally determined dependence of the vapor pressure on the surface temperature. Effects induced by surface tension variation and by buoyancy are also accounted for. The model, solved numerically by finite difference formulation, has allowed detailed flame structure, flow configuration and spread rates to be predicted [7,8,9,10]. In particular, the mechanisms controlling pulsating flame spread have also been determined, confirming the theory proposed by Glassman [21]. Some results of simulations performed by means of this model will be presented in the last section of the paper.

Models [1,2,7,20,39,41-43] briefly reviewed above have been applied essentially to simulate processes related to low initial liquid temperature. According to experimental observation, liquid phase momentum and energy transfer do become unimportant at high initial liquid temperatures and thus the analysis could be focused only on gas phase processes. However to date no detailed numerical simulation of flame spread in these conditions has been tried. Only very simplified analytical models have been formulated [16,26,29]. Theories [16,29] are based essentially on one-dimensional stream tube model and give, as main result, flame propagation rate, providing the thickness of the combustible layer is known. In the analysis [26] two-dimensional, inviscid, incompressible, isothermal flow field and constant spread rate are considered. The expansion of gas due to combustion of the flame front is replaced by an imaginary source behind the flame front. Based on these assumptions, predictions of velocity profiles, gas particle paths, fuel concentration distribution and flame front shapes are obtained.

As for laminar liquid fuel burning, most of the theories are based on simple analyses of experimental data and formulation of semiempirical formulae (see for example Ref. [5]). A detailed model is presented in [37] where two-dimensional momentum, chemical species and energy equations for the gas phase are solved. The level of the liquid surface is assumed not to change during the burning, and liquid phase processes are modeled only by prescribing boundary conditions at the liquid/gas interface. Forces due to surface tension are neglected and the liquid surface is isothermal at the boiling point. Heat flux balances allow the local liquid regression rate to be determined. This study, through a detailed prediction of flame structure and flow configuration and the investigation of the dependence of the process on fuel properties and enviroment conditions, has confirmed the importance of gas phase buoyancy forces on the burning process.

4. MATHEMATICAL MODEL

Essentially, only two models are available to date for a detailed simulation of combustion processes of laminar pool fires, that is the model of Ref.[7] for ignition and flame spread and the model of Ref.[37] for burning. Model [7] has been used to simulate ignition and flame spread at low liquid temperatures. Because of the use of approximate values of kinetic constants, no quantitative comparison with experiments has been tried. However the Boussinesq approximation for the gas phase could limit the model capability to give quantitative predictions. Presumably it cannot be used for qualitative predictions of high liquid temperature processes either, since flame structure characteristics and the high spread rates, observed for layered fuel/air mixtures, may be closely related to gas expansion. The other aspect to be improved in the modeling is the description of gas phase chemical reactions which has been afforded in a very simplified fashion. This remark is valid for the burning model [37] too, where a one-step reaction is still used to describe gas phase combustion. Furthermore, in this case, even if a rather complete description of gas phase processes has been used, the effects of the liquid phase processes on the burning are almost completely neglected. On the other hand, from experiments [44], they appear to be significant for small scale pool fires. Therefore, more advanced models must be solved in order to make a realistic description of the physical processes involved with the phenomena under study.

In this section a mathematical model for ignition, flame spread and burning of a liquid fuel from a pool is given. The model is comprehensive in that gravity effects, surface tension variation along the liquid/gas interface, fuel vaporization, chemical reactions, convection and diffusion of mass, momentum and energy for the liquid phase and the gas phase are described. The mathematical formulation here presented is based on the following assumptions:

a) the problem is two-dimensional and uses x as the coordinate parallel to flame spread direction and y as the normal one;

b) the flow is laminar;

c) flame and soot radiation are negligible; d) compressive work and viscous dissipation are negligible;

e) gas and liquid properties (heat capacities, viscosities, diffusion coefficients, thermal conductivities) are constant;

f) liquid density is constant except for variations induced by buoyancy described according to the Boussinesq approximation;

g) the depth of the fuel layer is constant.

Assumption b) is generally valid when small scale phenomena are considered [4,21]; also, in such a case, effects due to radiation are not important. Assumption d) is valid for low Mach numbers. Although the effects of variable properties can be very important for quantitative predictions, it is believed that assumption e) can be easily removed if the dependence of properties on environment conditions is known. Assumption g) is verified by most of the experiments.

In order to determine the controlling mechanisms of combustion processes of liquid pools, it is useful to introduce appropriate dimensionless groups measuring the influence of the different physical factors. Nondimensional equations for the gas phase are for:

- continuity

$$\frac{\partial \rho}{\partial t} + \frac{\partial(\rho u)}{\partial x} + \frac{\partial(\rho v)}{\partial y} = 0 \tag{1}$$

- momentum along the x direction

$$\rho\left(\frac{\partial u}{\partial t} + \frac{u\partial u}{\partial x} + \frac{v\partial u}{\partial y}\right) = -E\frac{\partial p}{\partial x} + \frac{1}{\mathrm{Re}}\left(\frac{\partial^2 u}{\partial x^2} + \frac{\partial^2 u}{\partial y^2}\right) \tag{2}$$

- momentum along the y direction

$$\rho\left(\frac{\partial v}{\partial t} + \frac{u\partial v}{\partial y} + \frac{v\partial v}{\partial y}\right) = -E\frac{\partial p}{\partial y} + \frac{1}{\mathrm{Re}}\left(\frac{\partial^2 v}{\partial x^2} + \frac{\partial^2 v}{\partial y^2}\right) - \frac{\mathrm{Gr}}{\mathrm{Re}^2}(\rho - 1) \tag{3}$$

- chemical species

$$\rho\left(\frac{\partial Y_i}{\partial t} + \frac{u\partial Y_i}{\partial x} + \frac{v\partial Y_i}{\partial y}\right) = \mathrm{Da}w_i + \frac{1}{\mathrm{ReSc}}\left[\frac{\partial}{\partial x}\left(\rho\frac{\partial Y_i}{\partial x}\right) + \frac{\partial}{\partial y}\left(\rho\frac{\partial Y_i}{\partial y}\right)\right] \tag{4}$$

$$i = 1, N$$

$$\sum_i Y_i = 1 \qquad i = 1, N$$

- energy

$$\rho\left(\frac{\partial T}{\partial t} + \frac{u\partial T}{\partial x} + \frac{v\partial T}{\partial y}\right) = \mathrm{Da}q + \frac{1}{\mathrm{PrRe}}\left(\frac{\partial^2 T}{\partial x^2} + \frac{\partial^2 T}{\partial y^2}\right) \tag{5}$$

- state equation

$$p = f1(\rho, T) \tag{6}$$

Nondimensional liquid phase equations are for:

- continuity

$$\frac{\partial u_l}{\partial x} + \frac{\partial v_l}{\partial y} = 0 \tag{7}$$

- momentum along the x direction

$$\frac{\partial u_l}{\partial t} + u_l\frac{\partial u_l}{\partial x} + v_l\frac{\partial u_l}{\partial y} = -\mathrm{E}_l\frac{\partial p_l}{\partial x} + \frac{1}{\mathrm{Re}_l}\left(\frac{\partial^2 u_l}{\partial x^2} + \frac{\partial^2 u_l}{\partial y^2}\right) \tag{8}$$

- momentum along the y direction

$$\frac{\partial v_l}{\partial t} + u_l\frac{\partial v_l}{\partial x} + v_l\frac{\partial v_l}{\partial y} = -\mathrm{E}_l\frac{\partial p_l}{\partial y} + \frac{1}{\mathrm{Re}_l}\left(\frac{\partial^2 v_l}{\partial x^2} + \frac{\partial^2 v_l}{\partial y^2}\right) - \frac{\mathrm{Gr}_l}{\mathrm{Re}_l}T_l \tag{9}$$

- energy

$$\frac{\partial T_l}{\partial t}+u_l\frac{\partial T_l}{\partial x}+v_l\frac{\partial T_l}{\partial y}=\frac{1}{\mathrm{Re}_l\mathrm{Pr}_l}\left(\frac{\partial^2 T_l}{\partial x^2}+\frac{\partial^2 T_l}{\partial y^2}\right) \tag{10}$$

In the equations (1-6), u and v are the longitudinal and normal velocity components, μ the viscosity, Y the species mass fraction, ρ the density, D the diffusion coefficient, c_p the specific heat at constant pressure, T the temperature and k the thermal conductivity. In the Equations (7-10), c is the liquid heat capacity, T_l the temperature, k_l the thermal conductivity u_l and v_l the longitudinal and normal velocity components, μ_l the viscosity, ρ_l the density and β the coefficient of thermal expansion.

The chemical production terms in the chemical species and energy equations (4-5) are defined once the kinetic model for combustion is specified. The simplest model is given by: $F+\nu_O O \rightarrow \nu_P P$. In this case, the chemical production terms, assuming a second order Arrhenius law, are expressed as:

$$\begin{aligned} w_i &= -\exp\left(\frac{-E}{T+P}\right) Y_O Y_F \rho^2 \frac{M_F}{\nu_i M_i} \qquad i=O,F \\ w_P &= -\exp\left(\frac{-E}{T+P}\right) Y_O Y_F \rho^2 \frac{M_F}{\nu_P M_P}, \qquad w_I = O \\ q &= -w_F \Delta H \end{aligned}$$

where A and E are the pre-exponential factor and the activation energy, ΔH the heat of combustion and M the molecular weight. The normalization procedure is given in Table 1 with the dimensional variables indicated by a prime. The dimensionless Reynolds, Euler, Grashof, Schmidt, Damkohler and Prandtl numbers are defined for each phase as:

$$\begin{aligned} \mathrm{Re} &= \frac{\rho_O v_* L}{\mu}, \quad \mathrm{Re}_l = \frac{\rho_l v_l L_l}{\mu_l} \\ \mathrm{E} &= \frac{p}{\rho_O v_*^2}, \quad \mathrm{E}_l = \frac{p_l}{\rho_l v_{l*}^2}, \\ \mathrm{Gr} &= \frac{g L^3 \rho_O^2}{\mu^2}, \quad \mathrm{Gr}_l = \frac{g\beta(T_l^* - T_{l0})L_*^3 \rho_l^2}{\mu_l^2}, \\ \mathrm{Sc} &= \frac{\mu}{\rho_O D}, \\ \mathrm{Da} &= \frac{w_* L}{\rho_O v_*} \\ \mathrm{Pr} &= \frac{c_p \mu}{k}, \quad \mathrm{Pr}_l = \frac{c\mu_l}{k_l} \end{aligned}$$

and the parameter $P = \dfrac{T_O}{T_f - T_O}$.

Fundamental for a correct description of flame spread process is the formulation of boundary conditions at the liquid-gas interface that couple the liquid phase equations (7-10) to the gas phase equations (1-6). Vaporization is assumed to occur only at the liquid surface. This assumption is valid until no boiling occurs and the vapor pressure of the liquid is lower than the total gas pressure. To describe the vaporization process, the Clausius-Clapeyron equation can be applied

$$p = f2(T_1) \tag{11}$$

or empirical expressions that relate the liquid vapor pressure to temperature [7].

The balances of fluxes of energy and mass for chemical species give:

$$-A\frac{\partial T}{\partial y} = -\frac{\partial T_l}{\partial y} + BmQ \tag{12}$$

$$\frac{1}{\mathrm{ReSc}}\rho\frac{\partial Y_F}{\partial y} = m(Y_F - 1) \tag{13}$$

$$\frac{1}{\mathrm{ReSc}}\rho\frac{\partial Y_i}{\partial y} = mY_i \quad i = O, I, P \tag{14}$$

where m is the vaporized mass flux and Q the heat of vaporization.

Conditions derived by continuity at the interface are:

$$m = Gv_l \tag{15}$$

$$\rho v = Gv_l \tag{16}$$

The balance of flux of momentum requires that the difference between the viscous stresses for the two phases must be equated to the variation in the surface tension, due to surface temperature variation:

$$\frac{\partial u_l}{\partial y} - S\frac{\partial T_l}{\partial x} = C\frac{\partial u}{\partial y} \tag{17}$$

Formulation of boundary conditions at the interface is completed by:

$$T = ET_l - F \tag{18}$$

$$u = Hu_l \tag{19}$$

Besides Schmidt and Reynolds numbers, further nondimensional groups appear in the boundary conditions at the liquid/gas interface:

$$A = \frac{kL_l(T_f - T_O)}{k_l L(T_l - T_{l0})}, \quad B = \frac{\rho_O v_* c L_l}{k_l},$$

$$S = \frac{d\sigma}{dT}\frac{(T_{l*} - T_{l0})}{\mu_l v_*}, \quad C = \frac{\mu v_* L_l}{\mu_l v_{l*} L}$$

$$E = \frac{T_{l*} - T_{l0}}{T_f - T_0}, \quad F = \frac{T_0 - T_{l0}}{T_f - T_0} \quad H = \frac{v_l}{v_*}, \quad G = \frac{\rho_l v_l}{\rho_O v_*}$$

A is essentially a ratio between the interface heat transfer by conduction for the two phases, B accounts for the relative importance of the energy flux by convection to heat transfer by conduction at the interface and has the form of a Peclet number. S is a measure of the surface tension force to a characteristic shear force and, when multiplied by a Peclet number, becomes a Marangoni number [1]. C is a ratio between characteristic shear forces for the gas phase and the liquid phase. Numbers E and F refer to characteristic temperature differences between the two phases, G is the ratio between the characteric mass fluxes for the two phases and H a measure of the relative importance of gas to liquid velocity.

Table 1

- Gas phase -

$$x = x'/L, \quad w = w'/A\rho\rho \quad m = m'/(\rho_O v_*),$$
$$y = y'/L, \quad T = (T' - T_O)/(T_f - T_O), \quad p = p'/(\rho_O v_*^2),$$
$$\rho = \rho'/L, \quad t = t'/(L/v_*), \quad v = v'/v_*,$$
$$u = u'/v_*, \quad \Delta H = \Delta H'/[c_p(T_f - T_O)],$$
$$E = E'/[R(T_f - T_O)],$$

- Liquid phase -

$$x = x'/L_l, \quad T_l = (T' - T_{l0})/(T_{l*} - T_{l0}), \quad u_l = u_l'/v_{l*}, 2v_l = v_l'/v_{l*},$$
$$y = y'/L_l, \quad t = t'/(L_l/v_{l*}), \quad p_l = p_l'/(\rho_l v_{l*})$$

Nondimensionalization procedure

In the normalization procedure here employed reference temperatures and velocities appear. In order to have a proper nondimensional formulation of the equations, these variables must be defined according to the particular physical phenomenon to be described. As for temperatures, it is relatively easy to choose the reference values as the adiabatic flame temperature for the gas phase, and flash or boiling points for the liquid phase. The reference velocity can be obtained by equating the coefficent of the dominant forces in the phenomenon to be modeled [1]. For instance, for the case of weak surface tension effects, the reference velocity is obtained by a balance of the buoyancy and inertia terms. The remaining boundary conditions as well as the initial conditions, can be assigned depending on the physical problem to be modeled and they are generally conventional.

5. A COMPUTATIONAL EXAMPLE

In this section an example of simulation of ignition and flame spread over methanol, obtained for values of the initial liquid temperature below the theorethical flash point and taken from Ref.[10] is presented. More precisely, an initial liquid temperature of -3°C, corresponding to experimental pulsating flame spread conditions, is considered.

The initiation of the spread process is shown in Fig.1 by means of the gas phase isotherms, the structure of the flow field for the two phases and the distribution of surface temperature for a time of .3s. The hot pocket of gas and the flame, after ignition, make the temperature to increase over a small portion of the liquid surface, with a maximum located about as the flame leading edge, whose position is assumed to correspond to the maximum of the heat flux from the flame to the fuel. Surface temperature gradients drive stream velocities at the liquid surface in the same direction as the flame propagates. A vortex in the liquid, whose center is located as the flame leading edge, is clearly visible from Fig.1B. Because of the boundary conditions at the interface, a recirculation in the gas phase ahead of the flame is also established (Fig.1A). However, being the small flame located well behind the maximum surface temperature gradient, the flame leading edge propagates in an opposed flow enviroment and, due to the very low fuel concentrations, moves very close to the liquid surface. In this stage of the process, the pre-heating of the liquid ahead of the flame occurs essentially through liquid phase convective and gas phase conductive heat transfer.

After ignition, the surface stream velocity increases because of the increase in liquid phase viscous stress at the liquid/gas interface. Then, the maximum in this quantity, initially located as the flame leading edge, and the center of the circulating eddy in the liquid phase move to the edge of the pre-heated region. Thus because of the enlargment of the recirculation zone, the flame leading edge meets concurrent flow velocities. This circumstance favours the flame propagation and the pre-heating of the liquid because gas phase convective heat transfer occurs in the same direction as the flame spreads.

For longer times, due to the enhanced convective heat transfer, surface tension driven flow velocities and surface temperature gradients decrease. On the other hand, the maximum gas temperature is still increasing as a consequence of the increase in the fuel vaporization. The heat flux from the flame to the fuel is also increasing with changes in the surface temperature gradient and consequently in the liquid phase viscous stress at the interface. The result of these two effects is clearly visible in Fig.2 from the stream function and temperature distributions plotted for t=.95s. Two circulation centers appear in the liquid phase, the first located in the neighbourhood of the flame leading edge and the second at the edge of the pre-heated region. The continuity of the stream velocity at the interface requires that two circulation centers should appear in the gas phase too. An eddy appears ahead of the flame, corresponding to that in the liquid located at the edge of the pre-heated region. However, at the flame leading edge, being the gas phase temperature gradients not negligible, buoyancy effects are important and a distorted eddy is observed.

As time increases further, the extend of the pre-heated zone and the concurrent flow velocities at the flame leading edge are continuosly reduced. As a consequence, a new change in the flow field structure is observed with only one circulation center in the liquid and the flame leading edge positioned in an opposed flow region (Fig.3). The conditions are thus qualitatively similar to those observed at ignition (t=.3s) and the whole cycle can repeat itself again.

The changes in the flame structure and flow field configuration are associated to

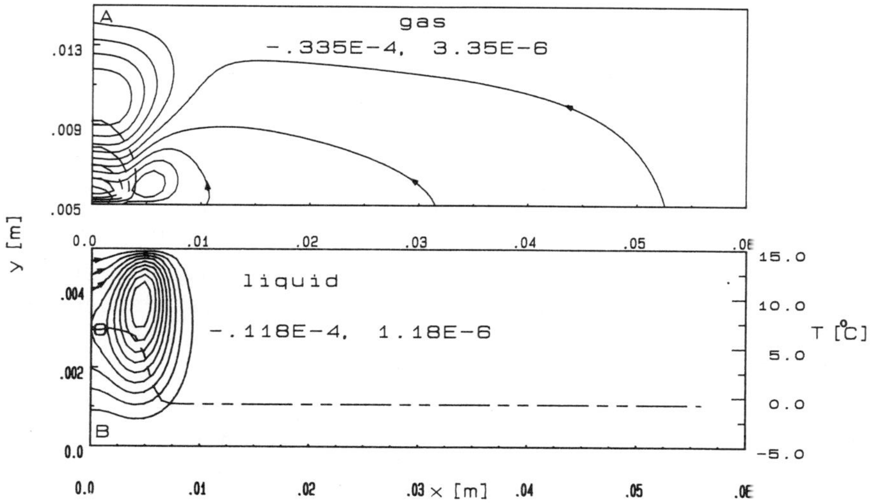

Fig.1 - Gas phase (A) and liquid phase (B) flow field configuration (solid lines), gas phase temperature (A) isotherms (dashed lines), surface temperature (B) distribution (dashed line) and flame leading edge position (B,marked) for t = .3 s. Values of the stream function [m^2/s] from the value (on the left) and with the step reported in figure, gas phase temperature [K] from 550 with step 250.

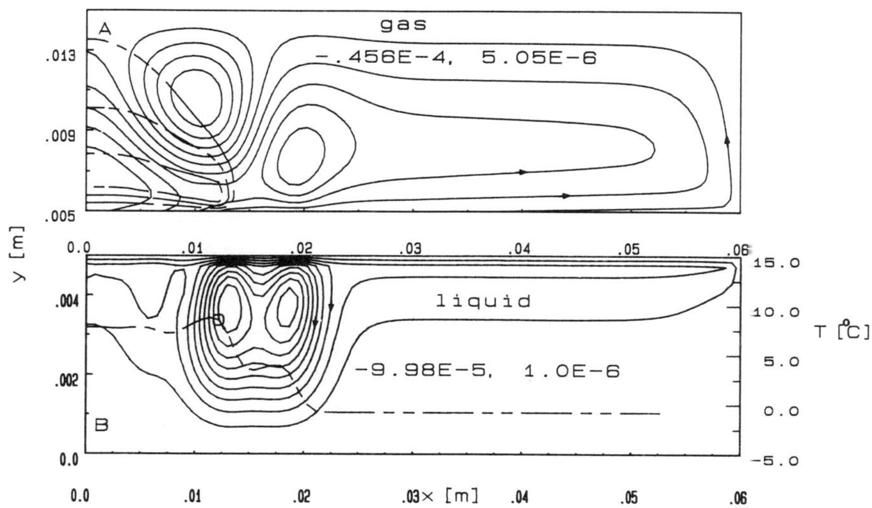

Fig.2 - Gas phase (A) and liquid phase (B) flow field configuration (solid lines), gas phase temperature (A) isotherms (dashed lines), surface temperature (B) distribution (dashed line) and flame leading edge position (B, marked) for t = .95 s. Values as Fig. 1.

a non steady flame propagation process. In agreement with experiments, oscillatory spread rates as a function of time have been predicted [9,10]. Concurrent flow at the flame leading edge and high fuel and oxygen levels in the recirculation zone make the flame to accelerate. As the extend of the gas phase recirculation zone is reduced, the flame leading edge is eventually positioned in an opposed flow region. This condition and/or the lowered fuel concentrations cause a deceleration in the flame spread rate.

It appears that the oscillatory phenomena of flame spread over a liquid, below the flash point, are the consequence of strong interactions between the liquid phase and the gas phase flow field. During the propagation process, changes in the flow field structure, species concentrations and temperature at the flame leading edge are indeed observed. The changes in the flow velocity at the flame front are caused by variations in the liquid phase flow field structure, consequence of the shift of the mechanisms controlling the pre-heating of the liquid from gas phase heat conduction to liquid phase and gas phase heat convection. The transition from opposed flow [11-13] to flow assisted [14-15] conditions at the flame leading edge causes variations in the characteristics of the spread process. In particular, variations occur in the chemical species values which are also observed to pulsate. This is a very important point because the flame propagation, under the above reported conditions, is a lean flammability problem and thus affected by chemical kinetics.

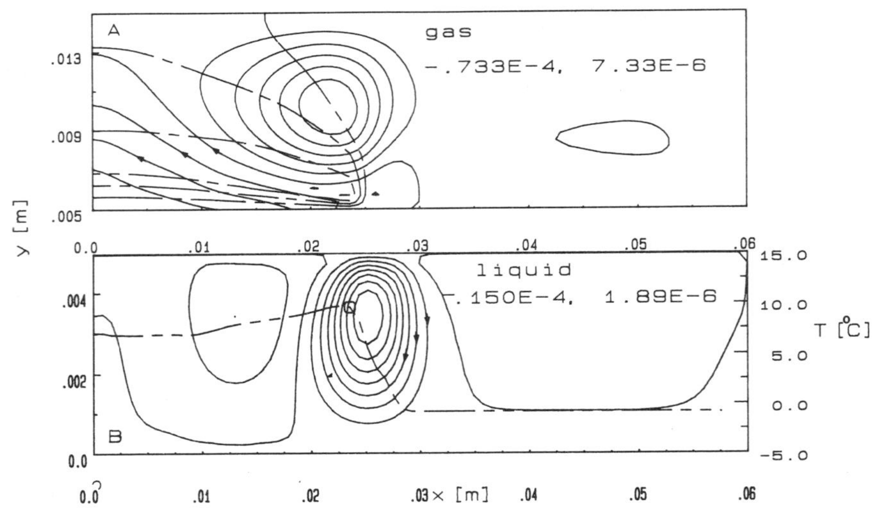

Fig.3 - Gas phase (A) and liquid phase (B) flow field configuration (solid lines), gas phase temperature (A) isotherms (dashed lines), surface temperature (B) distribution (dashed line) and flame leading edge position (B, marked) for t = 1.5 s. Values as Fig. 1. step 1.175-6.

In conclusion, the alternated flow conditions and, for the initial liquid temperature and viscosity of the liquid considered in the simulations of Ref. [10], the changes in the reaction rate at the front of the flame cause oscillatory spread rate.

6. CONCLUSIONS AND FURTHER DEVELOPMENTS

An analysis on the current state of art of experimental and theoretical research on ignition, flame spread and burning of a liquid from a laminar pool has been presented. It appears that, due to the complexity of the mathematical models, very few detailed simulations have been made in comparison with experimental measurements. Even though interesting information on controlling mechanisms of liquid fuel flammability has been obtained, the solution of more advanced mathematical models is needed.

To this end, the main physical processes of combustion of a fuel liquid from a pool (heat and mass transfer, hydrodynamics, liquid fuel vaporization and gas-phase chemical reactions) have been described through a rather comprehensive mathematical model, and the relative importance of each of them has been displayed by the introduction of some dimensionless parameters. Numerical solution of this model can be an important step to improve the understanding of laminar pool fire processes. The aspects of the problem which deserve further study are : a) quantitative predictions of ignition characteristics, spread rates and flame structure for a range of initial liquid temperature varying from well below the flash point to the boiling point; b) dependence of the flammability characteristics on fuel properties and enviroment conditions, particularly on forced flow conditions and gravity level; c) dependence of the burning process on liquid phase convective motions.

Numerical solution of the model here presented can also be important as first step towards the solution of more complex models accounting for turbulent flow conditions which are of great importance in practical systems. Finally, the formulation of efficient numerical techniques for the treatment of the present problem can be useful for the solution of similar mathematical models, such as for example the case of droplet dynamics.

REFERENCES

[1] *Abramzon B., Edwards D. K., Sirignano W. A..* Transient, stratified, enclosed gas and liquid behaviour with concentrated heating from above, J. of Thermophysics and Heat Transfer, Vol. 1, No. 4, 355 (1987)

[2] *Aggarwal, S., K., Iyengar, J. and Sirignano W. A.*, Enclosed gas and liquid with nonuniform heating from above, Int. J. Heat Transfer, Vol. 29 No. 10, 1593 (1986)

[3] *Akita, K.*, Some problems of flame spread along a liquid surface, Fourteenth Symp. (Int.) on Combustion, The Combustion Institute, 1075 (1973)

[4] *Akita, K., and Fujiwara, O.*, Pulsating flame spread along the surface of liquid fuels, Comb. Flame 17, 268 (1971)

[5] *Akita, K., and Yumoto, T.*, Heat transfer in small pools and rates of burning of liquid methanol, Tenth Symp. (Int.) on Combustion, the Combustion Institute, 943 (1965)

[6] *Burgoyne, J., Roberts A. and Quinton P.*, The spread of a flame across a liquid surface - II Steady state conditions, Proc. R. Soc., London, Ser. A, 308, 55 (1968)

[7] *Di Blasi, C., Crescitelli, S., and Russo, G.*, A model of combustion phenomena and of subsurface flows of laminar pools, Proc. of the V International Symposium on Numerical Methods in Engineering, Vol. II, 11, CMP Springer-Verlag (1989)

[8] *Di Blasi, C., Crescitelli, S., and Russo, G.*, Model of pulsating flame spread across liquid fuels, Proc. of the Twenty-third International Symposium on Combustion, The Combustion Institute, Pittsburgh, 1990, in press

[9] *Di Blasi, C., Crescitelli, S., and Russo, G.*, Numerical simulation of laminar pool fires, 14th Eurotherm Seminar - Heat Transfer and Major Technological Hazards, edited by L. Bolle, pp. 21-1,21-10, Lovain la Neuve (1990)

[10] *Di Blasi, C., Crescitelli, S., Russo, G.*, Model of oscillatory phenomena of flame spread along the surface of liquid fuels, Computer Methods in Applied Mechanics and Engineering, (1991), in press

[11] *Di Blasi, C., Crescitelli, S., and Russo, G.*, A finite difference approach of Navier-Stokes equations for heterogeneous combustion, Proceedings of the European Conference on Computer Application in the Chemical Industry, Dechema Monographs, 357 (1989)

[12] *Di Blasi, C., Continillo, G., Crescitelli, S., and Russo, G.*, Numerical simulation of opposed flow flame spread over a thermally thick solid fuel, Combustion Science and Technology, 54, 25 (1987)

[13] *Di Blasi, C., Crescitelli, S.,Russo, G., and Fernandez-Pello, A., C.*, Predictions of the dependence on the opposed flow characteristics of the flame spread rate over thick solid fuel, Proceedings of the Second International Symposium on Fire Safety Science, Hemisphere Pu. Co., 119 (1988)

[14] *Di Blasi, C., Crescitelli, S., and Russo, G.*, Numerical modeling of flow assisted flame spread, Computer Methods in Applied Mechanics and Engineering, 75, 481 (1989)

[15] *Di Blasi, C., Crescitelli, S., and Russo, G.*, Near limit flame spread over thick fuels in a concurrent forced flow, Combustion and Flame, 72, 205 (1988)

[16] *Feng, C., C., Lam, S., H., and Glassman, I.*, Flame propagation through layered fuel-air mixtures, Combustion Science and Technology, 10, 59 (1975)

[17] *Fernandez-Pello, A., C.*, Flame spread modeling, Combustion Science and Technology, 39, 119 (1984).

[18] *Fernandez-Pello, A., C., and Hirano, T.*, Controlling mechanisms of fire spread, Combustion Science and Technology, 32, 1 (1983).

[19] *Fisher, S., J., Harduin-Duparc, B., and Grosshandler, W., L.*, The structure and radiation from an ethanol pool fire, Combustion and Flame 70, 291 (1987)

[20] *Furuta, M., Humphrey, J., and Fernandez-Pello, A., C.*, Prediction of flame spread hydrodynamics over liquid fuel, Physico Chemical Hydrodymanics, Vol. 6, No. 4, 347 (1985)

[21] *Glassman, I., and Dryer, F., L.*, Flame spreading across liquid fuels, Fire Safety Journal 3, 123 (1980)

[22] *Glassman, I., and Hansel, J.*, Some thoughts and experiments on liquid fuel spreading, steady burning and ignitability in quiescent atmospheres, Fire Abstr. Rev. 10, 217 (1968)

[23] *Glassman, I., Hansel, J., and Eklund, I.*, Hydrodynamic effects in the flame spreading, ignitability and steady burning of liquid fuels, Combustion and Flame, 13, 99 (1969)

[24] *Hirano, T., Suzuki, T., Mashiko, I., and Tanabe, N.*, Gas movements in front of flames propagating across methanol, Combustion Science and Technology, 22, 83 (1980)

[25] *Hirano, T., Suzuki, T., and Mashiko, I.*, Flame behaviour near steps bounding layered flammable mixtures, Eigtheenth Symp. (Int.) on Combustion, The Combustion Institute, 647 (1981)

[26] *Hirano, T., and Suzuki, T.*, Theoretical simulation of gas movements in front of propagating flames through layered flammable mixtures, Combustion Science and Technology, 23, 215 (1980)

[27] *Hirsts, R., and Sutton, D.*, The effect of reduced pressure and airflow on liquid surface diffusion flames, Combustion and Flame, 5, 319 (1961)

[28] *Ito, A., Masuda, D., Saito, K.*, A study of flame spread over alcohols using holografic interferometry, Combustion and Flame (1991), in press

[29] *Kaptain, M., and Hermance, C., E.*, Horizontal propagation of laminar flames through vertically diffusing mixtures above a ground plane, Sixteenth Symp. (Int.) on Combustion, The Combustion Institute, 1295 (1977)

[30] *Liebman, I, Corry, J., and Perlee, H., E.*, Flame propagation in layered methane-air systems, Combustion Science and Technology, 1, 257 (1970)

[31] *McCaffrey, B., J.*, Momentum implications for buoyant diffusion flames, Combustion and Flame 52, 149 (1983)

[32] *Mackinven, R., Hansel, J., and Glassman, I.*, Influence of laboratory parameters on flame spread across liquid fuels, Combustion Science and Technology, 1, 293 (1970)

[33] *Murad, R., J., Lamendola, J., Isoda, H., and Summerfield, M.* A study of some factors influencing the ignition of a liquid fuel pool, Combustion and Flame, 15, 289 (1970)

[34] *Nakakuki, A.*, Liquid fuel fires in the laminar flame region, Combustion and Flame 23, 337 (1974)

[35] *Phillis, H.*, Flame in a buoyant methane layer, Tenth Symp. (Int) on Combustion, 1277 (1965)

[36] *Ross, H., D., Sotos R., G.*, An investigation of flame spread over shallow liquid pools in microgravity and nonair environments, Proc. of the Twenty-third International Symposium on Combustion, The Combustion Institute, Pittsburgh, 1990, in press

[37] *Sher, E.*, A theoretical study of the combustion of liquids at a free surface, Combustion and Flame 47, 109 (1982)

[38] *Sirignano W., A.*, A critical discussion of theories of flame spread across solid and liquid fuels, Combustion Science and Technology, 6, 95 (1972).

[39] *Sirignano W. A. and Glassman I.*, Flame spreading above liquid fuels: surface-tension-driven flows, Combustion Science and Technology, 1, 307 (1970)

[40] *Suzuki, T., and Hirano, T.*, Flame propagation across a liquid fuel in an air stream, Nineteenth Symp. (Int.) on Combustion, The Combustion Institute, 877 (1982)

[41] *Torrance, K., E.*, Subsurface flows preceding flame spread over liquid fuels, Combustion Science and Technology, 3, 133 (1971)

[42] *Torrance, K., E, and Mahajan, R., L.*, Surface Tension flows induced by a moving thermal source, Combustion Science and Technology, 10, 125 (1975)
[43] *Torrance, K. E, and Mahajan R. L.*, Fire spread over liquid fuels: liquid phase parameters, Fifteenth Symp. (Int.) on Combustion, The Combustion Institute, 281 (1975)
[44] *Yumoto, T., Takahashi, A., and Handa, T.*, Combustion behaviour of liquid fuel in a small vessel: effect of convective motion in the liquid on burning rate of hexane in the early stage of combustion, Combustion and Flame, 30, 33 (1977)
[45] *Williams F., A.*, Mechanisms of fire spread, Sixteenth Symposium (Int.) on Combustion, The Combustion Institute, 1281 (1976).

Colomba Di Blasi
Dipartimento di Ingegneria Chimica, Università di Napoli,
Piazzale V. Tecchio, 80125 NAPOLI, Italy